Environment and the City

'Cities' are now home for over half the world's population – the hubs
and gateways for capital, innovation, markets, resources and migration.
On current trends the future may be a 'planet of slums' – many cities
are places of poverty, crime and struggles for resources, where the
'environment' is polluted and hazardous. Other cities are sites of
conspicuous consumption, where the 'environment' is a commodity
seen from a car or hotel window. But even such extreme cases are
interdependent and connected in a global system of power and money,
resource flows and ideology.

Environment and the City is an introduction to the many layers of the
'human urban environment'. It examines the full range of issues and
elements that make up the urban environment, including the consumption
of resources, population pressures and the pattern of urban development.
These different issues and elements are examined through an inter-
disciplinary perspective, drawing equally on geography, sociology,
economics and political science, as well as the environmental and
resource sciences. As a consequence, the book is able to focus on the key
debates that are of critical importance for the cities of both the developed
and developing nations. The result is not a simple, single solution, rather
the book offers a set of directions and tools of enquiry that provide
a realistic and practical approach to understanding and managing
sustainable cities and regions.

This book is a concise and accessible guide for all students interested in
the environmental issues emanating from our urban societies. Primarily
written to aid student understanding, the easily navigable text features
boxed practical examples, discussion points, signposts to reading and
websites and a glossary. The book will also be useful to policy-makers
and practitioners in geography and environmental studies, urban planning
and development, urban politics and sociology, and others who think

the future of our urban civilization and its impact on the environment is pertinent.

Peter Roberts is Professor of Sustainable Spatial Development at the University of Leeds. He is an active researcher in the fields of urban and regional planning, development and environmental management, and has published widely on these topics, including *Integrating Environment and Economy*. From 2005 to 2008 he was Chair of the UK Government's Academy for Sustainable Communities.

Joe Ravetz is Co-Director of the Centre for Urban and Regional Ecology at the University of Manchester. He is the author of the landmark book, *City-Region 2020: integrated planning for a sustainable environment*, and co-ordinator of the Sustainable City-Region Programme.

Clive George is Senior Research Fellow at the Institute for Development Policy and Management in the School of Environment and Development at the University of Manchester. He is the co-author and editor of *Environmental Assessment in Developing and Transitional Countries*.

Routledge Introductions to Environment series

Environment and the City

Peter Roberts, Joe Ravetz and Clive George

Routledge
Taylor & Francis Group

LONDON AND NEW YORK

First published 2009
by Routledge
2 Park Square, Milton Park, Abingdon, Oxon, OX14 4RN

Simultaneously published in the USA and Canada
by Routledge
270 Madison Avenue, New York, NY 10016

Routledge is an imprint of Taylor & Francis Group, an informa business

Typeset in Times New Roman by Pindar NZ, Auckland, New Zealand
Printed and bound in Great Britain by TJ International Ltd, Padstow, Cornwall

British Library Cataloguing in Publication Data
A catalogue records for this book is available from the British Library

Library Congress Cataloging-in-Publication Data
Roberts, Peter.
  Environment and the city/Peter Roberts, Joe Ravetz, and Clive George.
    p. cm.
  Includes bibliographical references and index.
    1. Cities and towns—Growth—Environmental aspects. 2. Urban ecology.
  3. Sustainable development. I. Ravetz, Joe. II. George, Clive. III. Title.

  HT371.R56 2009
  307.76–dc22   2008046754

ISBN 10: 0-415-30246-3 (hbk)
ISBN 10: 0-415-30247-1 (pbk)
ISBN 10: 0-203-49592-6 (ebk)

ISBN 13: 978-0-415-30246-3 (hbk)
ISBN 13: 978-0-415-30247-0 (pbk)
ISBN 13: 978-0-203-49592-6 (ebk)

Contents

Series editors' preface
Environment and Society titles

The 1970s and early 1980s constituted a period of intense academic and popular interest in processes of environmental degradation: global, regional and local. However, it soon became increasingly clear that reversing such degradation would not be a purely technical and managerial matter. All the technical knowledge in the world does not necessarily lead societies to change environmentally damaging behaviour. Hence a critical understanding of socio-economic, political and cultural processes and structures has become, it is acknowledged, of central importance in approaching environmental problems. Over the past two decades in particular there has been a mushrooming of research and scholarship on the relationships between social sciences and humanities on the one hand and processes of environmental change on the other. This has lately been reflected in a proliferation of associated courses at undergraduate level.

At the same time, changes in higher education in Europe, which match earlier changes in America, Australasia and elsewhere, mean that an increasing number of such courses are being taught and studied within a framework offering maximum flexibility in the typical undergraduate programme: 'modular' courses or their equivalent.

The volumes in this series will mirror these changes. They will provide short, topic-centred texts on environmentally relevant areas, mainly within social sciences and humanities. They will reflect the fact that students will approach their subject matter from a great variety of different disciplinary backgrounds; not just within social sciences and humanities, but from physical and natural sciences too. And those students may not be familiar with the background to the topic, they may or may not be going on to develop their interest in it, and they cannot automatically be thought of as being at 'first-year level', or second- or third-year: they might need to study the topic in any year of their course.

The authors and editors of this series are mainly established teachers in higher education. Finding that more traditional integrated environmental studies or specialised academic texts do not meet their requirements, they have increasingly met the new challenges caused by structural changes in education by writing their own course materials for their own students. These volumes represent, in modified form which all students can now share, the fruits of their labours.

To achieve the right mix of flexibility, depth and breadth, the volumes, like most modular courses themselves, are designed carefully to create maximum accessibility to readers from a variety of backgrounds. Each leads into its topic by giving adequate introduction, and each 'leads out' by pointing towards complexities and areas for further development and study. Indeed, much of the integrity and distinctiveness of the Environment and Society titles in the series will come through adopting a characteristic, though not inflexible, structure to the volumes. Each introduces the student to the real-world context of the topic, and the basic concepts and controversies in social science/humanities which are most relevant. The core of each volume explores the main issues. Data, case studies, overview diagrams, summary charts and self-check questions and exercises are some of the pedagogic devices that will be found. The last part of each volume will normally show how the themes and issues presented may become more complicated, presenting cognate issues and concepts needing to be explored to gain deeper understanding. Annotated reading lists are important here.

We hope that these concise volumes will provide sufficient depth to maintain the interest of students with relevant backgrounds, and also sketch basic concepts and map out the ground in a stimulating way for students to whom the whole area is new.

The Environment and Society titles in the series complement the Environmental Science titles which deal with natural science-based topics. Together this comprehensive range of volumes which make up the Routledge Introductions to Environment Series will provide modular and other students with an unparalleled range of perspectives on environmental issues, cross referencing where appropriate.

The main target readership is introductory level undergraduate students predominantly taking programmes of environmental modules. But we hope that the whole audience will be much wider, perhaps including second- and third-year undergraduates from many disciplines within the social sciences, science/technology and humanities, who might be taking occasional environmental courses. We also hope that sixth-form teachers and the wider public will use these volumes when they feel the need to obtain quick introductory coverage of the subject we present.

David Pepper and Phil O'Keefe

1997

Series International Advisory Board

Australasia: Dr P. Curson and Dr P. Mitchell, Macquarie University

North America: Professor L. Lewis, Clark University; Professor L. Rubinoff, Trent University

Europe: Professor P. Glasbergen, University of Utrecht; Professor van Dam-Mieras, Open University, The Netherlands

List of boxes, figures and tables

Boxes

Figures

Tables

Preface

Sometime around the millennium, 'cities' became the habitat for over 50 per cent of the world's population. Meanwhile, the 49 per cent of the population who are not yet urbanized are more and more connected to cities – the hubs and gateways for capital, innovation, markets, services, resources and human population.

Cities can be divided crudely into those in 'developing' and 'developed' worlds – the 'South' and the 'North'. In very simple terms, the developing world includes the 80 per cent of the population who consume 20 per cent of world resources, while the developed world includes the 20 per cent consuming the 80 per cent remainder. At present, there are twice as many people in cities of the 'South' as in the cities of the 'North': by 2030 there may be four times as many. The future may indeed be a 'planet of slums' (Davis, 2005).

For many in the developing world, cities are often places of poverty, disease and crude struggles for power and resources – where the 'environment' is represented by polluted water, deforestation, or fear of landslides and floods.

For many in the developed world, the city can be a place of conspicuous consumption, leisure and culture, illusions and icons – where the environment is a commodity to be bought and sold, or seen from a car windscreen or tourist hotel.

In both worlds, the urban environment is obviously a physical system of water, air, land use, energy flows and other elements. It is also a 'human' system, with complex patterns and processes, which might be explained by economics, sociology, political science, psychology and so on. The causes, the impacts, and the solutions for physical environmental problems are generally found in these human processes and pressures.

That is why the term 'human urban environment' has been used to represent the interface between social, economic, political, and urban systems, and the environmental issues which confront them.

So why has this book been written? It's basically about making connections.

There are many kinds of connection between cities in the developed and developing worlds, through the complex linkages and supply chains of international trade, migration and global environmental resources.

There are also connections between the environment and the health of the city – public health, economic growth, social welfare, and the culture and community of the place.

In addition, there are also connections between cities and their hinterlands – traditionally cities were the hubs for information and capital, which drained their surrounding areas of resources and population.

Although such connections appear to be many and complex, one basic message is clear – there are alternatives to the conventional economic model and environmentally damaging growth path of the city. The many kinds of alternatives – economic, social and environmental – are generally grouped under the umbrella term 'sustainable cities'.

However, this means many things to many people. The core of the idea is that of a better balance between cities, their physical demands and environmental conditions, at the local, regional and global levels. This goes hand-in-hand with economic development, social welfare and human justice, and is essential for the future of cities themselves, and the planet as a whole.

Having said that, a first task is to get a better focus on what is 'urban', in the shape of settlements, from villages to mega-cities. A further task is a better focus on just what is the 'environment' – including local, regional, national and global levels; and with physical, human and financial and other resources in mind. These boundaries and definitions are explored in Chapters 1 and 2, which outline the concept and agenda for the 'human urban environment'.

This then points to the particular contribution of this book – a range of 'critical perspectives' on the 'human urban environment'. While it is easy to come up with simple solutions to individual problems – more greening, better regulation, smarter planning – it is often more difficult in a complex world to understand how issues interact, and why some things work and some don't. Experience demonstrates that this needs more than the traditional methods and tools of environmental management, planning, economics, sociology, political science and other disciplines. So this book follows some emerging critical perspectives – globalization, liberalization, risk, exclusion, consumption and governance. For each of the many issues in the 'human urban environment', it is essential to ask – what is the underlying discourse? Why don't tested solutions always work everywhere? And how is it possible to tackle complex problems with little money or power, when arguments over economics and politics, global and local, causes or effects, are all tangled up?

For such questions it is wise not to attempt to provide blueprint solutions, but rather to offer signposts and tool kits. What emerges is an agenda for 'integrated planning and management' – a set of pathways for putting together different sectors, different levels, different causes and effects, for steering towards more sustainable cities.

Another key feature of this book is the international perspective – some would ask, what has the 'green agenda' of the 'North' got in common with the 'brown agenda' of the 'South', separated as they are by massive differences in material wealth and social conditions? Most books look either at one or the other side, but for several reasons this book deliberately puts them together.

First, there is not a clear division between the 'South' or 'North', as the cities of many transitional economies, and those of Latin American and Asian countries, are somewhere between developed and developing worlds. Second, many of the contradictions and problems of the cities of the 'North' – in terms of divisions in their local economies, environmental quality and social cohesion – suggest there is much to learn from the cities of the 'South'. Third, each is clearly inter-dependent on the other – many of the consumer goods which pack the shopping malls and eventually the waste landfills of 'Northern' cities, are produced by low-cost labour in 'Southern' cities, often in extreme environmental and social conditions. It is impossible to attend to one set of problems without dealing with the other.

An additional feature of this book aims to look beyond the common view that cities have only recently gone off track, and that all that is required is the right policies to make them sustainable again. A more balanced view shows that most cities for most of history have hardly ever been in equilibrium – rather they survive many pressures and contradictions, while lurching between crisis and opportunity. There are always ways to improve, but it is essential not to be naïve about the reality of inter-dependence and competition, within and between cities.

So with these overarching ideas and themes, this book aims to review the evidence; provide signposts to critical thinking; and offer an agenda for future actions.

How is the book put together?

It is aimed at people without specialist knowledge, who are interested in making the 'connections' – the linkage between human, urban and environmental sciences. Each of the core chapters has a general structure including:

- an overview – introduction to the theme of each chapter;
- main narrative – outline of the case, the view from the South and North, and a discussion of critical perspectives;
- practical examples and case studies in boxes;
- signposts to further reading and useful organizations.

Writing this kind of book is a great opportunity to think through the issues, to gather together ideas from around the world, and to make them clearer and stronger. It is clear that the theme of *Environment and the City* is one that is urgent and essential for a more sustainable world in the twenty-first century. The authors hope that this enthusiasm is carried through the book, and is shared by its readers.

Acknowledgements

This project has taken a considerable period of time to come to fruition. Over the past six years the authors have benefited from many discussions and inputs, including the contributions made by their colleagues at Leeds and Manchester. A particular debt of gratitude is due to a former member of staff at Manchester, Joe Howe, who was initially a member of the project team, but withdrew due to pressure of work elsewhere. The authors wish to acknowledge his early inputs to this book.

Other inputs have come from friends and colleagues at Leeds, Manchester and elsewhere, including Stephen Cox, Andy Gouldson, John Handley, Colin Kirkpatrick, Norman Lee, Greg Lloyd and Fotini Papoudakis. Their willingness over the years to discuss difficult issues and to offer advice has been invaluable, as has their encouragement to persevere with what at times appeared to be an impossible task.

Particular thanks are due to UN Habitat who have provided many of the boxes and case studies through their excellent Best Practices and World Habitat Day databases.

Although the book as a whole is very much the product of a team effort – the authors have contributed to each other's drafts and thinking processes – each of the chapters has been led by an individual author. Joe Ravetz was lead author on Chapters 2, 3, 5, 6, 7 and 8; and initially mapped out the agenda for 'critical perspectives on the human urban environment'. Clive George was lead author on Chapters 4 and 9. Peter Roberts was lead author for Chapters 1 and 10, and editor for the entire manuscript. However, in the spirit of collective responsibility, this book is the product of collaboration and partnership – often contested and full of differences, but with a substantial core of agreement, common ground and mutual learning. This need for partnership in analyzing and identifying the future possibilities for the 'human urban environment' is perhaps an appropriate point on which to end.

Peter Roberts, Joe Ravetz and Clive George

Leeds and Manchester, UK

September 2008

1 Introduction

The city, man's noblest creation, hardly merits such an accolade in its late twentieth-century form (Eldredge, 1967, vii).

1.1 Environment and the city

Environment and the City looks at the evolution of cities in the developed and developing world, and the implications for resource consumption and environmental impacts, locally, regionally, nationally and globally. This is a huge topic and there is no way to put all the detail into one book. This is especially the case given that this book attempts to look at the city through two lenses: it looks 'inward' to the functioning of cities as individual entities, and it looks 'outward' to the operation of cities globally.

So this book focuses on the social science view of what is most relevant to the urban environment. This involves a cross-cutting and critical approach, with new thinking on multiple geographies of the city, including changing patterns of networks, exclusion, consumption, risk and expropriation. Then for each of the environmental themes – air, water, and others – there is a social, economic and political agenda, and for each activity sector – housing, employment, leisure, construction, transport and others – there is an environmental agenda.

The practical applications of this approach can be seen in the many methods and tools for urban environmental management, such as impact assessment, environmental taxation and citizen-based governance. From all this, the themes of the sustainable city and the sustainable community emerge – not so much as fixed menus, but more as a continuous learning process between all sections of society.

This continuous learning process is not restricted to one city or one group of individuals or professionals. Rather it is a universal need, set in the context of an urgent requirement for better understanding and action in order to prevent the exacerbation of known environmental (and social and economic) problems. Furthermore, it is evident that some of these problems are long-standing, including the difficulties encountered when attempting to reconcile the conflicting demands of, for example, economic growth and the conservation of scarce environmental resources.

It is for the above reasons that this book concentrates upon what Artur Glikson called the 'human environment', which he described as 'the space which surrounds human movement, work, habitation, rest and interaction' (Glikson, 1971, 1). This 'human environment' as the term suggests, transcends the normal definition of the environment because it deals with the transformation of biological and physical factors in space and the consequences, both negative and positive, of this process of modification. These are important matters which are considered in relation to the scope and methods that underpin the approach adopted in the book. For the purpose of the book Glikson's 'human environment' provides an organising concept for considering urban problems, thereby leading to the use of the term 'human urban environment' throughout the following text.

1.2 An urban world

Urban areas are now home to over half the world's population. They also represent the most significant concentration of global environmental challenges, including a range of major problems that are associated with the excessive consumption of resources; the generation of vast quantities of waste; the pollution of land, air and water; and a vast array of health and security concerns that would appear to be the inevitable collorary of dense urban living.

Cities have been a significant part of human existence for thousands of years, but for most of this history urban areas have played a relatively minor role in terms of housing the population and producing the goods and products necessary for survival. Although towns and cities have always exerted considerable power and influence far beyond that directly associated with their size – they have, after all, always been a focus for political, administrative, religious, cultural, educational, military and trading functions – in many parts of the world the presence and performance of urban settlements has not dominated human affairs as totally as is now the case. In the past the vast majority of the world's population lived in rural areas and their lifestyles required at least an elementary understanding of the environment upon which they depended for their survival. Nowadays the majority of the world's population live far removed from the natural environment.

Industrialization

This basic relationship between human settlement and the environment was disrupted with the advent of mass industrialization, which both stimulated the drift of population from the countryside into urban areas, and enabled the in situ growth of the urban population. Whilst there were, and still are, many advantages associated with urbanization, including, for example, the ability to support mass production, the provision of a wide range of services and the stimulation of research and innovation, many diseconomies and unaccounted externalities have emerged. Inevitably the attention of the public and many policy-makers is frequently focused on the negative environmental consequences of urban living. However, there are also many positive environmental attributes associated with urbanization, and it is essential to attempt to provide a balanced assessment of the environmental performance of towns and cities. A brief summary of some of the more important positive and negative environmental challenges facing cities is presented in Box 1.1.

The Aalborg Charter and the Bristol Accord

This book aims to provide a balanced assessment of the evolution and current condition of the human urban environment. In stating this objective, it is appreciated that any attempt to claim complete objectivity will be open to question, and that any assessment of such a massive topic can only ever be selective in terms of its geographic or topic coverage.

Box 1.1

Key urban environmental challenges

- The increasing size of many cities and the massive concentration of population in such areas.
- The massive direct and indirect consumption of finite and renewable natural resources, including water, minerals, food and other materials.
- The generation of substantial quantities of pollutants and wastes.
- The consumption of land for building and the degradation of many other areas of land, including brownfield and derelict sites and areas used for production and waste reception.
- The continued use of unsuitable and unsustainable modes of transport and the pollution and congestion which result from such usage.
- The absence in many cities of sufficient open space and the threat to biodiversity.
- The impact of urbanization on the climate at local and global levels.
- The excessive development of flood plains and the threat that this poses.
- The presence of conceptual and institutional barriers to the development and implementation of strategies to deal with urban environmental problems.
- The social and political barriers which prevent or inhibit the full engagement of all sectors and communities in city management.
- The absence of economic and other incentives to improve urban environmental conditions.

Nevertheless, in presenting the following analysis and discussion, the authors hope to be able to introduce what is now acknowledged as one of the most important issues confronting societies in both the more developed and less developed countries of the world. Although it is set in a European context, the overall nature of the challenge confronting cities was addressed by the Aalborg Charter, in the following terms:

> We are convinced that the city or town is both the largest unit capable of initially addressing the way architectural, social, economic, political, natural resource and environmental imbalances damage our modern world and the smallest scale at which problems can be meaningfully resolved

in an integrated, holistic and sustainable fashion. (Aalborg Charter, 1994, paragraph 1.3, cited in White, 2002, 4)

The challenge identified by the Aalborg Charter has subsequently been reflected in many other research reports and policy statements. However, the danger remains that the task of making cities sustainable will be fragmented with each individual organization or agency taking responsibility for a particular aspect of policy, such as transport, housing, social infrastructure or environmental management. Although as yet there is no global agreement on how best to take forward a comprehensive and integrated approach to the planning and management of the entire 'human urban environment', the member states of the European Union have agreed a broad approach to understanding and implementing a means of delivering sustainable development at the urban scale. This approach, known as the sustainable communities model, was agreed by the European Union member states at a meeting held in 2005 in Bristol (Office of the Deputy Prime Minister, 2005). The sustainable communities model, as agreed in the Bristol Accord, provides a practical means of delivering a fundamental objective of the European Union, that is the delivery of sustainable development.

The sustainable communities model brings together the theory and practice of sustainable development with the politics and practice of placemaking. It is a model that offers a response to the challenge set out in the Aalborg Charter and it aims to create and maintain better places 'where people want to live and work, now and in the future' (Office of the Deputy Prime Minister, 2005, 6).

An exploration of how to meet the challenges identified by the Aalborg Charter and the Bristol Accord provides a central theme for this book. Although the Aalborg and Bristol initiatives illustrate the actions taken in Europe, other broadly similar initiatives have been introduced in other countries and continents. As is demonstrated elsewhere, the intention of the book is to illustrate the ways in which cities have evolved and to identify pathways towards the future that offer them the opportunity to become sustainable communities irrespective of their locations in the 'North' or the 'South'.

1.3 The urban century and beyond

The scale of the problem facing cities in their attempt to become sustainable communities is considerable, and has become more severe over the past century. In 1900 about 150 million people, or less than 10 per cent of the world's population at that time, lived in cities; by 2000 the number of urban residents had increased 20-fold, and represented half the world's population. This drift to the cities, coupled with the ever-accelerating growth of the urban population, has resulted in a previously unknown concentration of people, problems and potentials in urban areas. This acceleration in the growth of the urban population shows little sign of slowing down. Indeed, the rate and scale of urbanization has continued to increase, generating problems in both the urban and rural areas. As the Brundtland Report (World Commission on Environment and Development, 1987) observed:

- in 1940 about one in eight of the world's population was an urban dweller and one in 100 lived in a city with a million or more inhabitants;
- by 1960, more than one in five persons lived in urban areas and one in 16 in a city with a million or more habitants;
- by 1980, almost one in three of the world's population was an urban inhabitant and one in 10 lived in a city with a million or more inhabitants;

Since then the trend of urban concentration has continued inexorably; as noted above, half of the world's population are now urban dwellers.

'South' and 'North'

The pattern of rapid growth and urban concentration is not confined to the more developed world – the 'North' as it is frequently referred to. Urban growth in less developed countries – the 'South' – has accelerated past the developed world. Satterthwaite (1996) notes that of the 286 million cities with a population of over one million in 1990:

- 119 were in Asia;
- 63 in Europe;
- 38 in Latin America;
- 36 in North America;

- 25 in Africa; and
- five in Oceania.

More significantly, of the world's 12 'mega cities', ten were in Asia or Latin America.

It is equally the case that the growth of cities, and of the world's urban population, shows little sign of abating in the near future. The estimated rate of population change indicates a further consolidation of the position of urban areas, with the most noticeable growth in urban population occurring in less developed nations. As Box 1.2 demonstrates, whilst the rate of urban population growth in the 'North' is expected to slow in future decades and to have increased by 20 per cent between 1994 and 2025, in the 'South' the urban population is projected to grow by 144 per cent over the same period.

Box 1.2

Urban population changes, 1994 to 2025

		Population (billions)		Change (per cent)
		1994	*2025*	*1994–2025*
'North'	urban	0.87	1.04	20
	rural	0.29	0.20	−31
'South'	urban	1.65	4.03	144
	rural	2.81	3.04	8
Total	urban	2.52	5.07	101
	rural	3.10	3.24	5

Source: United Nations (2005)

1.4 The environmental consequences of urbanization

In itself, the growth of the world's urban population need not be a cause for concern. However, much of the growth that has occurred has been unplanned, unregulated and unsustainable. As Box 1.2 indicates, the anticipated future pattern of the growth of population is likely to extend, and thereby reinforce many of the undesirable consequences of previous excessive urbanization. In stating that much of the urbanization that has taken place during the past half century was unplanned and unregulated, it is suggested that the future need not repeat the failures of the past. Equally, it is not assumed that urban growth need be undesirable or inherently unsustainable. Indeed, it can and has been argued that well-planned and well-implemented urbanization offers the best and perhaps the only chance of accommodating the rapidly increasing world population in a sustainable fashion. However, achieving this objective will require a significant change in attitudes and behaviours, including the introduction of what Hajer (1996) calls 'a new cultural politics' that can help to guide the future choices made by urban civic society.

So what has gone wrong in the past, and are there any lessons that can be identified and used in order to help inform the future planning and management of urban areas? Whilst the first element of this question is the legitimate concern of this and the following chapters, the identification and assessment of the lessons of good and best practice in urban environmental management are considered in the later parts of this book. This 'learning from doing' approach is now accepted as an essential element of the design and implementation of urban policy (Roberts, 2006a).

Cities and regions

Much of the urbanization that has taken place in the past two centuries has been accompanied by a radical redefinition of the relationship between urban areas and the natural environment. Many towns and cities are no longer functionally linked to their hinterlands, rather, they now depend upon a number of increasingly extensive and complex international supply chains. One direct consequence of this weakening of the urban–rural relationship is the dislocation of the city from the natural systems present in both the immediate city-region and the wider world. It is interesting to note the speed of the deterioration that has taken place in the relationship between cities and the immediate environment; even as late as the beginning of the twentieth century, it was still possible for Patrick Geddes to talk of certain small British cities as an integral part of a natural region where 'cities retain the essential character that is conditioned by their environment and occupation' (Geddes, 1915, 166). However, by the middle of the century the situation had changed dramatically.

The industrialised city

With the exception of certain small cities, such as those described above, by the early twentieth century the pattern of urbanization in developed countries had become indelibly imprinted upon the natural environment. From a position whereby an individual city could be considered to be a small but integral part of the overall pattern of human activity and a relatively insignificant intrusive element in a wider regional ecosystem, it rapidly became evident that the various processes of urbanization had caused the separation of town from country. It has been argued that this divorce of everyday life in urban areas from biological reality has resulted in the emergence of a situation in which 'the daily existence of urban dwellers is separated from natural ecosystems on which all life depends' (White, 2002, 5). One manifestation of the divorce of cities from the natural environment can be seen in the way in which urban areas consume vast quantities of resources: a typical European city of one million inhabitants consumes, on an average day, 2,000 tonnes of food, 320,000 tonnes of water and 11,500 tonnes of fossil fuels (European Environment Agency, 1995). Although the typical European city is a much more modest consumer of environmental resources than cities in North America, it is also the case that the residents of a typical European city consume a far greater quantity of resources than the global average per person.

Urban environmental metabolism

The concept of ecological cycles in environmental science examines how flows of substances, such as water and carbon, continuously circulate through the biosphere. Cities tend to disrupt such natural and self-organising cycles with a 'linear' metabolism – natural resources are sucked in, and pollution and wastes are pushed out along with products and services. A city or region which contains its own ecological cycles in a 'circular metabolism' would tend to be less vulnerable and damaging, in other words more sustainable. For instance, one key cycle would be food which is locally grown, digested and its nutrients returned to the soil within the hinterland of the city. Even where the cycle is at a global scale, it can still be sustainable, if its side-effects or risks are contained within acceptable limits and thresholds.

1.5 Ecological footprints

One method that can be used to measure and express the quantity and pattern of the consumption of resources by urban areas has been developed through the application of the concept of the 'ecological footprint', sometimes called the 'environmental shadow'. This concept, which allows for a meaningful and comparable measurement of the resources consumed by cities, has been defined as the 'land required to feed them, to supply them with timber products and to reabsorb their carbon dioxide emissions by areas covered with growing vegetation' (Smith, *et al.*, 1998, 20). The 'ecological footprint' is a helpful concept because it attempts to measure all the environmental consequences of urban living and consumption in a single consolidated statistic, usually global hectares of land per person. The component parts of the measurement include the direct consumption of energy, food, transport and other resources; the demand made upon land to grow and produce products; and the use made of other areas of land in order to satisfy human needs.

Footprints of 'North' and 'South'

Estimates vary of the 'ecological footprint' of individual cities, and research is continuing on the differences between urban and rural areas. However, at national level it is evident that the average North American's 'ecological footprint' is five times the world average, and 20 times the 'ecological footprint' of the city dwellers in the poorest nations. Box 1.3 provides some comparisons.

This measurement suggests that the more developed economies can only be maintained because of low consumption in less developed countries, and as a consequence, 'if everybody lived like today's North Americans, it would take at least two additional planet Earths to produce resources, absorb the waste or otherwise maintain life-support' (Wackernagel and Rees, 1996, 15). At present the total 'ecological footprint' of the world's population exceeds the available land area or bio-productive capacity

Box 1.3

Ecological footprint: a global comparison

	Population (millions)	Ecological footprint (global ha/person)
World	**6,148.1**	**2.2**
High-income countries	920.1	6.4
Middle-income countries	2,970.8	1.9
Low-income countries	2,226.3	0.8
Lowest: Somalia	9.1	0.4
Highest: USA	288.0	9.5
World available bio-capacity		**1.8**

2001 data. Source: WWF and Global Footprint Network, 2005.

by 20 per cent. This overshoot is likely to become much worse, and the environmental damage will escalate, as much of the rest of the world rapidly catches up with western standards of material affluence.

The reality at present is that the excessive consumption of resources and the profligate lifestyle of many urban dwellers in the advanced economies is only possible because of low levels of consumption in the less developed nations. Put simply, the 'ecological footprint' not only measures consumption, it also reflects the expropriation of resources, exploitation and the absence of social justice evident in many parts of the world.

The household footprint

Attempts have been made to apply the 'ecological footprint' model to the level of the individual household, or to a particular aspect of consumption. The use of the model at household level is of considerable value, because assessments show that indirect energy consumption, for processing and distributing food and other products, represents a significant element of the total account. In addition to this indirect form of consumption, the other main contributing forms of consumption include heating, other direct energy use, transport and household goods and services. Variations in consumption, and the size of the resulting 'ecological footprint', would appear to be associated with factors such as household size, social class, the availability of public transport and other lifestyle-related considerations (Roy and Caird, 2001). Particular forms and patterns of consumption also produce different 'ecological footprints'; evidence of these differences can be seen in cities which depend on private car usage as against those that are well served by public transport, and in cities where densities are higher or lower than average.

Standard methodologies are available to calculate the 'ecological footprint' of an individual city, or a household within a city, although the use of the methodology should be informed by the realization that local and regional circumstances and opportunities vary widely. However, the value of the 'ecological footprint' should not be underestimated; the message is clear 'humans – like all other species – must learn to live within the biological realities of the planet' (White, 2002, 16).

1.6 Sustainable development and sustainable communities

Especially in the case of cities, any discussion of the state of the environment has to be set within the context provided by the theory and practice of sustainable development. There is, however, a problem here which must be addressed at the outset; this is the knotty question of how best to define sustainable development. To some people it is a vision, to others it is a desirable and practical programme of direct action, whilst to many individuals and organizations it represents a challenge to the established orthodoxy that necessitates a radical change of direction. Whatever stance is adopted, the definition of sustainable development that has endured above all others is that put forward by the Brundtland Report:

> Sustainable development is development that meets the needs of the present without compromising the ability of future generations to meet their own needs. (World Commission on Environment and Development, 1987, 43)

Sustainable development and place

When the concept of sustainable development is applied to a given area of territory, such as a town or city, an additional dimension comes into play. This for the sake of simplicity is best described as the spatial dimension of sustainable development. Spatial considerations are of considerable importance in any examination of the practical application of sustainable development; this is because the carrying capacity of local environments and the distinguishing socio-economic characteristics of individual urban systems vary significantly from place to place and over time. Adapting the standard diagram that represents the triad of sustainable development concerns, Box 1.4 illustrates the interaction of the environmental, social and economic dimensions, and their influence upon the condition of an individual place. By definition, the condition of an individual place helps to shape the particular manifestation of sustainable development which is found there, on the one hand, whilst the working-out of the interactions between the three elements of sustainable development helps to shape the past, present and future of the individual place. As will be argued later, this addition of place and space to the sustainable development triad is equally as valid at supranational or national levels, as it is at the level of the individual city, locality or neighbourhood (Roberts, 2001) and is a matter of concern for everyone, now and in the future. In order to capture the issues of current (intra-generational) equity and equity between generations (inter-generational), Box 1.4 adds these dimensions, together with the need to ensure action through political choice and effective implementation.

Having introduced the essential notion of place and space, it is also important to recognize that sustainable development is not a static model, nor does it offer a fixed prescription. Rather it is a concept drawn from systems dynamics and it is concerned with the processes and outcomes of change, both in terms of the individual elements and overall. The general acceptance of this view has allowed and encouraged authors to describe sustainable urban development in many different ways: 'sustainable cities are cities where socio-economic interests are brought together in harmony with environment and energy concerns in order to ensure continuity in change' (Nijkamp and Perrels, 1994, 4), 'a sustainable community could be described as one which has an enduring integration of the social, economic and physical characteristics of our total environment' (Cook and Ng, 2001, 3), or 'urban sustainability projects cities onto the world scene of the future, … it is a dynamic concept bringing an emphasis to

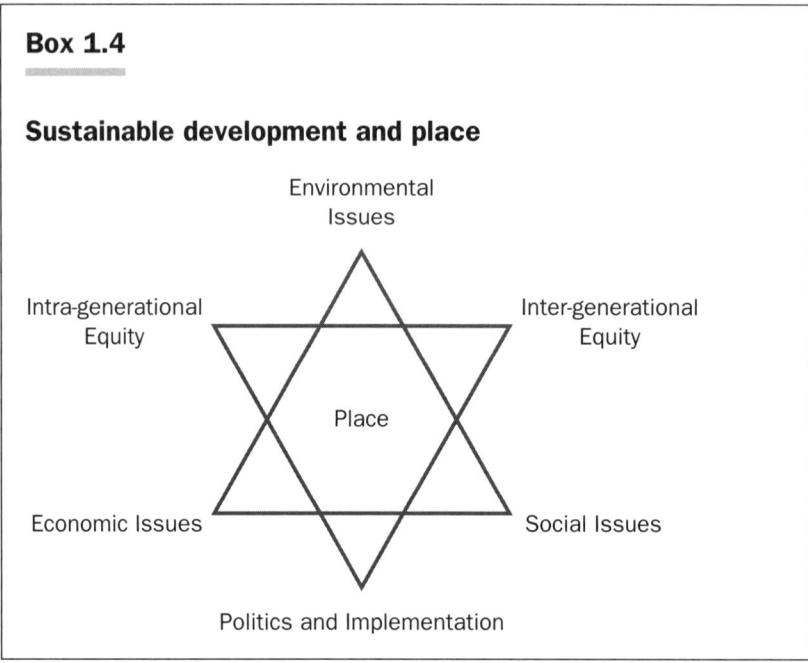

Box 1.4

Sustainable development and place

the quality of development, ... it has been defined more and more as a process and not as an end-point' (Mega, 2000, 227).

In considering the above ideas there are various approaches that seek to bring about integration between the different fields of action. One way is though the linking of three concepts:

- urban development – the evolution, adaptation and restructuring of urban systems in a changing national and international context; development is a dynamic process with no fixed goal other than the broad objective of improving quality of life – it may not be socially equitable or environmentally sustainable;
- sustainable urban development – any process of urban development consistent with an international development pattern that improves the quality of life of all the world's citizens in a sustainable manner;
- urban sustainability development – actions which steer the evolution of urban development towards sustainable urban development.

Another approach is to look at the role of cities in providing economic, social and environmental functions:

- environment – to reduce environmental impact and resource use to 'sustainable' levels, and to enhance environmental quality and safety;
- economy – to enhance long-term resilience, competitiveness, employment and the equitable distribution of resources;
- society – to ensure the equitable distribution of economic resources and to enhance health, education, security, equity, cohesion, diversity and quality of life.

Each of these roles is connected to the others. Actions that deliver on all three fronts are the win-win strategies that attract support from all directions. In reality there are many win-lose strategies – solving one problem while creating several others.

Sustainable communities

A further and more readily implementable attempt to provide a comprehensive and integrated response to the challenge of delivering sustainable development at all spatial scales – towns, cities, villages, and the countryside – has been introduced in the European Union (EU). As noted in section 1.2, in 2005 the member states of the EU agreed the Bristol Accord. This international agreement established the eight component characteristics of a sustainable community; in reality these characteristics are a fine grain expression of the three elements of sustainable development. European Union ministers agreed to utilize the sustainable communities model to guide both EU and national policies, to foster generic skills in place-making and management, and to guide the application of a funding mechanism known as the European Investment Bank.

The sustainable communities model provides a means of reconciling the requirements of sustainable development with the day-to-day business of national and local place development and management. The local activity system is discussed at greater length elsewhere in this book, but involves activities such as economic development, transport management, land use planning, housing provision, community development, social and economic infrastructure provision, and the management of cultural and leisure services. In essence, the sustainable communities approach offers a model of teamworking for the creation, development and management of places. Box 1.5 illustrates the model and the eight component elements that are brought together and delivered in an integrated manner through

Box 1.5

The sustainable communities approach

the ninth essential component of place shaping, which involves both the development and long-term management of places. This ninth component is essential to the provision of a sustainable community, as it represents the need for a 'full service' approach to the creation, development and management of places. Without the powerful integrating force exerted by the ninth component, the danger is that the dominant current model, whereby fragmented individual elements of provision are delivered to communities by separate public, private and voluntary sector bodies, will continue. Equally as important as thinking 'whole of place', is the need to introduce a means of thinking, 'whole of life'; places need to be continuously managed rather than subjected to infrequent bursts of intervention (Roberts, 2007).

1.7 Governance and engagement

Given the acceptance of the fact that sustainable development is a dynamic process which brings together all the actors present in a city in an attempt to plan and implement a comprehensive long-term strategy for the urban area, it is essential to recognize the role of governance in guiding and endorsing this process. Governance is more than government, and it moves beyond the realm dominated by elected members of national parliaments and city councils into the various communities present in a city. As Healey (1997b) has argued, in advanced urban systems the formal machinery of government can only achieve its objectives with regard to sustainable development if it works alongside the full range of formal and informal networks and groups present in a city. This collaborative working is essential in order to ensure both that a comprehensive analysis of urban problems is provided, and that the resulting strategies are realistic and can be implemented (Gilbert, *et al.*, 1996).

Stakeholder governance

Effective implementation of policy is the key to the resolution of many urban environmental problems, and this is unlikely to be achieved unless all actors and stakeholders in a city are involved throughout the processes of strategy building. The reason for this is associated with the realities of the control of resources and the many ways in which these resources can be mobilized. Moving beyond the traditional model, in which the central or local state exercised direct control over many aspects of funding and implementation, the management of a modern city essentially represents a partnership of stakeholders who collectively own, control and direct the utilization of resources. Because of this fragmentation of ownership and control, which in a typical 'Northern' city also reflects the privatization of many utilities and services, such as gas, water, electricity, telecommunications and public transport that took place from the 1980s onwards, it has become more rather than less difficult to control and co-ordinate the various component parts of public, private and voluntary sector policy and action.

In addition to this need to involve the providers of services, it is also essential to involve the ordinary members of the community in making and implementing decisions about how the city should be planned and managed. This active participation of citizens is essential to ensure the local ownership of policies, processes of delivery and the places themselves. An essential message in relation to many national and local systems of governance is the need constantly to restate the importance of representative democracy and accountability. This message reflects the

.portant role of non-elected actors and stakeholders, but it also asserts
.ne supremacy of elected representatives over private interests, including
representatives at the very local level of the individual neighbourhood or
community.

Complete urban governance

In order to plan and implement a strategy for the creation and
management of a sustainable city, it is essential to develop and adopt
a model of 'complete urban governance' which can help to ensure the
engagement and financial commitment of all relevant organizations and
individuals (Roberts, 2000). This 'complete urban governance' model
will, of course, differ in terms of its form and content between places and
it will involve the development of specific arrangements for the planning
and implementation of individual areas or aspects of city life.

As noted above, a much neglected aspect of the governance of
sustainable urban development is associated with the engagement of local
communities and their representatives. This is an essential element in the
analysis of urban environmental problems and opportunities and, as is
increasingly acknowledged by many academic authors and government
bodies, it also provides a firm foundation for the development and
implementation of appropriate strategies (Dobbs and Moore, 2002).
For the various reasons given above, it is essential to connect the local
communities present in an urban area to the various formal processes
of planning, management and implementation. Without such a link it
is unlikely that real and lasting change in the condition of the urban
environment can be achieved. As has been argued, whilst the slogan
'think globally act locally' is an appropriate means of dealing with
major policy issues, it is also essential that the highest level debates
and decisions are informed from below. An appropriate complementary
slogan 'think locally in order to act globally' has been suggested by
Gilbert, Stevenson, Girardet and Stren (1996, 17).

Although much of the local-global debate is very vague in terms of
the assignment of powers and responsibilities, it is possible to sharpen
the focus of the debate by reference to the principle of subsidiarity.
This helpful principle argues that policy decisions should be made and
implemented at an appropriate spatial level, for example, whilst it is
evident that the control of the global commons, such as the oceans,

requires international agreement, decisions regarding a small patch of derelict land in a city can best be made at a neighbourhood level. A further sophistication of subsidiarity provides guidance with regard to the appropriate level for the implementation of policy, for example, whilst the EU may decide on overall policies for waste management, it is appropriate for a city council to arrange for waste collection and processing to take place.

1.8 Communities, cities and regions

So far this chapter has used the terms 'cities', 'towns' and 'urban areas' as interchangeable expressions of the territory with which this book is concerned. There is no fixed or final definition of any of these terms which fits all circumstances. Cities vary in size from the global megalopolis, such as Tokyo or São Paulo with populations of 20 million or more, to a small historic city, such as St David's with a population of under 2000, which once represented a central point of power in a previous era of secular or ecclesiastical administration and organization. Equally, there is no fixed or universally applicable definition of what constitutes a city in terms of its physical form, or how an urban area can be distinguished from a surrounding rural region.

Cities, hinterlands and city-regions

For the purposes of this book, and reflecting much of the existing academic and practice literature, the primary concern herein is with larger urban areas and their surrounding regions. This is a flexible and appropriate scale of concern, and it has the merit of allowing for the environmental problems of individual cities to be set within the context of activities in the surrounding region. Geddes (1915) appreciated the importance of considering urban problems within the context of the adjacent region, and he coined the term 'conurbation' to describe the large sprawling cities of the early twentieth century. Indeed, in the case of the London region, he used the term 'conurbation' to summarise his description of the urban 'polypus' (rather than 'octopus') that like a coral reef has a strong skeleton and living polyps (Dickinson, 1967). Equally, in the modern era and referring to the newly emergent and rapidly growing cities of the less developed nations, Cohen (1993) argues that it is essential to adopt a regional or city-regional perspective in planning and managing the environment–city relationship. This requirement is all the more essential because individual administrative urban areas are frequently too small to cope with the trans-municipal incidence of environmental problems and, as a consequence, many such problems necessitate action at a regional level. Reflecting Cohen's assessment, the European Commission's Expert Group on the Urban Environment has suggested that there is a need to 'redefine the concept of urban, to take into account that cities and towns do not operate as closed systems, but are dependant on, and have a responsibility towards, their rural hinterlands' (European Commission Expert Group, 1998, 2), whilst it has also been argued that some polycentric urban systems would benefit from the introduction of planning and governance regimes at a regional or inter-regional level; the Pearl River Delta Region is an example of such a region (Hills and Roberts, 2001).

Cities, neighbourhoods and communities

A further consideration in examining the environment–city relationship is the issue of how the city can best be divided into smaller units for the purposes of analysis, planning and management. This discussion is linked to the earlier consideration given to the governance of cities, and it reflects the desirability of defining sub-areas, quarters, districts, communities, zones or neighbourhoods of a city that display both unity in terms of social cohesion and a degree of distinctiveness in relation to their natural and built environment characteristics. As was noted earlier, as well as emphasising the role of small areas in the implementation of global environmental policy, it is equally the case that neighbourhood and community concerns and analyses can help create new insights and policy perspectives. Such an approach is equally valid in cities in both the 'North' and the 'South', and would appear to be an appropriate solution for both poor and rich neighbourhoods and communities. In suggesting the need to emphasize the local it is, however, also important to recognize the need for the establishment of effective networks of neighbourhoods and communities in order to share experiences, disseminate good practice and provide mutual support through the identification and promotion of common concerns (Chatterton and Style, 2001). In some cities the effective mobilization of local people in their communities has been the catalyst for lasting change and it is this experience, with the confidence that accompanies it, that can help build capacity in their communities. A recent assessment of one such community in Liverpool illustrates the processes and consequences of this transformational effect (McBane, 2008).

1.9 Understanding and moving towards sustainable cities

The penultimate section of this chapter sets out some of the general issues that have to be considered in moving towards the planning, development and management of cities that are more sustainable. In addition to addressing the problems and opportunities that have been introduced earlier in this chapter – related to the governance, planning and management of cities – it is essential to develop an understanding of the overall choices and challenges that confront cities, including the desirability of identifying alternative pathways for the future development of the environment–city relationship.

Models for sustainable cities

The choice of models that can be used to guide the development of the sustainable city is extensive, both in terms of the analytical capacity of the models and their ability to develop scenarios from which future strategies can be prepared. At one extreme are the models that consider city systems as variants of ecosystems, in which human activities are part of a complex series of interactions between various natural elements (White, 2002). In the middle ground are models which attempt to integrate ecological factors with broader socio-economic and political aspects of sustainable development and, in some cases, to relate these issues to the use of land (Breheny and Rookwood, 1993). Other models focus on institutional capacity, governance and administrative issues (Gilbert, *et al.*, 1996).

Three city models

An alternative way of evaluating the scope, coverage and capability of the various models has been suggested by Haughton (1997) and elaborated further by Guy and Marvin (1999). Here the original analysis sought to identify how the use of resources in cities was linked to the wider zone of influence or 'footprint of the city'. Haughton's three models, whilst not mutually exclusive, offer alternative pathways. The 're-designing cities' model offers the prospect of reducing the call upon resources by re-designing cities as compact and energy-efficient urban areas; this model offers reduced resource flows and a lowered urban 'metabolism'. A second model is one which is described as the 'externally dependant city'. Here the urban area acts as a focal point for the input of additional resources from an increasing hinterland with little regard paid to the generation of wastes. The third model, the 'self-reliant city', is based upon the attempted internalization of economic and environmental activities, and emphasizes the establishment of a circular urban 'metabolism' whereby inputs and outputs of waste are more closely linked.

Evolution versus revolution

The value of these alternative models is that whilst they provide a helpful framework for the purposes of analysis and choice, they do not overly prescribe the processes of policy formulation and implementation. Indeed, what the models offer are generic guiding principles that can be shaped and applied to particular local conditions. Two additional points are of importance here. First, it is apparent the that 'evolution not revolution' approach to changing the nature and content of the environment–city relationship, which is represented by a number of ecological modernization formulations (Hajer, 1996), is likely to prove to be of particular value in mapping the future development pathway of cities. Second, it is important to realize and accept that there is no single model or vision that can be used to plan or develop the environmentally or socially sound city of the future. As Guy and Marvin observe, 'out of the mangle of the diverse practices, multiple rather than singular visions of the city are emerging' (Guy and Marvin, 1999, 272).

1.10 The purpose and structure of this book

The remainder of this book focuses on human–environment relationships and interactions in cities around the world, with the aim of providing a foundation for the establishment of a more environmentally and socially sound urban future. In other words, the intention is to examine the various issues and themes which define the 'human urban environment', and then offer some tools of analysis and policy formulation that can be used to help to plan, deliver and manage sustainable cities.

Managing complexity

This book is an introduction to a complex and often ill-defined subject. It crosses many boundaries: between the past and the future, between disciplines, between the academic and the practice worlds, and across the chasm that seemingly separates those who still cling to the view that 'business as usual' can continue unchecked from those who advocate a radical alteration to the way in which our cities function. Although these boundaries are infrequently acknowledged, they often serve to divide the various groups that are engaged in the planning, development and management of cities. As the OECD (1997) and many other inter-government organizations have emphasized, one of the major challenges facing urban policy-makers is to find ways of engaging the private sector, whilst at the same time encouraging the involvement and commitment of individual citizens to the cause of developing environmentally and socially sound cities. Without a sense of common purpose it is unlikely that progress can be made towards the sustainable city. In many cities of the 'North' the keys to successful schemes for the promotion of more sustainable urban development have been the establishment of a city-wide partnership, the engagement of citizens and the preparation of a vision and strategy that sets out the future pathway for a city (Burwood and Roberts, 2002).

Structure of the book

The early chapters of this book explore the general scope and agenda for the 'human urban environment'. Following this introduction, Chapter 2 sets out some basic ideas which help the reader to understand the complex and many layered themes of the 'human urban environment' and it also considers cities and regions as self-organising systems. Other topics considered in the first part are the transition of cities and the factors which drive change, and the urban environment in a global context, including a discussion of the idea of the global city-region.

The second part of the book considers the individual elements of the 'human urban environment' and examines the key drivers of change. Attention is focussed on the physical urban environment and the consumption of basic resources; the built environment, including an analysis of the various challenges to the future functioning of the city; the economic agenda of the city, both as an individual entity and as a part of a global system; and the social dimension of the urban environment.

In the third part of the book, the focus is on how best to deal with the challenge of planning, developing and managing cities in future. Emphasis is placed on the methods and tools that are necessary and available to analyse and deliver sustainable cities and communities. A final chapter offers a perspective on the future of cities as the home to the majority of the world's population; this chapter also provides some principles for the future of the 'human urban environment'.

In order to help to guide the reader through the book a number of key themes can be identified; these are presented in Chapter 2 and introduce a number of important features which relate to the issues covered in the following chapters. The book itself is structured around these key features:

- Critical perspectives: feeding the agenda for a 'human urban environment' are new ways of thinking about socio-economic-cultural-political issues; this involves topical, controversial debates and unfinished agendas, cross-cutting the social sciences (Chapters 1 and 2).
- Time dimension: every urban environment is at some point on a trajectory of development and modernisation: by understanding the past, and the transitions now in motion, we can better anticipate the future (Chapter 3).
- Global dimension: the trends of urbanization are increasingly driven by the pressures of globalization, where every city competes for a place in the hierarchy of urban systems; at other spatial scales, there are urban–rural relationships, urban infrastructure issues, and connections between local and global (Chapter 4).
- Environmental and physical dimension: this shows the physical environment through its various media – air, water, land, waste, energy etc. – each of these has local and global impacts, and a social, cultural, political and economic agenda (Chapter 5).
- Spatial and physical dimension: this concerns the physical structures and patterns of cities, and the physical fabric of buildings and streets, focusing on housing and transport, at different scales from local to the conurbation (Chapter 6).
- Economic dimension: urban growth and development is largely driven by urban economies, with key industrial sectors and workforce

pressures; there is an emerging agenda for environment-business management, supply chain management, sustainable economic development and so on (Chapter 7).

- Social dimension: this points to a multitude of issues in urban lifestyles – public services such as health and education, crime and security, cultural identities, gender divisions and the patterns of consumption (Chapter 8).
- Policy dimension: solutions for urban and environmental management and governance – investments, policies and plans, management systems and the role of new technologies (Chapter 9).
- Integration and forward thinking: the concept of integrated planning, delivery and management of communities, cities and city-regions underlies the goal of urban sustainable development – this is fundamental to the future of the 'human urban environment' (Chapter 10).

Further information

European Commission Environment   http://ec.europa.eu/environment/index_en.htm

Homes and Communities Academy   http://www.ascskills.org.uk

United Nations Environment Programme   http://www.unep.org/

Further reading

Blowers, A. (ed.) (1993) *Planning for a Sustainable Environment*, London: Earthscan.

Drewe, P., Klein, J. and Hulsbergen, E. (2007) *The Challenge of Social Innovation in Urban Revitalization*, Amsterdam: Techne Press.

Gilbert, R., Stevenson, D., Girardet, H. and Stren, R. (1996) *Making Cities Work*, London: Earthscan.

Organization for Economic Co-operation and Development (OECD) (1997) *Sustainable Development: OECD Policy Approaches for the 21st Century*, Paris: OECD.

World Commission on Environment and Development (1987) *Our Common Future*, Oxford: Oxford University Press.

2 The human urban environment

Scope and methods

2.1 Introduction

Everyone can think of an urban environment – cafes on the streets, shopping malls, housing complexes. Now picture a giant strip mine, with an army of migrant labourers in the remote bush of Southern Africa, producing aluminium for the cladding of office blocks in cities around the world. Is this another kind of urban environment? Following that trail, the minerals investors in New York, the engineers from Korea, the villages in Zaire who send their people to toil in the sun, these are each involved with the urban environment – in this case, the physical production of city structures. But does the 'urban environment' then mean just about everything on the planet? Some definitions and boundaries are clearly needed.

First, as in Chapter 1 this book focuses on the 'human' side of the urban environment – the social, economic, political and cultural patterns and pressures, which shape and are shaped by, the physical urban environment. The book leaves to others the technical details of air emissions or water supplies; global scale issues such as nuclear power or marine conservation, are also outside this book's boundary. Most importantly, the book explores urban environments in both the 'South' and the 'North' – otherwise known as the developing and developed nations – and what happens between them. Finally, the book keeps open the questions of 'what is a city', and 'what is an urban environment?'

Both these are many layered issues which cannot be taken for granted, as discussed in Box 2.1.

Box 2.1

What is the city anyway?

At the Global Forum 1994, I met Gary Lawrence, the deputy city planner of Seattle, who was well known for experiments with sustainability indicators and round tables. I tried to sound him out on climate emissions targets and urban transport models, which we were busy constructing for cities around the UK. But we didn't seem to get very far.

His angle was that Seattle was not actually a city as we know it, but a series of increasingly disconnected networks and sub-cultures, which happened to share the same road-space – some of these were very rich, others very poor; some local, some global. Only when we could start to re-connect these networks to each other, and to the 'city', whatever that was, could we talk about other more tangible goals.

Source: J. Ravetz, personal communication

In this chapter

This chapter is about the scope and methods of *Environment and the City*. It explores what is meant by these words, and where the boundaries are drawn. It then sets out some of the main methods which are used to explore them.

First the idea of mapping the 'human urban environment' is presented. This starts with basic concepts on urban development, and then extends it to environmental science. Behind both of these is the method of 'systems thinking' where the whole is greater than the sum of the parts.

This leads to the comparison of urban environments in the 'North' and the 'South'. This sets the city level in its context of a globalizing city system, and the global dynamics of centrality and dependency, wealth and poverty, growth and decline.

Turning to the human side of the agenda, there is a review of common social science methods as applied to the urban environment. Economics, sociology, anthropology, political science and information studies, each bring valuable insights and a body of evidence.

Following this trail leads to the central core of the book – a set of 'critical perspectives' on the urban environment. These are controversial debates, discourses, questions, campaigns cutting across academic disciplines, with many contradictions and much unfinished business, but which offer powerful tools for exploring the complex realities of the human urban environment. This chapter presents an overview, with further discussion following in Chapter 8.

2.2 Mapping the human urban environment

This quite complex bundle of ideas is summed up with the phrase 'human urban environment'. This means not only the physical environment of air and water; and not just the human experience of walking the streets. The human urban environment is about the relationships between the people, the economy, the city form and fabric, and the environmental qualities, flows, pressures and patterns which interact with these. Or, to summarize a multiplicity of causes and effects:

- Environmental pressures and impacts … on social and economic patterns and trends, local and global.
- Social and economic pressures and impacts … on environmental patterns and trends, local and global.

As for the response to a multitude of problems, local and global, this would look for a more sustainable relationship between cities, their societies and economies, and their local or global environmental impacts. Such a set of relationships could be titled 'urban sustainability'.

Mapping the metabolism

Figure 2.1 shows a typical city surrounded by its hinterland – a classic 'city-region'. This might refer to a freestanding city at the centre of a resource-rich region – cities such as Chicago, Birmingham or Osaka come to mind, as they are often quoted in urban studies. This city-region can be seen as if it was a giant material processor, with its metabolism fed by a constant flow of energy, water, food and other environmental resources. The question is, what kind of human activities are driving this flow?

If this cartoon is expanded into something more like a concept diagram or 'route map' for this book, some general features are visible, as shown in Figure 2.2. The first thing is a flow from left to right – from 'upstream' driving forces and inputs, to 'downstream' impacts and outcomes. This is based on a standard way of analysing environmental problems and solutions – the framework of 'driving forces, pressures, states, impacts and responses' – otherwise known as 'DPSIR' (European Environment Agency, 1996).

This route map also separates out environmental or physical factors from socio-economic factors (meaning social, cultural, ethical, political, institutional, economic and so on). The point is that a socio-economic cause, such as a dirty factory, can cause environmental problems, such as air pollution, and these in turn cause socio-economic problems, such as

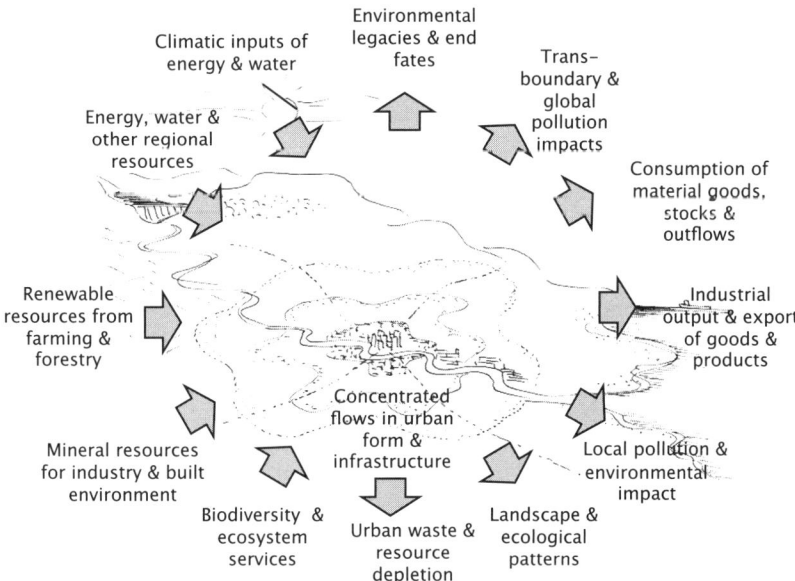

Figure 2.1 *Human urban environment – a resource flow metabolism*

poor health. In general, most environmental issues seem to start and end with human issues, shown in the diagram by the arrows.

Looking at the top part of the diagram, each of the boxes is linked to time and space dimensions. This is a way of highlighting the historic evolution of urban environments, and their pathways into the future. It also flags up their geographic space, from the neighbourhood level to a global network of cities.

Looking at the bottom right part of the map, the feedback arrows coming from the impact boxes lead towards a responses box. This contains a whole array of possible actions and solutions for human environmental problems, from government, businesses, communities or citizens.

Finally, the numbers on the left show the chapters of this book which focus on the different parts of the map.

It should be stressed that this is deliberately simplistic – readers should not get the idea that real-world environmental problems can be solved by

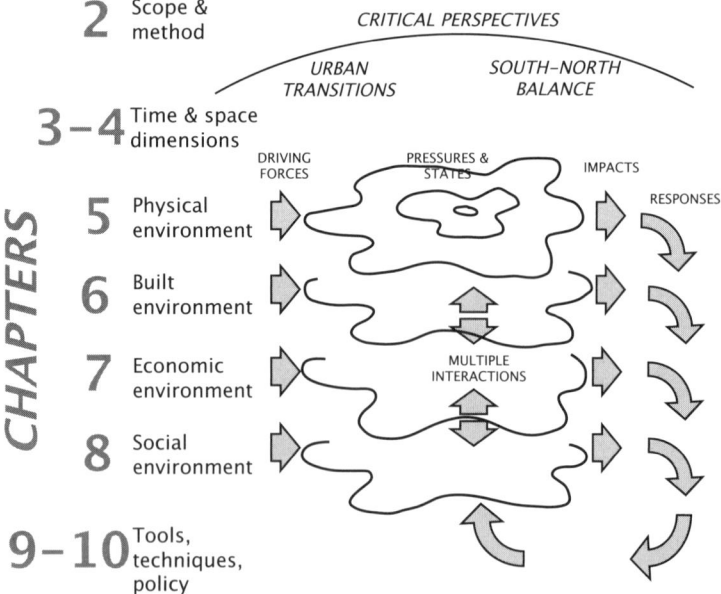

Adapted from DPSIR scheme, European Environment Agency 1996: Ravetz 2000

Figure 2.2 *Route map for the book*

a couple of boxes with an arrow pointing to the answer! However, this kind of route map can help in setting an agenda, exploring the nature of the problems, and unravelling complex lines of cause and effect between social, economic, urban and environmental factors.

Practical applications

There are plenty of examples which can be charted on this kind of route map. For instance, in Manchester, UK, economic growth generates transport demand which is fulfilled by more vehicles which create air emissions, lowering air quality and damaging public health, and widening divisions in quality of life. In this example, some alternative responses would be to regulate vehicle emissions, or set up public transport systems, or invest in 1 million urban trees, or relocate highways away from dwellings (Ravetz, 2000).

Another example, this time of a self-reinforcing 'feedback loop', concerns a conventional model of urban growth and decline. In a typical North American city, economic growth leads to increased car ownership, so fewer people walk or cycle, so street crime flourishes, which encourages people to move away, which lowers values and generates derelict land. In turn, this accelerates inner city decline and street crime, which reinforces the flight to the suburbs, and so encourages more car use, in a never-ending loop (Duany, *et al.*, 2000; Calthorpe, 1993). There are many policy responses which can start to address such a system-wide issue, but to be effective they must somehow get into this loop and start to turn it around.

What is a city? Defining urban systems

The very idea of a city is open to question, as introduced in Chapter 1. In the context of this book, cities are increasingly seen not as freestanding objects, but more as components within an 'urban system' – the question is, what is the nature and logic of that system? Such a question is increasingly challenging, as the economic and spatial functions of cities are shifting to other non-physical dimensions, such as cosmopolitan global networks and 'spaces of flows' (Soja, 2000; Sandercock, 2003; Wallerstein, 2004; Borja and Castells, 1997). It could be argued, for instance, that the executive lounges of airports around the world are more connected to each other, both functionally and culturally, than each airport with its nearest urban slum. However it is still possible to draw some simple lines around different levels in the urban system in Figure 2.2:

- first, there is an 'world urban system', external to any one city, describing the interactions between different sizes and types of cities as a kind of central spine of the global economy (Sassen, 2001);
- second, in larger countries there is a national urban system: this is both a geographic community of industrial, service and residential functions, and also the national policies and markets which provide the parameters for change;
- somewhere between national and local levels is the 'city-region system': this is a city together with its functional hinterland or bio-region, containing a labour market, water supplies, food supplies and so on (Sale, 1985);
- finally there is an 'internal urban system': a set of components and dynamics of cities and settlements within their built up areas, however these are defined. This could divide further into smaller and smaller districts, neighbourhoods, blocks and streets.

Within these layers, the scope of what is 'the city' is likewise open to many alternative views. There is a popular saying that 'the future is urban', which focuses on the mega-cities of over 10 million inhabitants, which are set to double in number in 20 years. In reality the bulk of the urban population will continue to live in settlements of between 25,000 and 500,000 people, although many of these are components of larger metropolitan areas, or 'polynucleated urban regions' (Hall and Pain, 2006). There are different ways to define such urban areas, in physical boundaries, economic activity or political units, and this is followed up in Chapter 3.

One crucial factor is the frequent mismatch and contradiction between different layers and types of city. Everywhere there are cities where most of the power and wealth is located outside the city boundary, and so pays no public taxes to the city. There is also the opposite case, where large populations live in peripheral suburbs or squatter camps outside the city boundary, working in the city but with few rights to property or access to urban services. There are many cities where the political structures are disconnected from economic functions and environmental patterns, and many other cities where social and cultural groups are also disconnected.

What is the environment? A systems view

The environment is clearly more than a collection of isolated pieces of soil or water – it functions and provides 'services' to human society, through the interactions and inter-dependency of all its components. This is the essence of the 'systems perspective', on which complex models can be built (Figure 2.3). Below are some key concepts, borrowed from ecological and bio-physical science (Kay and Schneider, 1994; Clayton and Radcliffe, 1996: Ravetz, 2000). These are illustrated with an example – a typical environmental programme in a northern city-region (Box 2.2):

- 'Stocks': the quantities of assets such as water or bio-mass, or the qualities of assets such as landscape value. Changes in such stocks could be measured with indicators, showing for example the gains or losses of urban green space.
- 'Flows': the throughput of resources such as materials, energy, people or money: for instance the flows of water or bio-mass.
- 'Patterns': the spatial or other forms of structure and self-organization, which determine functions and behaviours. For instance, the value of urban green space depends not only on its quantity, but on the pattern of its layout and design, its management, and to what extent it attracts crime or litter.

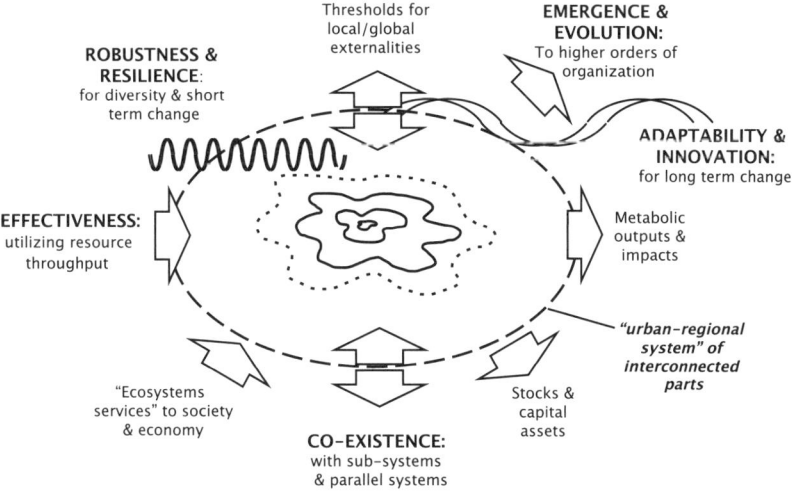

General characteristics of systems and their 'sustainability'.
Source: adapted from McLoughlin 1969: Bossel 1996: Ravetz, 2000

Figure 2.3 *Human urban environment – a systems view*

- 'Thresholds': certain limits on pressures which, when crossed, will change irreversibly the stocks, flows, patterns and function of the ecosystem or human-environmment system. For instance, if the urban open space is over-used by motor-cycles or ball games, it may lose its wildlife and vegetation.
- 'Resilience': the ability of an ecosystem or human-environment system to respond and adapt to shorter term pressure and external change, while maintaining essential structures and functions. The urban green space should be able to survive drought and other extreme weather, public events or vandalism.
- 'Adaptability': the ability of an ecosystem to respond and evolve to longer term change in its environment, while maintaining its integrity. Within the ecosystem there is 'succession' – the phenomenon of

---

Box 2.2

Greening of the Manchester city-region

Manchester, one of the world's first global industrial cities, created unprecedented levels of pollution and contamination in the winding river valleys which surround the city. A hundred and fifty years later, these were gradually transformed in a pioneering city-region-wide programme of reclamation and landscaping of country parks in the 1980s.

Since then, the strategic city-region authority was disbanded, and its 10 municipalities went through many stages of re-inventing ways to collaborate. Some country parks have been lost to business parks or motorway extensions. Some are in decline through neglect, vandalism or dis-investment, and there are problems with ecological invasions and adaptation to climate change. Others are highly prized by local communities and local groups, although there are conflicts between activities such as fishing, cycling, conservation, and between social groups. Some have become the focus of tension between the powerful property development lobby, and local residents who feel they are being pushed aside to make way for development and gentrification.

Source: Ravetz, 2000; Nicholson-Lord, 1987; recent material available on http://www.salfordstar.com

ecological cyclic change or transition from one kind of community into another. The urban green space in Manchester may be continuously changing and evolving, due to social change, climate change, urban pressures or environmental pressures, and its management – being part of that human-environment system – should be able to work with this.

What is the urban environment?

The 'ecosystems services' concept starts to make connections between the physical environment and its social or economic functions (Millennium Ecosystems Assessment, 2004). One approach is to look more closely at the environmental impacts of urban functions, at different scale levels of the city (further discussion is in Chapter 5):

● conditions and pressures on the local environment in urban areas: for instance, the quality of the urban green space and its ecology;
● conditions and pressures from urban activity on the surrounding city-region or 'hinterland' environment: for instance, downstream pollution as a result of the use of the urban green space;
● pressures on the external regional or global environment resulting from urban activity; for instance climate change impacts from urban energy consumption, which is influenced by the pattern of green space within the city.

A counterpart view looks at the urban and human impacts of environmental functions, again at different scale levels; again, following the urban green space example:

● environment and ecosystems as providers of resources for urban production or consumption; for example, the urban green space may provide water, shade or micro-climates;
● environment and ecosystems as providers of cultural assets: the green space may provide local amenity, aesthetic and cultural heritage for local communities;
● environment as a provider of absorption capacity for pollution and waste, and other environmental effects: the green space can provide space for floodplains and air filtration;
● environment as shaper of the urban social and economic structure of the city; for instance, the surrounding river valleys might be high amenity areas for the wealthy, or sink estates for the deprived.

What is the human urban environment?

The above issues produce a rather long list; the point here is not to build a catalogue for its own sake, rather to provide ways through for what are often tangled chains of cause and effect in the 'human urban environment'.

To summarize, the concept of the human urban environment covers all relationships and interactions between social, economic, political and cultural dimensions of the city; and the physical environmental conditions and pressures which affect them or are affected by them.

Following through the example above of the urban green space, there is a more complex narrative – about the use of the space for health or amenity, its value or maintenance costs, problems of vandalism on ageing trees, planning for climate adaptation, community involvement and urban regeneration, and many other issues.

If time allowed, then deeper themes would emerge – for instance, the political in-fighting which produced the green space; the ideological struggle between residents and the development lobby; the relationship of landowners and the planning system. This could also point towards local-global linkages, and the way in which outdoor leisure is driven by global media; or how Manchester's community allotments can draw inspiration from Asian market gardeners.

The implication is that the 'human urban environment' concept is a process of asking questions and exploring possibilities, not a fixed menu of problems and solutions. Local and global, physical and socio-economic, past and future, are all tangled up, in a rich picture of change and opportunity.

2.3 Global perspectives

Urban dimensions of the environment

Up to now the economic and social sciences have developed either with an urban focus, or with an environmental focus. There is surprisingly little which looks directly at the overlap and interactions between the two agendas – the particular issues, problems and solutions which relate to the human urban environment. So there is all the more reason for producing this book. Certain themes come to the foreground, which are different from those of conventional environmental studies or urban studies (Figure 2.3).

- Urban environmental impacts tend to fall mainly on the poor, while the wealthy generally have the means to relocate or guard against them; another view on this would see the poor being deliberately located in proximity to environmental problems. The 'environmental justice' agenda overlaps onto the urban poverty reduction agenda.
- Urban environments tend to concentrate and polarize environmental effects, such as traffic pollution or liquid effluent, to extreme levels which are not found elsewhere in nature. As above, such concentrations often impact directly on the most vulnerable communities and neighbourhoods.
- Many urban environmental impacts are the results of national or global supply chains, involving resource transfers, burden shifting, displacements and expropriations from elsewhere. For instance, industrial cities have struggled for centuries with the impacts of fossil fuel burning and human waste disposal; now, these burdens and impacts are often shifted elsewhere.
- Within the city and city-region there are also 'externalities', physical displacements and cumulative effects; for instance, deforestation in the river catchment upstream, magnifies the flood risk downstream, and the pressures of urban development and rapid population growth then locate communities directly at risk on the floodplain.
- Urban environmental quality becomes a major factor in the urban economy, influencing property values, regeneration potential, location choice, inward investment and competitiveness, as well as residents' local quality of life.

Urban environments in a global system

In the search for the causes of urban environmental problems, the trail often leads 'upstream' back towards the system of economic production, consumption and trade on a national or global level. This can be summed up with a simple picture of wealth and poverty – the poor generally suffer the bulk of environmental impacts, caused by the wealthy, although

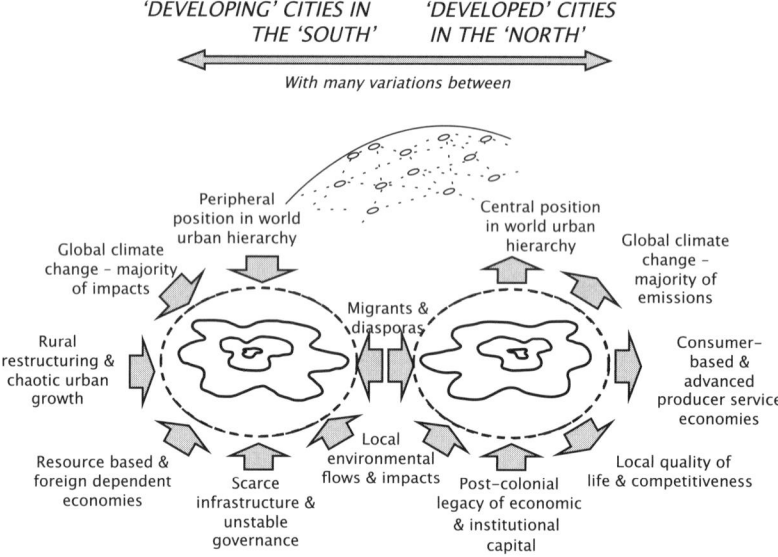

'DEVELOPING' CITIES IN THE 'SOUTH'   'DEVELOPED' CITIES IN THE 'NORTH'

With many variations between

Peripheral position in world urban hierarchy
Central position in world urban hierarchy
Global climate change – majority of impacts
Global climate change – majority of emissions
Rural restructuring & chaotic urban growth
Migrants & diasporas
Consumer-based & advanced producer service economies
Resource based & foreign dependent economies
Scarce infrastructure & unstable governance
Local environmental flows & impacts
Post-colonial legacy of economic & institutional capital
Local quality of life & competitiveness

A conceptual picture of dynamics of urban environmental change in North & South: based on Ravetz, 2006b

Figure 2.4 *Human urban environment – local and global*

the wealthy might not escape entirely. This can be further simplified to a caricature of the 'North' and the 'South' (in this book the 'North' is interpreted as the member states of the Organization of Economic Co-operation and Development [OECD]: see the notes in Section 7.2).

Generally the more affluent cities of the 'North' import increasing amounts of materials and products from overseas, where regulations, wages and labour standards are often lower. While the 'North' has more organized infrastructure, the sheer volume of consumption, together with new technologies, still causes urban environmental problems, such as transport emissions and food chain toxins. Meanwhile, standards and expectations are continuously raised, as environmental quality is seen as essential to competitiveness, property values, risk reduction and quality of life.

In contrast, the less affluent cities of the 'South' are generally driven by global markets to the lowest production cost and lowest environmental standards, with industries which are highly polluting and hazardous to their workforce and their surroundings. Public services and infrastructure are often disorganized, under-funded, struggling with overseas loan conditions, and in the grip of corruption or organized crime. As a result of chaotic change in rural areas, the city structure cannot cope with rural in-migration, which produces massive unplanned and un-serviced settlements, in locations at high risk of flooding, landslides and contamination.

In reality these caricatures are grossly simplified, and all cities contain both affluence and poverty in varying proportions; in particular the rise of the middle classes in rapidly growing nations such as India and China is changing the whole picture. However there is a structural difference between opposite ends of the spectrum. At one end is a typical North American or Western European city with a per capita income of $35,000, and, using the metric from Chapter 1, an ecological footprint of over 10 global hectares per person (details in Chapter 5). At the other end is a typical Sub-Saharan city, with a per capita income of less than $350, and a footprint of 0.5 global hectares per person. Ironically, the USA is both the richest and the most unequal large nation in the 'North', and the one with the lowest life expectancy.

Cities in the 'North' and 'South'

Some would ask, what has the 'green agenda' of the affluent 'Northern' cities got in common with the 'brown agenda' of 'Southern' cities, separated as they are by income differences of up to a hundred times? Most 'Northern' city dwellers have little experience of seeing children die from easily preventable diseases. Most of the literature is focused on either one side or the other, and this book talks simplistically about cities in the 'South' or 'North'. In reality this crude division is more of a spectrum, rather than two opposites.

First, many 'Southern' cities are on a trajectory which is heading rapidly towards the standards and conditions of 'Northern' cities. If they are not yet experiencing many of the problems associated with affluence, given the pace of global economic growth, it is likely to be a short time before they do. The concept of the 'urban environmental transition' (discussed in Chapter 3) describes this effect. While it took cities in the 'North' several centuries to make the current transition, in many cities of the 'South' the transition from the industrial revolution to the present is compressed into just a few years.

Furthermore, there is no clear economic or social dividing line between either end of the spectrum, as the cities of many transitional countries in Latin America and Asia are catching up rapidly with the affluent 'North'. In many ways it makes more sense to talk about a 'world urban system' than a clear split between 'North' and 'South'.

Meanwhile, there is increasingly direct interaction between 'North' and 'South' through immigrants and migrant workers, where kinship networks and business networks are maintained through international transport and telecommunications. There are many underlying contradictions and problems of the cities on both sides – such as social exclusion, cultural alienation, and community dis-investment – where exchange of experience and mutual learning is valuable.

Finally, each type of city is clearly inter-dependent on the other, if indirectly, through trade. Many of the consumer goods which pack the shopping malls and then the waste landfills of 'Northern' cities are generally produced by low cost labour in 'Southern' cities, often in extreme environmental conditions. The 'poverty of consumption' seen in the mental illness and community breakdown of the 'North' is indirectly implicated and responsible for the 'poverty of production' seen in the 'South', although this may not always be obvious to either side.

2.4 Social science perspectives

Following through this sketch of the 'world urban system', and its environmental agenda – the question arises, what have established disciplines such as economics, sociology, or political science got to offer? What is presented here is a sketch and overview, which provides a rapid tour of most of the relevant branches of the social sciences and their contributions to the human urban environment agenda. These are summed up in Figure 2.5.

Politics of the urban environment

City government, and its capacity for wider forms of 'governance', is the starting point for any response to problems in the human urban environment (as raised in Section 1.7 and explored further in Chapters 8, 9 and 10). Whatever their political colour, most city administrations are

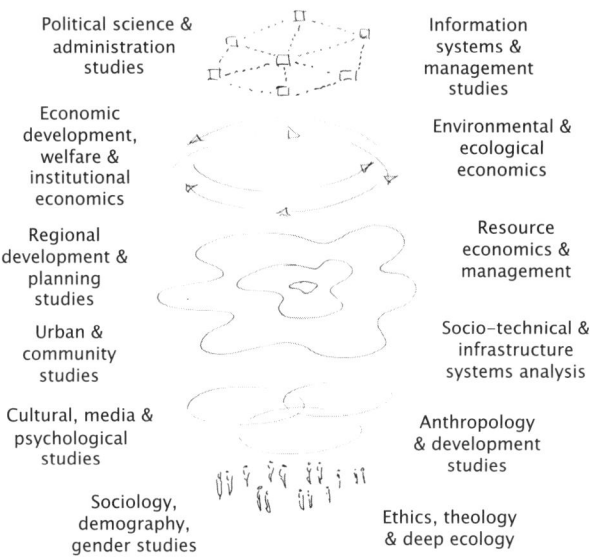

Political science &
administration
studies

Information
systems &
management
studies

Economic
development,
welfare &
institutional
economics

Environmental &
ecological
economics

Regional
development &
planning
studies

Resource
economics &
management

Urban &
community
studies

Socio–technical &
infrastructure
systems analysis

Cultural, media &
psychological
studies

Anthropology
& development
studies

Sociology,
demography,
gender studies

Ethics, theology
& deep ecology

Figure 2.5 *Human urban environment – social science approaches*

now in a transition to a new and uncertain world. Municipal governments were generally established on a Fordist model of centralized public services, with clear structures and channels for representation and decision-making:

- privatization or liberalization of essential city services such as water, drainage, electricity, transport and security: generally a shift of government roles from universal 'provider' to selective 'client' or 'enabler', working through diverse alliances and partnerships;
- a new agenda for economic development, with many agencies and alliances from business and others; together with new agendas for social welfare, using the private sector and third sector as service providers;
- new approaches to planning and management by objectives, performance monitoring and single-purpose agencies, and increasingly supported by information and communications technology (ICT), security and surveillance techniques;
- however in many countries there is a 'democratic deficit' in the effectiveness of urban managers, elected representatives and the

willingness of the population to vote for them. This may be due to alienation and distrust of politics in more affluent cities, or crude repression in military states.

Many city governments now have reducing control or influence on the essential services and levers of the urban system. At the same time, they remain the natural stewards of their territory, even while urban communities are increasingly globalized and networked. A wider set of institutions for urban governance would include the following:

- The resurgence of 'new social movements', citizen-based organizations, and non-governmental organizations (NGOs), generally under the umbrella of the 'third sector' (Etzioni, 1996; Castells, 1983). These organizations can often be more flexible and responsive in filling the gaps, mobilizing potential, revitalizing politics, and reaching parts of the population which traditional municipal government cannot. The rise of global environmental awareness is due in part to the actions of large NGOs, such as Greenpeace or the World Wildlife Fund (WWF), with businesses and governments following their lead.
- On a wider front, the role of the 'third sector' or civic society is seen as a counterpart to the public sector (government and policy) and private sector (business). The 'third sector' includes almost anything which is neither part of formal government or profit-making business. Depending on the context, universities and hospitals, trade unions and arts centres, as well as voluntary sector charities and community groups, may all be third sector organizations (Gorz, 1980; Offe and Heinz, 1992).
- 'Citizen participation' is a kind of meeting place between protest groups and the urban management regime now seen as the key to increasing the responsiveness and effectiveness of city government, particularly in urban planning and housing. It is also apparent in the local environmental agenda, where open space management, wildlife conservation, waste recycling and many others are all more successful and long lasting when embedded in local community action (Hamdi, 2004). The participation principle is growing rapidly as local governments see the benefits of agency structures with wide ranging partnerships (Gilbert, et al., 1996).

Underlying these broad trends are some parallel approaches from recent political science thinking:

- 'Regime theory' looks at how political growth coalitions are constructed and fought over (Logan and Molotch, 1987). This emerged from experience in cities in the USA, where there were particularly fierce conflicts around labour, housing, transport, urban renewal and race issues.
- 'Regulation theory' looks at how city administrations can unwittingly accelerate social divisions and segmentation, in terms of managing access to housing, jobs and services for poor or ethnic communities.
- 'New public management' emerged in the 1970s with an apparently scientific approach to public policy. This has been summed up in ways such as evidence-based policy, or zero-based budgeting, where financial spend has to be justified by its outputs and outcomes. There is also much management jargon around, such as the SMART approach – 'strategic, manageable, achievable, responsive and time-bound' (Stewart, 1989).
- 'Discourse analysis' looks at how environmental agendas, problems and solutions are often constructed as part of the 'hegemonic project' or strategic aims of an elite ruling coalition. This perspective comes from a wider concept of 'communicative action' which analyses how arguments are made, ideological power constructed, and decisions reached by different social groups (Hajer, 2003; Healey, 1997a; Habermas, 1992).

Political science perspectives are of course highly sensitive to the context – it is easy to assume the neo-liberal Western model of market democracy from a comfortable university in the 'North'. In reality, there are alternative approaches from other parts of the world; from less developed countries (LDC) and highly indebted poor countries (HIPC); transition countries and Newly Independent States of the former Soviet Union (NIS); India and China with 40 per cent of the world's population; from small island states and the First Peoples in former colonial territories; and from the Islamic and Arabic states (Said, 1978).

Economics of the urban environment

The economic aspects of the human urban environment are explored in Chapters 7 and 9, so again what follows here is a brief tour of methods.

'Welfare economics' looks at the distribution and re-distribution of benefits and costs between social groups, some of which may be measured by money. One of the first applications of welfare economics was on the question of pollution. A classic example is where one group's benefits from industrial production generates 'externalities' from a smoky chimney which cause another group's disbenefit in pollution and health impacts (Pigou, 1952). Welfare economics has looked at such urban questions as transport and housing, where there is a combination of public and private money involved, with winners and losers in each case.

'Institutional economics' has a wider focus – the economic interactions of organizations, management, labour, consumers and so on. One of the key concepts is 'transaction costs' or the friction of trade, another is that of property rights and responsibilities of ownership or stewardship (Williamson, 1985). This can be applied to problems such as the economics of poverty, where poverty is implicated as both cause and effect in the human urban environment. The applications to environmental questions are just now emerging (Jacobs, 1994; Paavola and Adger, 2005).

'Environmental economics' seeks to measure environmental resources and impacts through the medium of money, using various methods to construct shadow markets to assign money values for rivers, forests, landscapes and so on. It is generally based on neo-classical economic theory, with the logic that financial costs in cleaning up pollution need to be considered in that frame. However in dealing with non-monetary values, it often stretches the theory to its limit, and care should be taken to avoid spurious results (Common and Perrings, 1992).

'Ecological economics' also looks at the interactions of economy and environment, resources and institutions, but without trying to reduce all impacts to the common measure of money. It sometimes grapples with very challenging questions, such as the value of a human life, or the assessment of human happiness (Costanza, *et al.* 1997; Edwards-Jones, *et al.* 2000).

'Business-environment' issues look at the effect of industry and services on environmental impacts and resources: the environmental management systems, corporate responsibility, supply chains, technological innovation and the consumer culture which may support or hinder progress. This overlaps strongly with 'industrial ecology', which looks at clustering of industries so that one firm's waste is another's raw material. More widely it seeks the potential for reducing environmental impacts through managing product supply chains, both on supply and demand sides (Ayres and Simonis, 1997).

'Ecological taxation reform' (ETR) is the logical response to the argument that most countries tax labour at much higher rates than environmental pollution, and therefore that pollution taxes should be raised to correct 'market failures'. In practice every form of taxation generates winners and losers, and there are questions not only of who and when to tax, but of what to do with the proceeds, and of the many side effects for different social groups. On closer inspection many of the most polluting urban activities – energy, transport, agriculture and so on – are the most embedded in the politics and economics of the city, with a complex array of subsidies, tax breaks, quotas, concessions and others (O'Riordan, 1996).

The greening of international trade and finance concerns the conditions of trading and investment between cities and nations. This then informs the movement for Socially Responsible Investment (SRI), and the many arguments on the reform of world trade and financial flows – these may appear to be distant from the urban agenda, but in many ways they may be the most effective action for any city to take.

New economics focuses on the other end of the scale, on the local and social economy. This includes patterns of non-monetary trading or reciprocity in neighbourhoods, between social or kinship networks. It also looks at the localized level of economic activity, and observes how takeovers, buyouts and concentration of wealth in international firms can often generate unemployment and dependency in an urban economy (Douthwaite, 1995).

Sociology of the urban environment

The cities of Western Europe and North America provided the platform for the study of social groups and patterns, the starting point for sociology as a discipline (Weber, 1922; Veblen, 1899). The impacts of the urban environment on people and communities are influenced by the divisions of class, gender, race, ethnicity, age, disability and so on. Similarly the urban environment is partly constructed and maintained through public services, such as health, education, and crime prevention, which also define the identities of urban communities. Many of these issues surface in the critical perspectives in the next section, and others in the lifestyle and community agenda of Chapter 8, so below is a brief summary.

Poverty issues are the starting point for the human urban environment. In the cities of the 'South' the poor die earlier due to factors such as the cost of public water supplies; while in the cities of the 'North' the poor die earlier due to traffic emissions and other externalities. In the 'South' there has been growing awareness of the multiple and compound factors of urban poverty, and of the coping strategies of urban dwellers, and of the pressures in rural areas which make cities more attractive, and of the complex reciprocal roles between urban and rural areas. In the 'North' urban poverty is seen as an object for regeneration, which may have the effect of improving the place while dispersing the people; while there is emerging across Europe a more comprehensive approach to fostering 'sustainable communities' (further discussion in Chapter 10).

One of the most far-reaching global trends is the re-balancing of gender relations, with many implications for environmental issues – for women who cook in unsafe kitchens, or who are cottage workers with no security. Children and youth issues surface in the 'South', where in some areas half the population is under the age of 20, and where children are most at risk from pollution, waste, poor sanitation and household hazards. In the 'North' there is an opposite problem where many children are over-protected, obese, driven to school and entertained by ICT. Meanwhile there is growing fear and paranoia about 'environmental security' and civil disorder exacerbated by a widening generation gap. In both 'North' and 'South' younger people with few prospects tend to construct their identities through drugs, fundamentalism, gang warfare and other forms of disorder.

In contrast are the issues of demographic change in general. Across the 'North' there is an emerging 'third age' – a bulge in the post-retirement population – a result of lowering rates of mortality, improved healthcare and increasing personal incomes. This is significant for urban environment campaigns as the elderly are often less transient, and more able to engage in social movements, voluntary activities and lobby groups. At the same time they are more vulnerable to urban environmental hazards such as air and water pollution, noise and disorder, and the extreme effects of climate change.

The themes of 'environmental justice' and 'environmental property rights' are powerful drivers of change, influencing governance, the corporate social responsibility of business, and changes in personal lifestyles and consumption patterns (Westra, 2008; Roszak, 1993). This can affect controversial issues such as the siting of roads and landfill sites, or the conversion of 'First Peoples' reserves to tourism parks. Underlying this are the fundamentals of property rights and responsibilities, and tenure systems in land and buildings, which identify in legal terms the environmental responsibility between neighbours, and between individuals and the city. These are then tangled with ethnic cultures and alternative debates on modernization: the current tension between the Islamic world and the neo-liberal Western consensus is possibly a marker of deeper schisms in a post-colonial world (Said, 1978). These surface in the issues of the urban environment, where problems and values are differently perceived, where civil society is differently organized, and where external neo-liberal solutions tied to international aid may be counter to local cultural practices.

Cultural issues in the urban environment

Cultural issues cover an even wider agenda – patterns and systems of thought and belief, which underpin human activities and relationships. These are often concentrated and magnified in cities, the meeting points and melting pots for new cultural movements.

Ethical issues and a moral philosophy approach underlie much of the modern environmental movement. Environmental justice campaigners show regularly that the urban poor live next to landfills and major roads, and many residents become involved with political activism through local environmental campaigns. The movement of Local Agenda 21, which followed the World Summit on Sustainable Development in Rio in 1992, depended on an ethical foundation which highlighted the links between local actions for global benefits.

However, few things are simple: the ethical approach can often become very tangled, for instance when dealing with nature conservation and wildlife management. Many of the national parks in Africa and Latin America have become attractions for wealthy urban tourists, which generates the income needed for protection and management but at the cost of driving out native inhabitants and their livelihoods.

The Corporate Social Responsibility (CSR) ethos which drives (some) businesses towards environmental and social improvements throughout their supply chains depends very much on an ethical foundation. There is increasing awareness that organizational change is not only a matter of management structures, but of corporate ethos and culture which works at all levels. It also sees markets and consumers as driving this, at least as much as producers and distributors.

Psychology and cognitive science approaches start with the formation of behaviour, and the potential for change at the individual level. This might consider, for instance, the phenomenon of car addiction in transport policy; the need for recycling in waste management; or the definition of thermal comfort in housing. The 'hierarchy of human needs' looks at behavioural desires and aspirations, finding that people often consume material goods or mobility as a substitute for more meaningful personal fulfilment (Ekins, 1994). This is very useful in highlighting the scope for more sustainable production and consumption, as it seems to suggest that reducing the material demands of affluence might follow from increasing social and cultural cohesion in the city.

Urban anthropology is more concerned with the combined patterns of belief, experience, incentives and perceptions which underpin the patterns of choice and behaviour (Hannerz, 1980). In western cities this can be interpreted as 'structures of feeling' or analysis of the patterns of everyday life and experience among different social groups in the city (Taylor, *et al.*, 1996).

Finally the role of religions and spiritual belief systems may underlie each of the above. As some traditional forms of religion decline, there is a resurgence of other more fundamental Christian-based and Islamic-based faiths in many parts of the world. Churches and mosques can play a powerful role in urban areas, in regeneration of the urban environment, and in the faith-based frameworks which can motivate corporate social responsibility and social change.

Socio-technical issues in the urban environment

There is an emerging agenda for studies of 'socio-technical systems', meaning the combined interactions of consumers and producers, technology and management, markets and regulations, and innovation or transformation (Randles and Green, 2006; Ravetz, 2006b). This has obvious significance for urban environments, as shaped by pipes, wires, networks; and also as shaped by the flows of information and intelligence which are made possible by such technologies.

- Infrastructure and networks as institutions: this institutional approach looks at the provision of services such as energy, water, transport and telecommunications. These were traditionally an area for engineers and economists, but not explored by other social scientists. Now, there are new insights on the effect of private or public networks on the cohesion or fragmentation of cities and urban spaces. The issue of 'spatial exclusion' due to inadequate public transport, is both an engineering and a social science agenda (Graham and Marvin, 2001; Borja and Castells, 1997).
- Knowledge management, and information and communications technology (ICT). The influence of ICT continues to be profound, with the advent since 1990 of the internet and mobile telephony as major forces in the economy and civil society (Castells, 1996; Mitchell, 1996). While there appear to be boundless possibilities for organising and communicating data, there are also huge implications for privacy and surveillance, for misuse and sabotage, and for a widening gap between the insiders and the 'digitally excluded'.
- 'Science in society' is an agenda for examining the role of science and scientific evidence for policy and business. Many of the most important scientific topics – climate change or genetically modified organisms – are deeply contentious, with many competing claims by lobbyists, NGOs, the media and intermediaries. Even apparently straightforward questions, such as the impact of urban air pollution on human health, turn out to be tangles of evidence and interpretation. For such scientific agendas, where the risks and stakes and uncertainties are all very high, a new approach of 'post-normal science' is needed which actively builds social awareness, transparency and informed consensus (Funtowicz and Ravetz, 2002).

2.5 Critical perspectives

This leads to a central feature of this book – the *critical perspectives* on the human urban environment. Critical perspectives come in the form of controversial debates, unfinished agendas and theoretical battlegrounds. They tend to cut right across conventional textbook fields of economics, sociology, political science and others.

Why? Because it is easy to produce simple analysis – for instance of economic price and demand functions – but often not simple to relate these to real-world situations. Likewise there are many simple solutions around – more green space or better planning – but such solutions often don't seem to work. Almost every good idea to improve the urban environment seems to come with a host of problems and barriers – lack of funding, corrupt government, faulty technology, tax evasion and others. The net result can be seen in the overall indicators of environmental performance of cities in the last few decades – any improvement in efficiency is outweighed by increases in material throughput and consumption. Experience shows that while such real-world problems need good economics, sociology, political science and so on – they also need more cross-cutting and responsive approaches.

Contested logic and conflicting discourse

So, the discussion throughout this book follows some of the key critical perspectives, from globalization to social exclusion. This is by no means a complete list, but more like a 'starter' set, setting up debates, mapping the landscape, and fitting some of the pieces of the puzzle together. Further discussion is in Chapter 8.

Each of these themes is 'critical', in the sense that they do not accept existing structures of power and ideology as inevitable and fixed – rather, they may critique and argue for change (Calhoun, 1995). They are also critical in the sense that they do not provide simple ready answers – rather they offer a lens to look through, a way of asking relevant questions. Each is built upon underlying themes of ideology and power; and each revolves around the practical issue of working out what is really going on, and what can be done about it.

For almost every urban environmental issue there are profound questions. Why are simple technical problems so difficult? Why does pollution continue, and land fall into dereliction, and the poor continue to breath the emissions of the wealthy? Why is so much urban environmental policy constructed for a fictional world where taxes are paid and decision-makers govern wisely? Why do the curvaceous graphs of environmental economics ignore the reality of exclusion and expropriation? And how can society tackle real conflicts, where the economics, politics and sociology are all pointing in different directions?

For such real-world questions there are no easy formulae, blueprints or textbook answers. Instead, critical perspectives help to explore the questions as a live and urgent debate between policy-makers, academics, businesses, NGOs and so on. There are also deeper underlying agendas – explorations of the nature of power and ideology in a capitalist system – which inform and stimulate these critical perspectives. It is not easy to visualize such perspectives, but the diagram shows some pointers (Figure 2.6).

Key critical perspectives

Globalization is a dominant force in cities around the world. It takes economic forms in the structure of business and finance; political forms in the regulation of trade; and cultural forms through media and information and communications technology (ICT). Globalization also affects workforces, family and kinship networks, tourism and leisure. There is also a counterpart strand, sometimes termed 'localization', where the cultural identities of people and places are re-interpreted and re-constructed at the local level. For the urban environment, analysis of globalization is essential to understanding not only the conditions in cities, but the forces which produce them (Sassen, 1994; Soja, 2000).

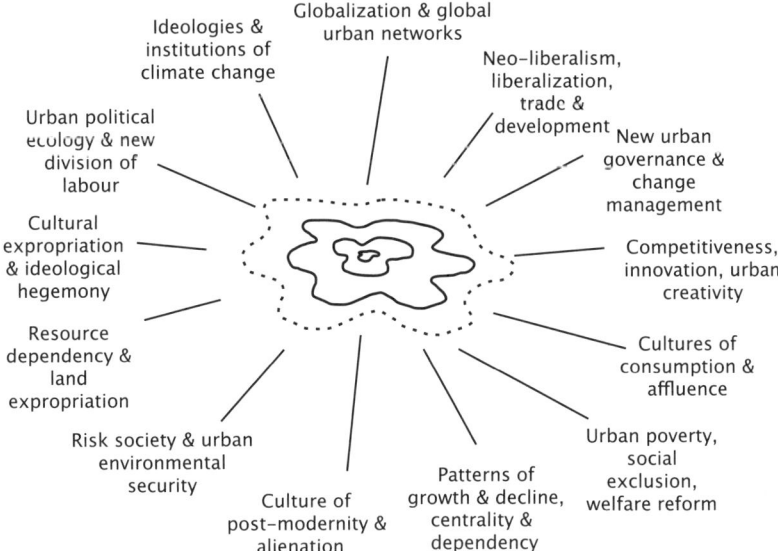

Figure 2.6 *Human urban environment – critical perspectives*

Alongside this is the paradigm of 'liberalization' of trade, finance and public services. In many cities this often involves privatization, franchising, cost recovery, and market segmentation into profit-making and other sectors. Liberalization has been forced on many developing nations and cities through the process of structural adjustment, driven by loans and investment-funding packages (Harvey, 2005; Stiglitz, 2002).

A culture of consumption and affluence in both wealthier and poorer cities is a major driver of individual and cultural identity and community. This then drives new perceptions of urban environments, particularly through leisure and tourism, and thereby the investment or dis-investment which takes place. 'Sustainable consumption' is a response to this, an agenda for reducing the environmental impacts of affluence to more responsible levels, while aspiring to increase quality of life.

This leads to the emergence of a 'risk society' which is driven and structured by uncertain risks from environmental change and different social groups. This produces both the need for security against crime, civil disorder and perceived risks of migrants; and the importance of trust between citizens, organizations and governments (Beck, 1995). Often the response is a defensive partitioning of the cities into gated communities with video surveillance. But in turn there is a new positive emphasis on the 'liveable city' and the 'sustainable community': the public realm and the priority for open space, safe public transport and a multi-cultural community (Roberts, 2008).

The agenda for the human urban environment starts and ends with poverty; the urban poor receive the majority of environmental impacts, while even the rich do not entirely escape. In reality the problem of poverty is not so much a simple lack of income, but a cumulative causation of stress coming from employment, housing, education, health, as well as environmental factors, all adding up to what is now termed 'social exclusion'. In response there are many solutions in both 'North' and 'South', from community participation and empowerment, to asset and livelihood approaches, and micro-finance schemes (Yunus, 2008).

The cycles of growth and decline, and the geography of 'hubs and peripheries' are well known. It is now emerging that the quality of the urban environment is instrumental in the competitiveness of places, the investment flows and values of different parts of cities; the technological

Box 2.3

Managing change in mega-cities

Every notion we may have about planning and architecture evaporates here. What do you do about cities with over 10 million inhabitants? What do you do about cities that threaten to swell into metropolises of 25 million inhabitants (Sao Paulo and Rio de Janeiro)? What do you do about cities that were planned for a few hundred thousand people but within a few decades have 2 to 3 million inhabitants? You cannot do them justice with "normal" planning or "normal" architecture. That would suggest that the contemplative slowness of the plan or design would work here. In Brazil, action is chronically overtaken by events. No time for consideration, no time for reflection. That's a European luxury, but here every municipal organization is powerless against the proliferation of the city. All that can be done is to keep things under control. Urban planning becomes a matter of policing rather than a political or cultural discipline.

Source: Bosch *et al.*, 1999

innovation of certain sectors and the links to different forms of urban networks and infrastructures.

In response to the agenda for 'change management', new modes of urban governance are emerging in cities around the world. This can be seen with rational management approaches, alongside more entrepreneurial governance, with partnership between municipalities and other stakeholders. There is also priority given to citizenship and civic or corporate responsibility, and more intelligent and responsive ways of doing environmental planning and management. The example from São Paulo in Box 2.3 shows the challenge of managing change in a dynamic situation, which forces a rethink of the nature of city planning.

Underlying agendas – power and ideology

There is a rich seam of debate in the social sciences, particularly in the 'post-structuralist' vein from the 1960s onwards, which explores the social construction of reality, power and ideology. There are many applications of this to urban studies, and many to environmental issues, but few as yet which explore the crucial meeting point of the urban environment. The urban environmental policy agenda in cities around the world, apparently well-meaning, can be deconstructed in terms of ideology, institutions, power, and other more subtle forms of influence and expropriation (Hajer, 2003; Brand and Thomas, 2005; Heynen, et al., 2006).

The underlying agenda of power, and its applications in reproducing ideology and ideological dominance, is explored by Michel Foucault (1991). This debate sees ideological power as an often tacit and invisible guiding of social structures and institutions, in contrast to previous 'top-down' political or military power structures. This helps to understand how the urban environmental agenda is formed, by which power coalitions, and for what reasons, and how it may then be an arena for ideological conflict.

A contrasting approach is focused on communications, and the 'communicative rationality' which is possible to construct between different social groups with different values and worldviews. This approach is followed through into urban issues; the 'communicative planning' approach aims at practical applications based on local democratic debate and deliberation (Habermas, 1992; Healey, 1997a).

In practice cities are not abstract things but highly complex structures in space and form. The role of physical space in reproducing capitalist structures of power and ideology is explored by Henri Lefebvre (1991). There is a powerful argument on the nature of centres and peripheries, and on the fragmentation and reconstruction of urban space to suit a capitalist ideology and social hierarchy.

The context to all this might be summed up as the condition of 'post-modernity' – a kind of umbrella phrase which contrasts with 'modernity', or the generalized trust in political institutions, technological progress and social order. In contrast, post-modernity describes a world of individualism and privatization, breakdown in social order, and paranoia from environmental risks (Harvey, 1995; Amin, 1994). The former British premier Margaret Thatcher famously said 'there is no such thing as society', and in many ways she was reflecting the mood of the times (*Woman's Own*, 31/10/1987).

For instance, the case study below shows how urban questions – 'how can city-dwellers get their water?' – can intersect with environmental questions – 'why is the groundwater so depleted?' – as well as economic questions of who pays; political questions of who decides; and social questions of who benefits.

Example of critical thinking

This example is drawn from a topical and universal issue – the provision of water supplies for urban and rural areas (Box 2.4). As in later chapters, we can see how critical perspectives can help to highlight a complex problem and challenging solution.

- A 'globalization and neo-liberalism' perspective observes the problems caused by imported engineering, and by pressure for cost recovery.
- A 'consumption' perspective observes the benefits of integrating consumers into the supply chain to facilitate demand management and responsible use.
- A 'risk' perspective focuses on the problem of falling resource levels, and aims to rebuild the citizenship and social capacity needed for increased security.
- An 'exclusion' perspective focuses particularly on the problems of the poor and marginalized communities, and designs a system around their ability to pay and be involved.
- A 'governance' perspective sees the problems caused by a technocratic management and the opportunities in a more decentralized and participative mode of organization.

Towards a theory of the human urban environment?

Is there an overall theory which brings all these together? The classic theories of urban geography focused on economic and social interactions, the urban environment agenda introduces another layer of complexity, and the critical perspectives take it to a higher level of debate. It is not the job of this book to write such a theory, even if this was possible. But we can make some guesses at the direction it might follow, to be explored in later chapters.

Much current thinking comes with a sense of fragmentation, division,

---

Box 2.4

Urban and rural water – critical perspectives

The public utility system of Tamil Nadu in Southern India, providing water to 32 million people, had many problems – failed monsoons, over-abstraction by farmers, and illegal connections – and the results were groundwater scarcity and financial crisis. A major review was held with full community consultation, including input from women and marginalized groups. A new strategy set out to change the employees' attitudes and incentives, consult client communities, and safeguard or revive traditional water sources. This programme was piloted with several hundred villages, with the result that costs were reduced by 40 per cent, while raising funds for further investment.

This example shows a typical problem caused in part by previous 'technocratic' solutions based on heavy engineering and over-use of natural resources. Solutions can often come through more sustainable forms of governance, communications and management. Although this example concerns water in a developing nation, there are lessons for developed nations in other sectors such as local transport or waste management.

Source: Nayar, 2006 (UNEP, 2002a)

---

disconnection and divergence – between cities around the world, between social and cultural groups, between public and private sectors, and so on.

At the same time there is a resurgent theme of integration, re-connection and universality, in terms of the one world, local-global, networked village and citizenship agenda, for sustainable cities, multi-level governance and the ideals of Agenda 21. For instance, the Make Poverty History campaign of 2005 – whether or not it is seen as a real success or a media stunt – linked together events located in several hundred cities around the world, with over 1 billion viewers in real time.

The scope of an underlying concept is under development at the time of writing. It focuses on the cross-cutting themes of connectivity and inter-dependency for sustainable development, which can be traced in many aspects of governance, markets, technologies, communities and cultures (Ravetz, 2008b).

2.6 Conclusions

This chapter has aimed at a challenging task – to map out the mental landscape and a practical set of tools to tackle the human urban environment. This involves looking at cities from the 'North' and 'South', and at scales from the local to the global. It focuses on the interface between the physical urban environment, and the socio-cultural-political-economic patterns and processes which shape it, and are shaped by it.

Naturally the human urban environment agenda is complex and many-layered. There are problems and opportunities to be found in every community, in every city, in every sector, for almost every environmental effect, at every scale in time and space. Generally these problems are compound bundles of causes and effects, cutting right across economics, politics, engineering, biology and so on. They also raise some controversial debates and unresolved questions, which are highlighted by the critical perspectives.

Overall, this book is but a brief outline of a large and growing territory.

Who will use this knowledge, and for what? This depends on the role and task in hand, whether as researchers, policy-makers, entrepreneurs or other stakeholders in the urban environment. As explored in chapters 9 and 10, the key is the way in which such knowledge can be mobilized, for better understanding and better actions.

Further reading

Badcock, B. (2002) *Making Sense of Cities: A Geographical Survey*, London: Arnold and New York: Oxford University Press.

Benton-Short, L. and Short, J.R. (2007) *Cities and Nature*, London and New York: Routledge.

Clayton, A. and Radcliffe, N. (1996) *Sustainability: A Systems Approach*, London: Earthscan.

George, C. and Lee, N. (eds) (1998) *Environmental assessment in a developing world*, Edward Elgar.

George, C. and Kirkpatrick, C. (eds) (2007) *Impact Assessment and Sustainable Development: European Practice and Experience*, Cheltenham, UK and Northampton, MA: Edward Elgar.

Girard, L.F., Forte, B., Cerreta, M., de Toro, P., Forte, F. (eds) (2003) *The Human Sustainable City: Challenges and Perspectives from the Habitat Agenda*, Aldershot: Ashgate.

Harvey, D. (1995) *The Condition of Post-Modernity: An Enquiry into the Conditions of Cultural Change*, Oxford: Blackwell.

Pacione, M. (2001) *Urban Geography: A Global Perspective*, London and New York: Routledge.

Ravetz, J. (2000) *City-Region 2020: Integrated Planning for a Sustainable Environment*, London: Earthscan. (Chinese version: Chan's Publishing, Taiwan, ISBN 957-705-276-2).

Roberts, P. and Sykes, H. (Eds) (2000) *Urban Regeneration: A Handbook*, London: Sage.

Soja, E. (2000) *Postmetropolis: Critical Studies of Cities and Regions*, Malden, MA and Oxford: Blackwell.

United Nations Human Settlements Programme (2004) *The State of the World's Cities 2004/2005: Globalization and Urban Culture*, Sterling, VA and London: Earthscan.

Sources of information

UN Habitat or UN Centre for Human Settlements: the main international agency of the UN concerned with human settlements   http://www.un-habitat.com

Best Practices Database of the UN Habitat   http://www.bestpractices.org

Urban Indicators Programme (UIP) and Global Urban Observatory of the UNCHS   http://ww2.unhabitat.org/programmes/guo/guo_indicators.asp

World Urban Forum: biennal meeting of local governments   http://www.unhabitat.org/categories.asp?catid=41

European Sustainable Cities and Towns Campaign and Aalborg Charter   http://www.ndparking.com/sustainable-cities.org

Allianz Knowledge Partnership   http://knowledge.allianz.com/

Homes and Communities Academy   http://www.ascskills.org.uk

3 Future city
Urban environments in transition

3.1 Introduction

The story of cities is in many ways the story of material civilization. Early cities were shaped and formed by their environment, as providers of food, water, shelter and transport. The industrializing city then began to widen that marketplace to a global scale, and the post-industrial city begins to change the marketplace itself.

This is the theme of the 'urban environmental transition', the changing structure and metabolism of cities as the combined result of economic, technological, social and political changes. With the growing impact of urbanization on the global environment, this is as much a global problem as a local and urban one.

Looking to the future, there are alternative paths for urbanization and the urban environment. It seems very likely that cities will be warmer, more crowded, more connected, more exciting and possibly more stressful, in both 'South' and 'North'. What is up for debate is the quality of cities – will they be segmented into zones of safety and danger? Will clean air and green grass be for the rich, or for everyone? Will the cities of the 'South' struggle against the invisible stranglehold exerted by the corporate giants of the 'North'? To explore such questions, there are future projections and scenarios, which can be linked back to choices and decisions for the present. The goals of the 'sustainable city' are then not just a utopian dream, but also a practical menu for action.

In this chapter

This chapter looks at the dynamics of change in urbanization and urban conditions; and the implications for the human urban environment. It starts with a brief history of the environment in cities from their emergence to the present day.

Then, the 'urban environmental transition' concept sums up the trends, pressures and dynamics of change for cities in different parts of the world. There are two parts to this – the socio-economic context and the environment metabolism, i.e. the throughput and transformation of resources, energy and materials.

Finally, there is the question of where cities are heading from here – an outline of future projections and alternative scenarios for the human urban environment.

3.2 Brief history of the urban environment

The pre-industrial city

In the early stages of urban development, the nearby hinterland supplied most of the agricultural and mineral produce. Permanent settlements began to grow at strategic points such as the confluence of rivers and coasts. The internal spatial structure of the city was based around its original role in military defence and trading, on which later were overlaid functions in government and civic society.

Water supply systems were built at an early stage in most cities, although often available only to the wealthy. Although covered drainage was installed in the first recorded city, Mohenjo-Daro in Pakistan, in about 2600 BC, underground drainage generally comes later. Other planned infrastructure and urban technology was generally minimal, and most ordinary housing was un-serviced by either water or sanitation.

Housing conditions were primitive, and the air was polluted by wood or charcoal fires, and the uncontrolled operation of industries such as food, leather and textiles. The use and recycling of resources was generally as far as the available transport could carry. However, certain commodities, such as salt or precious metals, were traded on an international scale. For imperial systems, such as those centred on Rome, Beijing or Tokyo, there were extended supply lines of food and textiles for capital cities and their burgeoning populations of service workers.

Such cities generally adopted some form of grid or axial planning, which could be organized and funded by a powerful state, acting in combination with a wealthy religious institution. The rectangular layouts could then be extended indefinitely and enabled a variety of urban forms such as parks, piazzas and terraces.

Such planning was accompanied on a conceptual level by various images and theories of the 'ideal city'. Modern forms began to emerge in the colonial age: cities as far apart as New Delhi, Buenos Aires, Paris, Philadelphia and Washington DC set out an agenda of modern city planning, with built-in services, street and building design patterns, and public open spaces, or 'green infrastructure' (see discussion in Chapter 6) (Mumford, 1961; Kostof, 1999).

The industrializing city

With the growth of manufacturing through the industrial revolution, a more complex spatial structure begins to emerge, dominated by transport infrastructure – by water up to the eighteenth century, railways in the nineteenth, highways in the twentieth and by air travel in the present day. Other public functions such as government, health and education also emerge as key elements in the urban form. With the rapid increase in population to meet the labour requirements of the factories, the city water and sanitation systems were under severe stress, with mass outbreaks of infectious diseases before the investment was made in urban infrastructure. Energy for housing and industry was mainly via coal, and air pollution was severe and widespread.

In the earlier industrial revolution, the exploitation of resources for industrial production was on a regional scale, although national and global supply lines soon followed. In the absence of controls, cities rapidly grew in a landscape of mining waste, industrial pollution, derelict land, deforestation and rural poverty. Cities such as Manchester and its hinterland, which combined a global trading function with local manufacturing in textiles technology, showed the worst excesses of the industrial revolution, and paved the way for radical political thinking (Engels, 1845).

In the later industrial revolution, environmental health became one of the defining forces in urban growth, and a key priority for the emerging municipal governments in new industrial cities. In Europe, the USA and elsewhere the fierce competition for investment led to civic 'boosterism', where services such as municipal water supplies were a major selling point. This was enabled by the generation of new kinds of capital through the issue of municipal bonds, and such large-scale investment funded city-wide infrastructure in drainage, electricity, gas, transport and, later, telephone systems. Such processes took several decades in the North, but are now greatly accelerated in other parts of the world, as illustrated in Box 3.1.

Box 3.1

Self-financed urban infrastructure, Angola

Luanda Sul is a trend-setting model for urban infrastructure, based on a self-sustaining programme which valorizes public assets through careful land use management and planning. This was a self-financing process following initial investment of $30 million in 1995: this has provided 70km of water supply pipes, 23km of drainage, 12km of power lines, 2,210 houses and adequate shelter for 16,700 people.

The programme operates from an Achievement and Management Fund, mobilized through: (i) the sale of concessions (or land tenure rights) derived from the allocation of public land for private development; (ii) taxes and tariffs perceived on the exchange of goods and services; and (iii) investments made by the private sector. The government, by issuing guarantees for private investments, provided the basis for the self-financing of the programme.

Source: UN-Habitat Best Practices, http://www.bestpractices.org

---

Modelling the urban environment

In the classic theory of urban development, the environment has two crucial roles: as the provider of resources, such as food or timber; and as a location with trading potential, with rivers, coast or other transport corridors.

As urban-regional economic analysis developed, it built in the cost of food production and cost of transport to market, to the economics of land values, construction costs, population migration and so on. By following the mathematical logic towards the optimum layout, the 'central place' theorists generated an idealized hexagonal spatial pattern. This assumed that towns and cities were laid out on an endless flat plain, with major hubs, minor hubs and connecting corridors, so that each settlement has its appropriate scale market (Christaller, 1966).

It is interesting to compare the assumptions in this idealized locational model with the realities and nuances of an urban environment dimension:

- environmental impacts of industry: many cities show an asymmetric development, where the wealthy live upwind and upstream of industry, while the poorest generally live in the most polluted areas;
- environmental infrastructure, such as water supply, sewage or energy, shapes the pattern of spatial development, and that in turn depends on the technology and the political economy of investment and return;
- environment and landscape, amenity and quality of life factors: these are the most volatile, but in later service-sector and post-industrial cities, often the most crucial factors.

The city of services

In the later stages of the industrialization process, large cities began to re-shape themselves into a more extended structure. This was first identified by the Chicago school of human ecology, as a set of concentric urban rings, then as segments (Burgess, 1925; Hoyt, 1939). New transport technologies, first the railway, then urban rail and tram systems, and then the private car, provide the physical capacity for mass movement. On the demand side, the rising affluence of consumers, and availability of finance for development, enables a voluntary migration of the more affluent to suburban locations, in parallel with the thinning out of formerly dense inner city and industrial areas. At the same time, further increases in transport speed and availability enable the functional integration of groups of settlements into extended conurbations, which subsequently urbanize a wider zone of the city hinterland, and transform formerly rural settlements into urban satellites.

In the last half-century many of the problems of industrial production have gradually been brought under control by a combination of enlightened local government and commercial interests. However the problems of consumption are mounting, not least because in a cleaner and more affluent city, environmental standards are continually raised. While many local environments within and around many cities in the 'North' have never been cleaner, at least on the surface, many environmental problems are simply displaced to other regions or other nations, and the expropriation of environmental resources takes place on a global scale (McGranahan, 2006).

For traditional air quality problems there are basic controls on coal burning and factory emissions, but traffic emissions are a growing problem for all cities, especially those where micro-climates facilitate urban photo-chemical smog. The continuing diffusion of population, workplaces and other facilities means that public transport is less viable, and so the entire urban economy and society rapidly becomes motorized or 'automobile dependent' (Newman and Kenworthy, 1999).

The long-range transfer of acidifying emissions, or 'acid rain', was the first focus for international environmental policy. For water supplies, many cities are constructing inter-regional or international transfer systems. While sanitation is universal at least in the 'North', there are new artificial substances in food chains and industrial effluent, and waste water treatment becomes ever more costly.

With rising affluence, many urban environmental pressures now become the result of consumption rather than production, which is increasingly displaced to other parts of the world. While much of the population in wealthy urbanized countries aspires to a quiet rural life, the majority are contained in larger urban systems, often dominated by congestion, noise, crime and a host of environmental risks.

One result is an expropriation of environments on a global scale for leisure and tourism, where quality of location becomes a commodity in itself, often over-riding the needs of native communities. Another result is the displacement of environmental problems on a global scale, of which the foremost case is that of climate change (White and Whitney, 1992). For this and other issues, there is much irony in the fact that the wealthiest nations of the 'North' cause direct climate change emissions many times the rate of the 'South', and even more indirect emissions through the import of products – and yet it is the cities of the 'South' who will suffer the greatest impacts from climate change.

The post-Fordist city

The city type now emerging in many parts of the 'North', and in many diverse ways in the 'South', does not have a single name or definition. The 'post-Fordist' city is one way to characterize a more fluid city system, beyond the conventional economic and political structures of the twentieth century (summed up by the standardized mass production method of the Ford motor company) (Amin and Thrift, 2002). The 'space of flows' is another way to describe a city system which is more identified and dependent on its global connections than its local economy and population (Castells, 1996).

In economic terms, such a city may be the locus for global networks in media, finance, education or advanced technology. In population terms, such a city may be dominated by a global elite of mobile entrepreneurs and professionals, at the same time as being a hub for a multitude of migrants and nationalities. In spatial structure, such a city may be part of a functional city-region, conurbation, agglomeration or other geographical unit, without clear boundaries or definition. In political terms, such a city may not have a single clear identity or government structure, being a complex set of overlapping and competing units at various scales, from the public, private, civic and voluntary sectors – a fragmented and many-layered vision of an urban community (Soja, 2000).

What then becomes of the human urban environment, in such a fluid and fuzzy city system? This relates to the critical perspectives introduced in Chapter 2. It seems that physical environments are increasingly segregated and privatized, so that the wealthier will buy their way into environmental quality, bypassing the risk and insecurity represented by other social groups. Such environments are increasingly globalized, in the sense that high-quality leisure and cultural environments become globally accessible commodities, marketed for the culture of consumption.

There are huge implications for urban environmental management in such a post-Fordist city, a theme which recurs in later chapters. While many of the world's cities are still trying to establish democratic and effective public government and administration, the agenda has in many ways moved on, for cities in the post-Fordist transition, into a more complex, fluid and entrepreneurial situation. Elected city governments have to deal with neighbourhood and regional stakeholders, from public, private, civic and community sectors, in transient communities and professions. As most city services are privatized or franchised, partnerships and network connections may be the key to power and wealth. But even the city of the technocratic elite contains its own contradictions, with conflicts over space and resources between local and global, or between modernization and conservation – for instance, where city airports expand, or where regeneration displaces local businesses.

Colonial cities

Naturally there are many variations on this highly condensed history of human urban environments. Most cities show patterns of growth and change with various stages of the above overlapping, especially in cities of the 'South' with very rapid industrialization.

Colonial cities were integral to the age of Northern European imperialism, and were set up for specific military, trading and government functions. Some colonial cities were created on virgin sites, such as Mumbai and Hong Kong. Others were more like 'dual cities', where the European city was adjacent to an existing centre which provided services and population, such as in Delhi and Shanghai. In Africa, colonial cities such as Kinshasa or Maputo were set up as transit points for the commodities of the interior, and were deliberately kept short of investment and modernization. Generally, colonial environments could be completely different in each part of the city: for affluent Europeans there would be gardens and boulevards, while for servants and workmen there would be overcrowded tenements, pollution and disease.

Squatter cities

Surrounding many former colonial cities, and some cities in the 'North', are informal settlements, containing in total over 700 million people and a large proportion of the world's slum dwellings (UNCHS, 2001). Many cities of the 'South' are characterized by *ad hoc* squatter settlements, generally on the outskirts or on pockets of vacant or industrial land. These have local titles – *barriadas* in Peru, *favelas* in Brazil, *bustees* in India and *bidonvilles* in former French Africa. The environmental conditions of such settlements can be extreme; without formal tenure or planning, usually without essential services, often subject to severe industrial pollution and in situations prone to flooding or landslides. The definition of informal settlement can be stretched to street dwellers, without even the comfort of a home-made shelter, of which there may be over 100 million, mainly in Asia and Africa (Neuwirth, 2005). Ironically, 'Northern' cities also contain squatted dwellings, more often in inner-city vacant properties which are caught in the cycle of obsolescence and re-development.

New towns and planned settlements

Around the world are settlements which have been planned and built directly by the state or through government-backed programmes. Some were built in earlier ages of direct monarchy, such as Addis Abbaba in Ethiopia, or St Petersburg in Russia. Many of the more recent were based on the Garden City concept originating from the Town and Country Planning Association, and were built in the UK, elsewhere in Europe and in North America, in response to the severe overcrowding and environmental stress of industrial cities (Howard, 1898; Hall and Ward, 1998). There are many smaller settlements or extensions, such as the Garden Suburbs, but the most notable larger new towns are those built for civic purposes, such as Brasilia, Canberra and Ottawa.

In various parts of the 'North', the new town concept now shows a resurgence under the banner of 'new urbanism', a combination of urban design co-ordination, mixed uses and higher densities. One variation is that of the leisure or dormitory town, such as Seaside in Florida (Duany, *et al.*, 2000); another is the insertion into an urban structure of a 'transit-oriented development' (Calthorpe, 1993). Elsewhere there are larger scale political and economic urban projects; for instance the Korean capital

Seoul is moving its government functions 150 miles south, to rebalance the national spatial development, and also to remove the government from the North Korean border. The urban environment in new towns is generally of higher quality, but there can be problems with economic and social vitality; new towns such as Chandigarh in India were not designed around local economic activity, and there are now many self-built shops and workplaces on the sides of the monumental boulevards.

One of the most topical forms of planned settlement is the 'technopole', with a concentration of research, high technology and education, often built on a digital communications spine (Castells and Hall, 1994). Other forms of specialized settlements occur often as coincidences of politics, geography, industrial structure and proximity to natural resources. Such specialized roles show great variety:

- Company towns: for example, Levittown, USA; Port Sunlight, UK;
- Mining industry cities: for example, Magnitogorsk, Russia; Broken Hill, Australia;
- Trading or port cities ('entrepots'): for example, Tijuana, Mexico; Rotterdam, the Netherlands; Hong Kong, China;
- Science cities: for example, Novosibirsk, Russia; Kuala Lumpur, Malaysia;
- University cities: for example, Cambridge, UK; State College, USA;
- Government and civil service cities: for example, Brasilia, Brasil; Canberra, Australia; Abuja, Nigeria;
- Imperial cities: for example, St Petersburg, Russia; Versailles, France;
- Retirement settlements: for example, Seaside, USA; Bournemouth, UK;
- Cultural cities: for example, Venice, Italy; Madurai, India; Llangollen, UK;
- Tourist and leisure settlements: for example, Las Vegas, USA; Cancun, Mexico; Marbella, Spain;
- Sectarian or sacred cities: for example, Salt Lake City, USA; Jerusalem, Israel; Mecca, Saudi Arabia; Varanasi, India.

One of the effects of global urbanization may be to encourage such specialization, while at the same time the flattening of corporate structures means that the integrated company town is in theory less essential. Each of these urban types has a particular urban environmental profile and policy agenda.

Finally it should be stressed that these are very generalized sketches. The sheer ingenuity and resourcefulness of settlement and city builders in fitting into the most unlikely environments should never be underestimated (Box.3.2).

Box 3.2

A new generation of cave dwellings, China

Over 1,000 environmentally sustainable dwellings have been built in the Yaodong cave area of the Loess Plateau in China using traditional energy-saving methods and vernacular housing design. The low-cost houses are built through self-help construction and the use of innovative solar energy systems and natural ventilation methods help to reduce energy consumption to a minimum.

Source: World Habitat Day Awards, http://www.worldhabitatawards.org

---

3.3 Urban Environment Transition – context

Urban environments around the world are clearly in a state of flux and structural change, in parallel with urban economies and societies. Such processes of change are referred to as the 'Urban Environment Transition' (UET). This reflects the idea that urban environmental problems are both effects and causes of the dynamics of development and urbanization on a global scale (McGranahan, *et al.*, 1996). This section looks at the global urbanization projections in the context of socio-economic transitions. The following section looks at the UET itself.

Urbanization projections

At present, 3.3 billion people live in urban centres across the globe. By 2030 this number is predicted to reach five billion, with 95 per cent of this growth in developing countries. Over the next three decades, Asia's urban population will double from 1.4 billion to 2.6 billion; Africa's city dwellers will more than double from 294 million to 742 million;

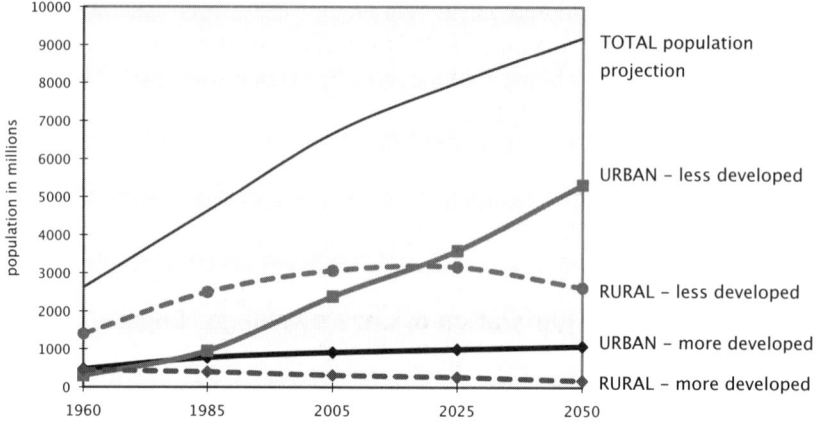

Source: UN Department of Economic and Social Affairs, 2007

Figure 3.1 *World Urbanization Prospects*

while Latin America and the Caribbean will see a slower rise from about 400 million to 600 million (UN Population Fund, 2006). While mega-cities dominate the development agendas, overall growth in urban centres of 10 million or more inhabitants is expected to level out; over the next 10 years, cities of less than 500,000 will account for half of all urban growth.

The 'business as usual' or default projections for urbanization at the international level are constructed in detail to 2030 by the United Nations, in the World Urbanization Prospects series (UN Department of Economic and Social Affairs 2006). Summary tables for these projections are shown in the appendix.

The projections show the expectation that current levels of urban growth will continue almost unabated. They also show a remarkable turning around of historic trends, to a reduction in the rural populations in both More Developed Countries (MDCs) and Less Developed Countries (LDCs). The urban–rural balance is of course dependent on pressures and policies in both rural and urban areas: development aid, agriculture, water, climate change, healthcare and so on are some of the issues which affect the urban migration trend. In terms of location, there is a very clear shift in the geography of the largest urban agglomerations or 'mega-cities' of over 5 million population:

- up to 1950, half of all urban agglomerations were in Europe;
- as of 2000, the more recent agglomerations were spread around most continents;
- projected to 2015, the majority of new agglomerations are expected to be in Asia.

However this headline focus on the largest agglomerations can easily miss the many smaller urban systems in the ½ million to 5 million band, where the rate of agglomeration is proceeding more rapidly in many instances (Bell and Jayne, 2006). As to what kind of urban environments can be expected, on current trends, the proportion of slums and/or informal settlements could increase to over half of the world's urban population by 2030. So it is arguable that the majority of future urban dwellers will be in quasi-temporary, low-tech shacks, short of fixed systems such as water, sanitation and electricity, even while mobile phone and other transient networks thrive, in a more spontaneous and fluid urban environment (Webster and Lai, 2004).

Urban environment transition

The brief history of the urban environment above shows how the development of industrial production, the evolution of urban form, and the creation of infrastructure and technology, each move in parallel and inter-dependent processes. Each of these has implications for the human urban environment.

A very topical question is raised by the relationship between economic growth and changes in environmental quality. In some countries, some environmental conditions such as water quality deteriorated in parallel with rising GDP in the early stages of industrial growth, and then improved as incomes continued to rise. In such cases a graph of historical pollution levels against national income has followed an 'inverted U' shape known as the Environmental Kuznets Curve. In many other cases, and particularly for global impacts such as climate change and biodiversity loss, environmental degradation has continued to rise closely linked with economic growth (Kuznets, 1955; Ekins, 1997; World Bank, 2001b; McGranahan and Marcotullio, 2006).

This can be seen by charting the dominant urban socio-technical factors over several centuries (Figure 3.2):

- a 'brown agenda' – focus on poverty, including human sanitation, water supply and air pollution;
- a 'grey agenda' – focus on production, including industrial pollution and urban transport impacts;
- a 'green agenda' – focus on consumption, including global climate emissions, consumer supply chains and corporate responsibility.

One obvious example of a transition is that from the horse-drawn city of the 1850s, with all the economic, technological and cultural apparatus of horses, to an automotive city in 2000, with the equivalent apparatus for the private car. This 'technological transition' was then a key factor in the 'spatial transition', that is the shift from a denser centralized city to a more diffused polycentric city. There followed an 'environmental transition', which shifted the resource flow from animal feed to oil, and the pollution agenda from animal wastes to traffic emissions (Newman and Kenworthy, 1999).

The UET curves in Figure 3.2 also say something about the differences between cities in the 'South' and 'North'. With accelerating

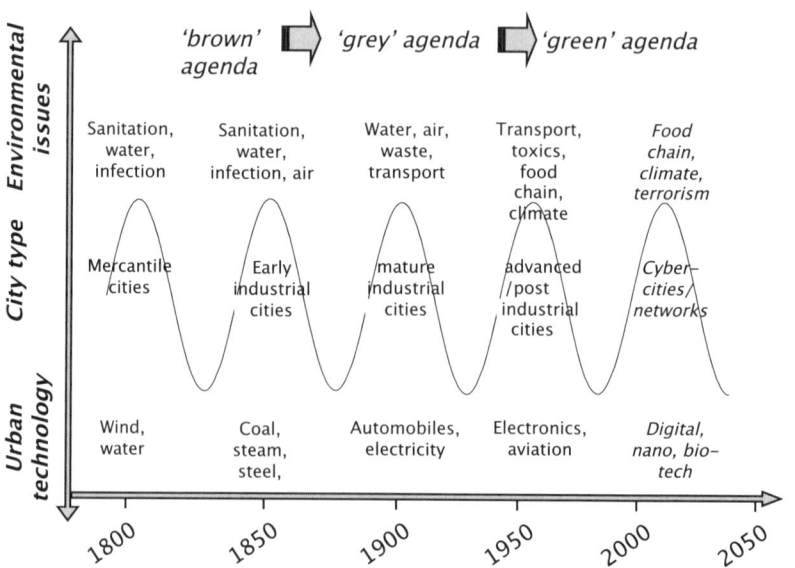

Overview of the historical trajectory of urban development, technology and environment: based on Marcutullio & McGranahan, 2006: Ravetz, 2006

Figure 3.2 **Urban environmental transitions**

industrialization and globalization, many cities in the 'South' have overlapping and simultaneous kinds of environmental pressures, from squatter settlements to airport growth – contributing to the state of fluidity, disorganization and fragmentation.

The UET can also be seen through the lens of a critical perspective, as instrumental to 'capital accumulation' and economic division on a global scale. This applies not only to capital and ownership, but to informational goods and positional goods, such as network connections, high-quality locations and so on. The downside of the UET can be seen with new patterns of segmentation, deprivation and exclusion.

Finally, it is clear that there is no fixed trajectory, but rather many alternative options for the future. With an understanding of the UET it is possible to see the simple aspirations for the sustainable city and sustainable communities, and relate them to the more complex reality of the inter-dependent trajectories of cities and regions, in both 'South' and 'North'. The question of the future of the city and its communities returns at the end of this chapter and in the final part of the book.

Socio-technical transitions

Exploring the dynamics behind the projections, one starting point is to observe the long waves of economic development identified by Nikolai Kondratiev (Barnett, 1998). Kondratiev observed that approximately every 50 to 60 years since the late eighteenth century there had been a severe economic downturn followed by a period of stagnation and then a sharp upturn. Joseph Schumpeter explained these long waves in terms of cycles of major technological innovation, capital investment and associated institutional change (Schumpeter, 1939; Brotchie, *et al.*, 1987). The first industrial wave started in Britain in the mid eighteenth century: a later one focused on the military-industrial complex in the USA in the 1970s. Each of these economic and technological waves then influenced changes in urbanization, urban form and urban infrastructure. The advent of the railway, electricity, telephone, motor car, elevator, jet aeroplane, personal computer and now the internet, have generated new patterns in urban forms and structures.

For most of the industrial city period, economic specialization and its competitive advantage could be defined in terms of the city or city-

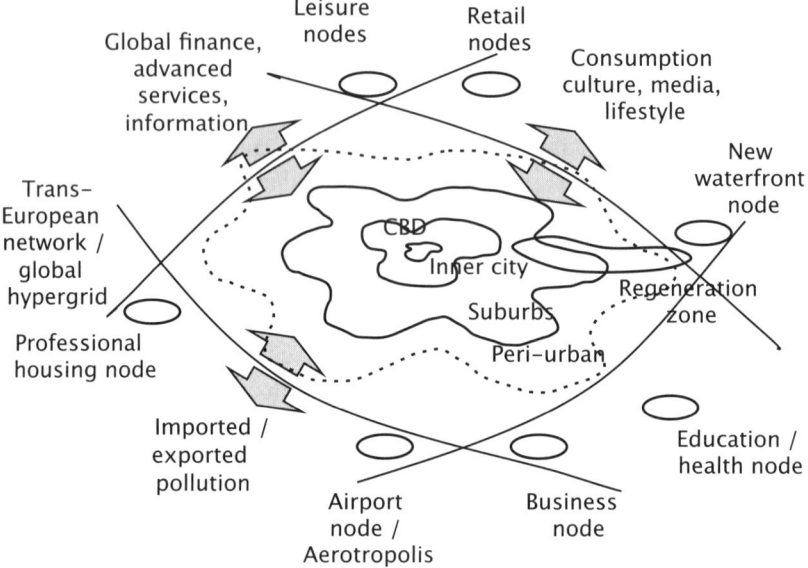

Flows of capital, information and cultures in a post–industrial globalizing city–region

Leisure nodes

Retail nodes

Global finance, advanced services, information

Consumption culture, media, lifestyle

Trans– European network / global hypergrid

New waterfront node

CBD

Inner city

Suburbs

Regeneration zone

Professional housing node

Peri–urban

Imported / exported pollution

Education / health node

Airport node / Aerotropolis

Business node

Figure 3.3 *Urban informational transitions*

region's location and resources in terms of material flows (Solow, 1970). That general model is now in transition to a more post-Fordist city of cultural and informational 'spaces of flows' (Castells, 1996). The city and its region now tends to function more as a node in a global 'hypergrid' – networks of motorways, railways and airports for movement of people and goods, and networks of satellites and wires for movement of information and capital (Figure 3.3). Many patterns of urban activity and urban form are turning inside out, as the growth nodes of production and consumption migrate to the urban fringe or 'edge city' – a new landscape of retail, leisure and business parks with easy links to the hypergrid (Garreau, 1991). The city itself, and its people's reason for being there, centre on services and consumption, and the city's cultural *cachet* aims to compete in a global hierarchy. Such cities often place large sections of their population into a form of 'spatial exclusion', especially those without access to cars or disposable income.

There are certain paradoxes in such a transition – in many cities nineteenth, twentieth and twenty-first century cultures and economic systems exist side by side. Meanwhile as production and consumption becomes globalized, there is a countervailing trend of localization – a new kind of 'place advantage' gained through cultural tourism and local environmental quality (Girard and Nijkamp, 2009; Brand and Thomas, 2005). In physical terms, edge cities are counter-urbanized, while historic centres are re-urbanized and industrial areas regenerated (as explored in Chapter 6). In socio-economic terms, 'uneven development' creates clusters and sinks of unemployment and exclusion (Savage and Ward, 1993).

Each of these changes has huge implications for the urban environment, and the environmental impacts and displacements caused by urban activity. The impacts of climate change, resource extraction and marine pollution may now be caused mainly by 'Northern' cities, and experienced mainly in the 'South' (see Chapter 5). Many urban material flows now come and go through the global hypergrid, which is increasingly privatized and deregulated, and beyond the control of city or even national governments. In this new context, urban environmental management may be an even greater challenge than ever before.

Economic and business transitions

Whatever happens in the cities of both 'South' and 'North' is increasingly driven by the global economy, and the several trillion dollars of volatile capital which now circles the world every day (Hertz, 2005). The result is a rapid restructuring in almost every level of every urban economy:

- global and regional integration – the dominant trend of liberalized free trade, carried out through frameworks such as the Single European Market, North American Free Trade Agreement, or World Trade Organization;
- deregulation and privatization – blurring the boundaries between state and market, reducing market barriers and business controls;
- international division of labour – where the globalization of production pushes lower value activity towards lower cost countries, who then are locked into relationships of dependency;
- 'flexible specialization' – large corporate firms shift towards downsizing and out-sourcing of previously in-house activities; together with rapid obsolescence of industrial infrastructure, technology, skills and occupations;

● reshaping the urban hierarchy – new structure of hubs and peripheries, control and dependency, between cities and nations.

The last point illustrates that such economic transitions are rarely straightforward or neutral, but rather instrumental in conflict and competition between cities, nations, corporate empires, sub-cultures, social groups and so on. There is a seemingly inevitable pressure for cities to attract investment, build airports and motorways, and dominate their hinterlands, no matter what the environmental cost.

Informational transition

One of the most potent factors for the future city or city-region is almost invisible on the surface. Information and communications technology (ICT) is rapidly changing the nature and location of production and consumption, and also provides a new global nervous system for any human activity or organization (Mitchell, 1996). ICT also contributes to several other technological revolutions now in full swing – not only in digital, but in nano-technology and bio-technology, which have the potential to transform not only communications but life itself (Kaku, 1998).

The implications for the physical city alone are huge – daily commuting for instance can be reduced by tele-working, but is often replaced by a more resource-intensive demand for more flexible and long-distance travel. Energy demand in smart houses could be greatly reduced, but new applications and comfort standards often then draw in further demand. Artificial climates for outdoor spaces, for instance, may help to increase the vitality of urban spaces at the cost of huge energy flows. Digital participation and decision-making could bring new life to urban governance, although not simple in practice, and risking the exclusion of many sections of society. The convergence of technologies – geographical, social networking, and distributed computing – opens up boundless possibilities for organizing future urban life (Hudson-Smith, 2008).

In the wider picture, ICT is instrumental to a post-modern perspective on cities and urban life as a 'space of flows' (Castells, 1996). The ICT networks themselves are one layer of construction of such space, and the nodes and hubs are another layer. A further layer is the potential for

'mass collaboration' or co-production of economic, cultural and social
resources on a global scale (Tapscott and Williams, 2006). Over-hanging
all this are the shifting patterns of dominance and hegemony, which
may be enabled and reinforced by the ICT system, and managed by the
new technocratic elite, using cities as the primary arenas (Sardar and
Ravetz, 1996). On the other hand there are many examples of positive
developments (Box 3.3).

Box 3.3

Telecenters: changing the way people work, USA

Municipal energy consumption, climate change, air pollution and the
rising cost of fuel are forcing cities to look for responses. Chula Vista in
San Diego County, USA, is currently dealing with serious traffic, air
quality and environmental concerns, by making use of the 'information
superhighway', to counter the effects of the 'concrete superhighway'.
Residents can drop in to their 'Neighbourhood Telecenter' where they
can use computers, modems, telephones and other office services. These
telecentres reduce automobile trips, traffic congestion, energy consumption
and air pollution, promote a better quality of life and productivity by
working closer to home.

Source: UN-Habitat Best Practices. http://www.bestpractices.org

---

Demographic transitions

In parallel with the economic dimension is a 'demographic transition'
(Caldwell, *et al.*, 2006). This reflects the rapid changes in population
structure, health and mortality over a period of generations as in the
industrialized 'North', or more rapidly as in the 'South'. It can be seen as
evolving through several phases:

- in the pre-industrial phase birth rates and death rates are high,
  depending on fertility as much as mortality, with a population size
  subject to natural disasters, wars and famines;
- as more modern healthcare spreads, death rates fall rapidly, while

birth rates are still high, leading to rapid population growth and urbanization, as seen in some cities of the 'South';

- as urban life becomes the norm, and particularly as women are increasingly educated, economically active, and able to access birth control, birth rates fall towards the lower levels of death rates. The result is almost zero population growth and a demographic ageing effect, as seen in many cities in the 'North';
- there are many localized issues which are cause or effect of such a transition. There may be a rapidly falling population in rural areas, large inertia in the housing stock in urban areas, the localized ageing of the population in retirement areas, or the problems of restructuring and urban shrinkage.

Overall, the demographic transition is closely linked with urban environmental factors. Industrial pollution and workplace hazards, poor housing and lack of infrastructure all raise the death rate.

Social and cultural transitions

In parallel to the demographic trend, there are equally fundamental social and cultural dynamics, at both local and global scales:

- other types of demographic trends – changing gender balance, family structure, disposable time and income, kinship networks and household organization;
- cultural trends – differentiation and a shift from former patterns towards individual or community self-identity, empowerment and alternative states of consciousness;
- social trends – polarization and segmentation of communities into those with or without work, opportunities, networks and norms: further diffusion of the 'moral economy';
- psychological trends – increasing aspirations for community, affluence and fulfilment, alongside stress, depression and alienation.

Overall, such trends might appear to lead to an ageing population with rising disposable time and money, chasing diminishing job opportunities and expanding lifestyle activities. Parents and children in particular may struggle with the diffusion of career structures, family structures, and the increasing alienation of the public realm. The overall effect for urban dwellers could be an acceleration of social and cultural fragmentation

– bringing the opportunities of diversity and empowerment, in parallel with the many problems of exclusion and stress.

Transitions in 'South' and 'North'

Clearly the cities in 'South' and 'North' are at different stages of transition. As for why such disparities exist, there are clearly relationships of inter-dependency and expropriation between one and the other. Managing this global imbalance may be the largest transition challenge of all.

In many cities of the 'South', there is a typical transition in function from a regional hub for primary production (agriculture, forestry, minerals), to a site for globalized manufacturing. In that transition there is a typical problem of 'over-urbanization' and disorganization, where the city grows more rapidly than the capacity to manage it or invest in it. The result is often seen in the huge communities of unserviced or illegal households, and gross problems of air, water and ground pollution. However there is also evidence that urbanization generally raises opportunities, incomes and living standards, above the more diffused but severe poverty and deprivation of rural areas.

In counterpart, in cities of the 'North', there is a typical transition from a regional manufacturing base to a globalized service-based economy. This can also accelerate the polarization of wealth and poverty, in both inner cities and surrounding hinterlands. In turn this transition sets the context for urban environments, their problems and potential solutions.

However the diversity of urban environmental conditions does not fit neatly on a 'South'–'North' spectrum. For instance, in the former industrial cities of the Russian Federation and New Independent States, there has been a very rapid process of re-structuring and de-industrialization. This has helped to reduce the direct industrial pollution of air and water, but has left a legacy of contaminated and derelict land, together with a lack of funds to deal with it. In the newly industrializing countries and cities around the Asia-Pacific Rim and elsewhere, there is rapid growth in consumer affluence and demand for housing, transport, waste management and other services. China is now seeing the world's most rapid urbanization and modernization, with environmental problems on an equally huge scale. In contrast, Asian city states such as Singapore provide models of stable and highly centralized environmental management (McKinsey Global Institute, 2008a).

Overall, the conclusion might be drawn that the majority of urban environments are on the edge of chaos, except for a minority of relatively affluent, well-organized cities in the 'North'. The conclusion might also then be drawn, that these 'Northern' cities are clean and green, largely because their physical resources and supply lines are expropriated from overseas, and the bulk of their external impacts are also exported back. Each of these conclusions might be too simplistic in dealing with a great diversity of problems and solutions (Ravetz, 2006a).

Each of these socio-economic transitions has distinct effects on urban environmental issues; the range of effects is summarized in Table 3.1.

3.4 Urban Environment Transition – effects

The urban transition

One of the most alarming global trends is the phenomenon of urbanization, particularly as concentrated in developing countries. Most of the world's 30 largest conurbations are in the 'South', each of them larger than some nations, and containing the most extreme environmental problems (UN-Habitat, 2005b). Meanwhile in the more industrialized parts of North America or Western Europe, up to 90 per cent of people are already in urban areas, and there is an opposite trend of 'counter-urbanization' or outward migration – threatening the viability of both urban and rural areas, and dependent on high-impact lifestyles and technologies. Urbanization and counter-urbanization can be seen as two sides of the same coin, reflecting different stages in a general urban development path, where cities are both the engines, and the dustbins, of economic growth.

This has been set out as an 'urbanization curve', which shows different development stages for cities within their nations (Champion, 1999). This seems to work in parallel to similar curves for the economic transition from primary to tertiary activity; and the demographic transition from high to low mortality. It also identifies the position of different world regions on different parts of the curve:

Table 3.1 Key drivers and effects in the Urban Environment Transition

ENVIRONMENTAL EFFECTS	Individual-household environmental health	Local environmental conditions	Urban-regional environmental impacts	Global environmental impacts	Supply chain & resource flows
TRANSITION TYPES					
Economic production transition	Basic domestic infrastructure	Rising efficiency of utility & infrastructure	Displacement of regional resources	Growing impact on climate, resources	Reorganization to global systems
Productivity/ business transition	Customized business models	New institutions for infrastructure	New institutions for regional resources	Growth in corporate responsibility	Material & energy efficiency
Consumption/ lifestyle transition	Consumer choice & access	Segmentation of housing & neighbourhoods	Segmentation of urban & regional communities	Growth & specialization of leisure & tourism	Increasing complexity & material turnover
Informational/ structural transition	ICT in personal profiles, diet, healthcare	ICT in monitoring & management	Counter-urbanization production & consumption	Functional specialization of global system	Advanced ICT in supply chain management
Demographic/health transition	Rapidly reducing mortality	Rapid reduction in pollution & communicable disease.	New patterns of urban-regional migration	Rapid increases in material consumption	Increasing specialist services for ageing population

Adapted from Ravetz, 2006a

- in the initial stages the population is mainly rural and engaged in primary industries;
- in the accelerating urbanization stage there is mass migration from rural to urban areas;
- in the third stage the urban population stabilizes and may begin to decline, as a more affluent and mobile population counter-urbanizes over a wider area;
- a possible fourth stage might involve wider urban diffusion and global networking.

Spatial development transition

At national and regional levels, the 'spatial dynamic' of urbanization – that is, the growth and changing structure of cities and settlements – is the result of interactions between demographic change, economic change, social structures, housing, transport, public services and so on. Chapter 6 looks at this in more depth, but here it is significant that there is complexity at every scale – general population shifts, concentration in hubs of different scales, thickening or thinning of existing urban areas, and migration or displacement of social groups. This is why statistics on urban growth and change need to be taken with great caution; depending on the boundary definitions, for instance, a large metropolitan urban region such as New York or London, might be classed as shrinking or expanding.

In most developing nations, there is a general migration from remoter rural areas into extended metropolitan areas, although within this general pattern there are often flows of migrant workers who are based both in the city and in rural areas. There is also much movement of local urban populations to gravitate towards sub-centres, new industrial zones, new trading hubs and so on. Some agglomerations, such as Manila or Kuala Lumpur, also evolve new kinds of urban–rural interfaces, where urban functions are inter-woven with intensive agriculture (Allen and Dávila, 2002).

In both 'South' and 'North', there is an over-arching trend towards 'metropol-ization' – an economic and socio-cultural restructuring of larger and larger areas, combining conurbations, smaller settlements, mixed urban-rural areas, and transport hubs and corridors (UN-Habitat, 2004). Also described by the unwieldy term 'functional agglomeration', with the result of 'mega-urban regions' (MURs), this then drives spatial development patterns on the ground. The largest MUR in the world is debatable, but candidates would include Tokyo-Osaka-Kyoto-Kobe-Nagoya with 60 million; Hong Kong-Shenzen-Guangdong MUR with 120 million; and Greater Shanghai-Nanjing MUR with 83 million (Hu, 2003; Lacquian, 2006).

Environment and resource transitions

Each of the spatial development dynamics above has effects on environmental flows, conditions and impacts. The summary in Table 3.2 shows some key linkages between spatial and environmental transitions.

- transport is the first most obvious impact from physical agglomeration; shopping, education and leisure are now carried out at greater distances with greater specialization in the extended motorized city;
- production and servicing of the built environment; as residential and workplace density reduces, and locations diversify, the sheer volume of buildings goes up, together with their energy and water demand;
- technological infrastructure – the range of appliances and media equipment grows rapidly, with the increase in smaller households and workplaces; each with their own supply chains, energy requirements, waste management problems and so on;
- land use change; directly in the conversion of open land to built land and gardens; there are also indirect effects on the pattern of land use and land-based activity in previously freestanding settlements and rural areas;
- restructuring and diversification of socio-economic activities and patterns; not only travel on a daily or weekly basis to access shops, leisure etc. It may involve more complex patterns of partial migration, due to factors such as dual-career families, distant college courses, specialized leisure activities, and extended social networks and diaspora, at every scale from the urban to the global.

In general, both the physical urban area and its hinterland shows a changing pattern of risks and opportunities. Many common environmental pollutants are being replaced with the more insidious hazards of modern production and consumption, such as genotoxics, carcinogenics, and bio-accumulation in the food chain. As heavy industry migrates to other locations which offer lower costs, the clean-up of the

Table 3.2 Spatial dynamics and environmental flows

ENVIRONMENTAL EFFECTS	Environmental flows	Environmental stocks	Environmental conditions	Environmental impacts	Environmental benefits
URBAN TRENDS					
Urbanization	Direct increase in urban metabolism	Direct land use change	Intensification of urban conditions	Transport demand growth	Transport & energy efficiency
Suburbanization	Shift in metabolism to suburban	Rapid land use change: shift of biodiversity	Outward spread of urban conditions	Transport/energy demand growth	Increase in domestic green space
Counter-urbanization	Shift to long range commuting/networking pattern	Rapid activity & community changes	Displacement of urban conditions	Transport/energy demand growth	New rural-urban fringe landscapes
Re-urbanization	Shift to affluent urbanist metabolism	Intensification of urban land use	Intensification of urban conditions with greater affluence	Gentrification with loss of biodiversity on derelict land	Land & water reclamation
Functional agglomeration	Increasingly complex metabolism due to specialization	Specialization of land use & activities	Polarization of conditions due to increased fragmentation & segmentation	Transport demand growth	Increased investment due to economic growth

Adapted from Ravetz, 2006a

urban environment shows gradual improvement, while rising affluence generates consumption of imported goods.

On the ground, local territorial conflicts mount over positional goods, where assets and property values are based on exclusive locations. Sectors such as housing, transport and waste management are each entangled in controversies over environmental risk and justice, and these conflicts also help to define new social groupings and sub-cultures. Such consumption-based environmental agendas are driven by rising affluence and aspirations for identity-creating goods, places and lifestyles.

In terms of environmental processes the picture is increasingly complex. Material movements are increasingly 'trans-boundary' with long distances from origin to destination; 'trans-media', with many reactions between gases, liquids and solids; and 'inter-generational', displacing impacts and responsibilities from present to future. Assessment of hazard and risk depends on how the system boundaries are drawn – even detailed life-cycle analysis of products or processes can easily underestimate total system impacts. Standards for such impacts can be set with pollution thresholds and critical capacities, for resource demands, industrial emissions, environmental quality standards, and human health risk. However the increasingly lengthy supply chains mean that causes and effects may be on opposite sides of the world.

Resource flow transitions

As well as concentrating pollution, urbanization consumes massive amounts of materials and energy. The evidence on resource flows has been aided by recent work on the material flow and ecological footprint of cities and nations around the world (Eurostat, 2000; Chambers, *et al.*, 2000; Barrett, *et al.*, 2006). The impacts of consumption are highly uneven – the richest nations and cities use a footprint of over 10 global hectares per capita, and the poorest use less than a tenth of this. However, the impacts of production are more evenly spread – implying a growing transfer of resources from the poor to the rich.

A critical perspective sees this as part of the global pattern of displacement and expropriation of resources by the rich, which then transfers environmental impact and depletion from the rich back to the poor in reverse. This starts from the basic imbalance where commodity-based economies are often maintained in a state of peripherality and dependency, while advanced service-based economies are able to invest, innovate and dominate markets and supply chains (Stren, *et al.*, 1992). There are different patterns of resource flows involved:

- direct expropriation of resources and impacts: importing of materials and commodities from poor to rich nations, and transferring pollution back;
- indirect expropriation of resource and impacts: induced effects in poor countries of disruption and pollution due to industrial production, with smaller material transfers to rich countries;
- extended supply and use chains with impacts at many stages and locations;
- structural and institutional displacements, as both cause and effect of the supply chain effects of expropriation; for instance the phenomenon of developing country debt.

This pattern of resource dependency is generally pictured as a national or regional issue, and it is only recently that the urban agenda has emerged. However, as above, it is the cities which are the major consumers of resources, the hubs for extraction and manufacturing, and also the residual destination for in-migration from failed rural development.

Cities are also the locus for new technology, for energy and utilities. The 'urban energy transition' is hugely important in shifting cities from a supply based on centralized fossil fuels, towards decentralized renewable energies and distributed heat and cooling (Droege, 2008) (details in Chapter 5).

Environmental management transition

The final part of the UET picture concerns the policy responses. The context here is the modernization and liberalization of governance and regulation, and so in parallel there are transitions in environmental policy and management practice. These can be seen again on a historical path (Murphy and Gouldson, 1998):

- in pre-industrial cities, the first concerns were with fire protection, and later with drinking water and sanitation: after London's 'great stink' of 1856, where the river pollution was so severe that Parliament was

suspended, construction began on what was then the world's largest
sanitation system;

- there followed several waves of formal regulation of fuel sources and
emissions, as in the UK's Clean Air Acts of 1956, or the Pittsburgh
Smoke Control Ordinance of 1941;

- as environmental science developed, and governance was modernized,
there emerged a more responsive mode of negotiated dialogue with
polluters, as in the BPEO (Best Practical Environmental Option)
principle;

- co-ordination between all environmental pressures is then achieved (in
principle) through frameworks such as the EU system of Integrated
Pollution Prevention and Control;

- in parallel, the scientific and management resources are taking shape,
with ICT as a catalyst. These include integrated datasets and analytic
models, covering pressures, conditions, benchmarks, supply chains,
product life-cycle analysis, and underlying environmental processes.

What is now emerging is a more comprehensive approach to 'integrated
assessment' and integrated environmental management (Bailey, 1997;
Rotmans and van Assaelt, 1996). As well as drawing on better evidence
and modelling than ever before, this takes a more open discursive
and participative approach to complex problems, involving a range of
stakeholders, looking at upstream and downstream issues, with alternative
perceptions of risks and values. This 'post-normal' approach is suited
to problems of high uncertainty, controversy and urgency – which are
typical of the global challenges of the twenty-first century (Funtowicz and
Ravetz, 1993).

3.5 Alternative futures

If the parallel transitions above are more or less clear, is the future also
clear? Not at all. Many changes are fuzzy, contradictory, specific to each
city and community, and driven by unexpected events. To deal with
such uncertainty, scenarios are more useful than projections or forecasts.
Scenarios can raise a structured series of 'what if' questions, with a range
of tools which can help to envision, adapt and influence change as it
emerges (Schwartz, 1991; Ringland, 2002).

One kind of scenario is based on forecasting, which projects forward
from current trends as far as they can be taken – the 'business as usual'

scenario. Another kind is based on a more creative envisioning of desirable future goals and possibilities – in this case, the complex bundle of aspirations, scientific targets, political discourse and planning policies known as sustainable development. Using a process of 'back-casting' (the counterpart to forecasting) the policies and actions to achieve this can be explored in detail (Dreborg, 1996).

World scenarios

The concept of sustainability emerged with the Limits to Growth arguments of the 1970s, anticipating a future collapse of the global resource base. Many global scenarios backed up with systems models were pioneered by both NGOs and transnational companies (Meadows, *et al.*, 1992). The World Business Council for Sustainable Development explored a range of global scenarios with three clear dynamics as the starting point – a world population in the region of 10 billion; a wave of technological innovation in the early 21st century, and the transformation potential of ICT (World Business Council for Sustainable Development [WBCSD], 1998).

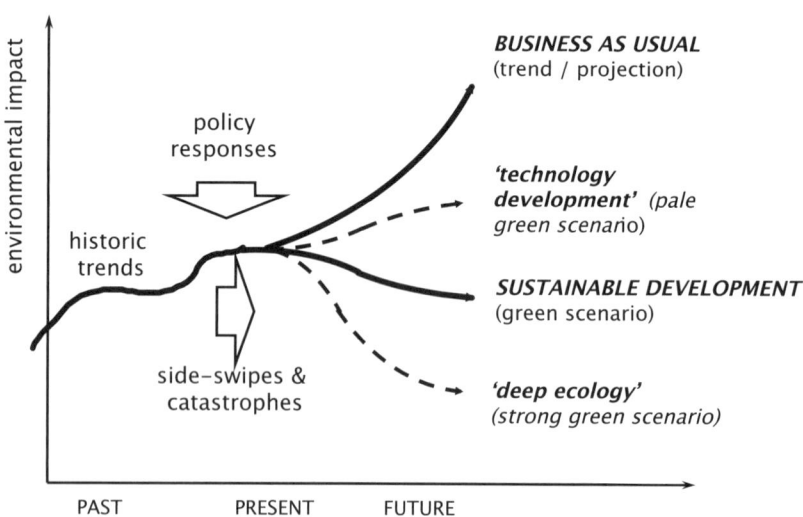

Figure 3.4 *Urban environmental scenarios*

The simplest form of scenario analysis, as in this section, compares the business as usual with the sustainable development scenarios, although there are many possible variations. These can be linked to simplified measures of global environmental impact, as summed up by climate emissions and/or the ecological footprint. In this case, 'Factor 1' refers to a future based on current growth trends, and 'Factor 4' refers to a fourfold or 75 per cent reduction in the ecological footprint: this can in principle be achieved by doubling resource efficiency and halving resource use, as in the book *Factor Four* (von Weizsacker, *et al.*, 1997).

There are also some variations – a 'pale green' scenario might rely on technology for solutions, while existing patterns of economic and social development continue. In contrast, a 'deep green' scenario would look across the board at political, economic, social and cultural change, in order to achieve more radical environmental changes. More sophisticated scenario frameworks start to explore such possibilities. The Special Report on Emissions Scenarios of the Intergovernmental Panel for the Scientific Assessment of Climate Change (IPCC), constructed four main families of scenarios. These were based on a range of governance systems (globalized or regionalized), and a range of societal values (private enterprise or public responsibility); these were then used as the basis for modelling future climate emissions (IPCC Working Group III, 2000).

A topical question is why so few published scenarios relate to urban development and urban conditions, beyond the basic ratios of population and urbanization – perhaps cities are too complex, too individualized, or too unpredictable. These are some key scenario variables which relate particularly to the human urban environment:

- urban spatial development – clustered or diffused;
- urban form and fabric – rapid or slow change;
- urban economic development – globalized or localized;
- urban social development – polarized or collective;
- urban environmental management – from public-based regulation, to market-based trading systems.

'Business as usual' for the urban environment

The UET curves shown in the last section show a dotted line disappearing into the future, with destination unknown. One interpretation is that beyond the cities of poverty, of production and of consumption, lies a further and more positive phase in the transition – the ecological city or sustainable city (Bai and Imura, 2000). Unfortunately many analysts are not so optimistic, noting the numerous signals of new environmental hazards alongside the old ones, and the many barriers to a more sustainable city model.

This shows, above all else, how environmental problems are constructed and mediated in terms of competition, conflict, coercion, domination and dependency – between social groups, political elites, economic communities, cultural networks and so on. Even in a situation of rising affluence for most urban dwellers, the underlying environmental agendas are likely to remain (European Environment Agency, 2000). With the expectation of being proved or disproved, some likely trends and projections for the human urban environment can be put up for discussion:

- climate, water, desertification – the projected rise in global temperatures will hit urban eco-systems in unpredictable ways. Water will be in greater demand and shorter supply than ever, and may replace oil as the main agenda for geo-political strategy. As access to water is increasingly privatized, the grass of the rich will literally be greener than that of the poor (Scheumann, *et al.*, 2008);
- energy systems – as oil and gas reserves begin to decline, and as world demand escalates, other energy sources (including many renewables) with their environmental effects will come to the fore;
- risk and hazard distribution – from gross environmental pollution to more subtle risks of food chains, bio-accumulation, occupational risk, and the new bio-technologies and nano-technologies;
- digital darkness – the potential of ICT continues to unfold, with new aspects of surveillance and repression, social engineering, cyber-realities, fundamentalist cults, sabotage and counter-terrorism;
- positional goods – i.e. space, territory and environmental quality, will continue to be appropriated by communities with power and wealth, using cities and regions as the arena;
- new forms of social/cultural conflict – the advent of the race riot and the suicide bomber may bring new forms of insurgency and asymmetrical conflict, and this will transform urban spaces and public facilities;
- migration and mobility – accelerating global mixing of communities

and cultures, due to the specialization of production and consumption, and the enhanced transport and communications;

- slums and informal settlements – on current trends, the proportion of slums is increasing and could contain over half the global urban population by 2030, and two thirds by 2050 (Neuwirth, 2005);
- urban transport – congestion continues in most cities as the mode of choice shifts to the private car, even while vehicle emissions are gradually cleaned up and the city grapples with car dependency;
- urban land uses – alongside the new urbanism café culture, real mixed uses are likely to see workspace, leisure, culture, education and retail overlapping in new configurations, with increasingly privatized security control;
- segmentation and fragmentation – the partitioning of urban spaces into safety and danger, privilege and deprivation is likely to increase despite the public aspirations for a 'city for all';
- regional spatial development – the shift will continue towards larger city-regional agglomerations, competing for prominence in global networks, with continuing pressure for edge city business and retail parks;
- local spatial development – partly in reaction to globalizing forces, partly to encourage global investment, there is a re-invention of neighbourhood-scale planning and environmental action.

Sustainability in the urban environment

The many strands of a sustainable development scenario for the urban environment run throughout the rest of the book. This scenario would aim to bridge the gap between ideals and practicality, and the barriers of inertia and materialism. It would also draw from the UET approach and the critical perspectives above, and look beyond the often simplistic calls for greening, towards the practical realities of sustainable communities (Ravetz, 2000; Roberts, 2008).

In this light there are major challenges for a 'sustainable' human urban environment, in both the 'South' and 'North', to be explored in the following chapters:

- extended local–global urban supply chain systems – the implications of the global urban hierarchy and international trade patterns are that many conventional boundaries are superseded. Environmental planning and management will need to extend to a whole systems approach for supply chains and urban environments;

- multi-level and multi-agency governance – how to coordinate environmental management across many social groups, vertically and horizontally, with changing needs and desires;
- social and cultural polarization and exclusion, characterized by a fractured and fragmented human urban environment – where the power of information-driven market forces is overwhelming, but the outcome may yet be a dysfunctional and divided city.

3.6 Conclusions

This chapter has presented a rapid tour of 5,000 years of urban history, with a glimpse of some alternative paths for the twenty-first century. At the centre is the concept of transition and transformation – social, economic, cultural, urban and environmental. This is a challenging and complex theme – not only for academic interest, but for the practical task of improving urban environments around the world. In other words, to avoid fixing tomorrow's problems with yesterday's solutions, it is important to be clear about the changes in motion in different kinds of cities, and what they mean for the way forward.

Further information

UNEP Millennium Assessment   http://www.millenniumassessment.org/en/index.aspx

The Future 500   http://www.future500.org

Tellus Institute Boston   http://www.tellus.org/

UK DTI Foresight   http://www.foresight.gov.uk

UK Strategy Unit   http://www.cabinetoffice.gov.uk/strategy.aspx

European Commission Thematic Strategy on the Urban Environment   http://europa.eu./comm/environment/urban/home_en.htm

UN-Habitat   http://www.un-habitat.com

World Future Society   http://www.wfs.org/

Worldwatch Institute   http://www.worldwatch.org/node/913

Further reading

Blowers, A. (ed.) (1993) *Planning for a Sustainable Environment*, London: Earthscan.

Ayres, R. & Simonis, U., (eds) (1997) *Industrial Metabolism: Restructuring for Sustainable Development*, New York: United Nations University Press.

European Environment Agency (2000) *Cloudy Crystal Balls: An Assessment of Recent European and Global Scenario Studies and Models*, Environmental Issues Series 17, EEA, Copenhagen (available on http://www.eea.europa.eu).

Knox, P.L. and Taylor, P.J. (1995) *World Cities in a World-System*, New York: Cambridge University Press.

McGranahan, G. and Marcotullio, P.J. (eds) (2006) *Scaling the Urban Environmental Transition*, Tokyo: United Nations University and London: Earthscan.

Martin, J. (2006) *The Meaning of the 21st Century: A Vital Blueprint for Ensuring Our Future*, New York: Penguin.

Hall, P. (1998) *Cities in Civilization*, New York: Pantheon.

Ponting, C. (1992) *A Green History of the World*, Harmondsworth: Penguin.

Ringland, G. (2002) *Scenarios in Public Policy*, Chichester: Wiley.

Raskin, P., *et al.*, (2003) *Great Transitions: The Promise and Lure of the Times Ahead*, Boston, MA: Tellus Institute (available on http://www.tellus.org/index.asp).

Schwartz, P. (1991) *The Art of the Long View*, New York: Doubleday.

Strategy Unit (2002) *The Future and How To Think About It*, London: Cabinet Office (available on www.cabinetoffice.gov.uk/strategy.aspx).

UN-Habitat, (2001) *Cities in A Globalizing World: Global Report on Human Settlements 2001*, Nairobi: UN-Habitat (available as of Dec 2008 on www.un-habitat.org/categories.asp?catid=555).

World Business Council for Sustainable Development (WBCSD) (1998) *Exploring Sustainable Development: Global Scenarios 2000–2050*, London: WBCSD.

Worldwatch Institute (2007) *The State of the World: Our Urban Future*, Washington DC: Worldwatch Institute.

4 Urban environments in a global context

4.1 Introduction

Globalization and modernization are like tidal waves, driving the development of cities both in the 'South' and the 'North'. This reflects not only the law of the market and international trade, but the policy and ideology which surrounds it – trade liberalization and deregulation in the 'South', and the drive to maintain competitiveness in the 'North'. Such global forces often serve to expropriate natural and human resources from the 'South' to the 'North', while leaving a trail of pollution and disruption. The result is often seen in environmental problems at every scale – smoky chimneys, urban road traffic, regional desertification and water crisis. Ironically, waste from the cities of the 'North' is increasingly sent to the 'South' for disposal and recovery.

Although problems are encountered in the 'North' as well as the 'South', this chapter focuses on the cities of the 'South' and the brown agenda of urban development, as a means of understanding the global impact or ecological footprint of cities (an idea introduced in Chapter 1). Rapid urbanization, shanty towns and sweat shops are the popular images of 'Southern' cities, whilst land tenure problems and post-colonial dependencies may stymie their progress.

This might suggest that a global perspective is a key to understanding and improving the human urban environment all round, in that the problems of affluence are inextricably linked with the problems of poverty. For such improvements it is clear that new ways are needed of combining market forces with more effective environmental governance, at local, regional, national and global level.

Naturally, in this it is necessary to generalize about the 'South' and 'North' as representing opposite ends of a spectrum – the reality is of course much more varied, both between and within cities. Every city and region has a complex story to tell, and potential to realize, in the international urban order.

In this chapter

The following sections of this chapter deal with many of the above issues, ranging from discussions of population growth and urbanizing forces, through considerations of the city edge and the challenges of planning cities, and then to an analysis of pollution and its consequences. A final section examines the global city-region.

4.2 Growing populations

International divisions

The backdrop to urbanization in 'North' and 'South' is the international division of wealth and income.

The poorest 20 per cent of the world's population has seen its share of global income decline from 2.3 per cent to 1.4 per cent in the past 30 years, while the share of the richest 20 per cent increased from 80 per cent to 85 per cent (UNCHS, 2001). On that basis the upper 20 per cent are on average 60 times more wealthy than the poorest. The 250 richest individuals control more wealth than half the world's population.

Urbanization worldwide

As we saw in Chapter 2, defining a city is not straightforward. Its population may be defined as the whole metropolitan region, a municipality or just the city centre. Even so, the recorded statistics indicate the major changes. In 1900 the world's largest city was London, with a population of 6.5 million (World Bank, 2001b). Next came New York, Paris and Berlin, followed by Chicago and Vienna, each with 1.7 million. Tokyo was number seven on the list, at 1.5 million. A hundred years later Tokyo was at the top, with 26.4 million, followed by Mexico City and Bombay, each with 18.1 million, and then São Paulo (17.8 million). Developing countries are expected to contribute 24 of the world's 30 biggest cities by 2010, with Bombay (forecast to be 23.6 million) close to passing Tokyo (stuck at 26.4 million), and Lagos (20.2 million) close to overtaking both.

Environmental health in the industrialising 'North'

By 1900, London and the other industrialised cities of Europe and North America had got reasonably close to dealing with the problems caused by cramming several million people into a confined space. The turning point came in the 1830s and 1840s. The great cholera epidemic of 1831, among other concerns, prompted Friedrich Engels to write his *Condition of the Labouring Classes in England*, and to join Marx in drawing up the Communist Manifesto. Other less strident voices had greater effect. The Poor Law Board's 1842 report on The Sanitary Condition of the Labouring Population of Great Britain, prepared by Edwin Chadwick, and the 1845 report of the Parliamentary Commission of Enquiry into The State of Large Towns and Populous Districts, detailed much the same concerns: 'cellar dwellings crammed with people; vast refuse heaps in the midst of towns; rivers foul with sewage; swamps draining into water supplies' (Hill, 1985). Action followed. Improvements in housing conditions, water supply, sewerage and waste disposal began to make a difference. In 1849 Dr John Snow identified the causal links between leaking sewers, contaminated drinking water and cholera, thereby laying the scientific foundations of the massive infrastructural investment programme which eventually overcame London's most pressing environmental and health problems. Much the same took place in New York and other industrialising cities in Europe and North America (UNCHS, 2001).

Environmental health in the 'South'

The problems faced in London and New York two centuries ago now confront many developing countries. Although these countries' total population growth is expected to slow in the coming decades, urban areas will continue to absorb an increasing proportion of the rising total. The provision of housing, water supply, sanitation and waste disposal remain sorely inadequate (Brown, 1996; World Bank, 2001a). Only 18 per cent of low-income city dwellers have their own water supply. Communities of 500 people or more may be served by a single tap, perhaps functioning for only a few hours a day. Only 8 per cent have a sewer connection, while the rest may share a toilet with a hundred or more other people. It has been estimated that over 90 per cent of sewage is discharged untreated into rivers or lakes as well as coastal waters. Between 20 and 50 per cent of solid waste is uncollected. Over a billion people live in urban areas with poor air quality, arising mainly from motor traffic and industry. In many major cities, air pollution exceeds World Health Organization guidelines for at least two key pollutants. Meanwhile, many people rely on wood or dung for cooking and heating, leading to indoor as well as outdoor air pollution.

Cities in the transitional countries of central and eastern Europe suffer from different problems. Here, socialist planning systems created relatively effective urban infrastructure, but highly polluting industries. Transformation of their economies has tended to reduce inefficient industrial production, and hence industrial pollution. However, in many areas the old industries have yet to be replaced by new wealth creating activities, which can create difficulties in maintaining the infrastructure. Rapidly rising car ownership has introduced a new source of atmospheric pollution and generated congestion.

City Development Index

As an aggregate measure of these effects, the United Nations has devised a City Development Index (CDI), which combines the Human Development Index (HDI) with other factors that are directly related to the urban environment:

- City product: GDP per capita or the nearest equivalent;
- Infrastructure: access to water, sanitation and energy supply;

- Waste: a measure of the relative eco-efficiency and environmental quality;
- Health: as in the HDI;
- Education: as in the HDI.

Inhabitants of cities in the high CDI group (the top 20 per cent in the world) have a life expectancy at birth of 75 years. For those in the bottom 20 per cent, it is 50 years (UNCHS, 2001). Bad as this is, it is not as bad as it was in industrializing England. In Macclesfield near Manchester, a textile town at the time, the average age at death in 1847 was 24 (Hill, 1985).

The CDI is a good index of urban poverty and urban governance. Its health, education and infrastructure components are particularly good variables for measuring poverty outcomes in cities. Similarly, its components for infrastructure, waste and city product are key variables for measuring the effectiveness of city governance in creating a clean and prosperous environment. The CDI correlates strongly with the city's economic performance. Generally a high-income city will have a higher CDI, although some poor cities such as Ulaanbaatar and Hanoi have relatively better environmental and social standards. Cities which are poor but have even worse environmental and social standards include Lagos and others in Africa.

The CDI of a city reflects its progress towards meeting target 11 of the Millennium Development Goals (MDG): 'By 2020, to have achieved a significant improvement in the lives of at least 100 million slum dwellers' (United Nations, 2002, 31). The United Nations system has assigned UN-Habitat the responsibility to assist member states monitor and gradually attain MDG 11. Future steps are envisioned to assist member states with activities such as technical cooperation on slum upgrading and urban management.

4.3 Urbanizing forces

Urbanization – blight or blessing?

Contrasting perceptions of urbanization regard it either as a blight on rural development and the environment, or as the driving force of human progress. One view sees it as 'siphoning private and public resources from rural areas into urban, leaving the former impoverished and further fuelling out-migration'. The other sees it as the principal mechanism 'underlying technological innovation, economic development and socio-political progress'. A more balanced view argues that the 'utilisation of surplus rural labour in other (usually urban) activities is a prerequisite for raising rural incomes and living standards', and that 'the concentration of human, technical and financial resources in cities has become an increasingly important asset in the more internationalised world economy of today' (DFID-UN, 2002, 17).

Urbanization via surplus labour

The utilization of surplus rural labour played a key role in the urbanization of the 'North' in the eighteenth and nineteenth centuries. Then as now, the surplus arose from rapidly increasing agricultural output per worker, which enables larger total populations to be supported by smaller rural ones. In Britain's case this was achieved partly through the new machinery emerging from the industrial revolution, but more strongly from the programme of enclosures and clearances which restructured the country's agricultural organisation (Hobsbawm, 1969; Hill, 1985). The surplus labour either emigrated to Britain's colonies or sought a living in the cities. Similar mechanisms are occurring in developing countries now, through the 'green revolution' of agricultural technology, the introduction of modern machinery, and the restructuring of agricultural organisation to increase productivity. With fewer opportunities for emigration, the bulk of the surplus population goes to the cities in search of a new economic role there.

Urbanization and the wealth of nations

Why the city? The alternative view would prefer to see the rural economy take advantage of its own increased productivity, eat more of the food it grows, expand its own economy, and create its own new high-earning opportunities for productive employment. This has sometimes happened to a degree, for example in Japan and Korea (Ward, 1988; Edwards, 1992), but it tends not to. The city has two big competitive advantages over rural areas; its concentration of wealth, which creates a demand for service labour and the ability to pay for it, and its concentration of commerce. It is the second of these which gives the densely-packed city its status as the main driver of a country's economic development. The process has been described as one of 'complex symbiotic relationships' formed among the city's various producers, in which

> city markets – whether of consumers or producers – are at once diverse and concentrated. These two qualities of the local market make production of many kinds of goods and services economically feasible that would not be feasible in rural places (Jacobs, 1986).

Limits on the city's size

Given time, even in rural areas, the people and small firms that between them embrace the wide range of skills that are needed to produce an innovative product might make contact across the distances involved, provided that a market for the product can be found. A market is found more easily in the city. It can also be satisfied more cheaply from the city. Transport costs and the time taken to communicate militate against rural areas being able to compete with the city in both innovation and production. Even if the product is for rural use, its purchasers can buy it more cheaply from the city.

Under normal market forces the city will expand its population until this competitive advantage gets cancelled out by counteracting factors (Henderson, *et al.*, 2001). There are several such effects. If land becomes scarce, costs may rise; if the supply of new labour diminishes, wages may rise; if supply of the city's products approaches demand, prices may fall; if congestion within the city slows its internal communications, it begins to lose its competitiveness. In developing countries, the agricultural output per worker is still far below that in developed ones, resulting in a continuing supply of surplus labour. In an increasingly international marketplace with low external transport costs, the city's market can be expanded beyond national borders. The city may therefore continue to grow until choked off by congestion costs or land prices. In cities with no natural geographical boundary such as mountains or sea, the price of land may not be an issue, leaving congestion as the ultimate constraint.

Urbanization and the service economy

Displaced subsistence farmers and surplus agricultural workers do not migrate to the city to offer their special skills in this diversely innovative process on which the city thrives. Their skills are more likely to be in tending goats or seedlings. Their options are generally limited to unskilled factory work, or more commonly, tapping into the demand for service labour that the city's wealth creates.

In Chapter 3 we saw how the city has changed from a place where people exchanged agricultural produce, to the networked city-region of a global economy and society. In some ways, however, the changes are smaller than the continuity. Services have always been the prime focus of the economy of the city, as a financial centre, a marketplace, a centre for the arts, education and government, a construction site, supplying buildings for other service activities, plus varying amounts of manufacturing. England is the only country where manufacturing has ever employed the majority of the workforce. Even there, employment in services rose faster than in manufacturing even when the growth of manufacturing was at its height in the eighteenth and nineteenth centuries (Kumar, 1978). The same applies in developing countries now, where services still absorb the bulk of surplus rural labour (Selya, 1999). Some new migrants, such as market traders, may find a way into intermediate levels of the hierarchy of service delivery through which the city's wealth is spent and trickles down. Most go in at the lower levels, such as construction labour, cleaning or hawking, where specialist skills are not needed. If they fail to find an entry point any further up, they or their children may eke out a living at the bottom, picking over the city's rubbish dumps.

In an increasingly integrated global economy, the main difference between a developing country city and an industrialised one arises from what they sell to each other. Goods and services with high added value per employee go in one direction, in return for low value ones going the other way. As well as growing as a proportion of employment, services have also grown as a proportion of GDP, to the extent that they now dominate high income countries' economies. This has raised the prospect that all future economic growth might occur in services, thus 'decoupling' it from material consumption and direct pollution. For this to be a practicable proposition, consumption in high income countries would have to fall, such that the rising service component of their own economies is not merely an indication of increased imports of material goods from countries that bear the associated environmental costs.

'South'–'North' convergence

At some point in the future, the scenarios discussed in Chapter 3 for cities of the 'North' and the 'South' may converge. If they do, this is likely to entail greater availability in the 'South' of the technologies on which high-income economies are based, and lower dependence in the 'North' on material consumption. The possible mechanisms by which this might happen include changes in the global trade regime, the behaviour of corporations, and issues of governance and development strategy in developing countries. In the meantime, the cities of the 'South' will continue to face much the same environmental issues as they do now.

As noted already, the human and environmental problems associated with urbanization, while bad, are little if at all worse than they were in industrializing Britain. In Europe and North America urbanization was a process of transition, which led to their current levels of wealth. One of the biggest issues of sustainable development is whether cities in developing countries can make a similar transition now.

4.4 The urban–rural interface

Cities in developing countries rarely grow through a series of planning decisions to extend their boundaries. They just grow, partly by their own high fertility rates and falling mortality rates, but mainly because of declining employment opportunities in rural areas. Boundaries are fuzzy, and tentacles of urban activity extend out into the fields. As the city grows, the tentacles reach further, while the inner regions of the fringe become increasingly urban. The fringe is the peri-urban interface, where urban and rural activities meet. Its economic, social and enviromental characteristics are a mixture of the two. Small farms coexist with industrial workshops, middle class commuters' houses and urban workers' informal or illegal settlements, all with different and often competing interests, practices and perceptions (Allen and Dávila, 2002).

Land tenure and urbanization

Differences arise in the first place over land. The tenure rights of the rural poor are often informal and traditional, not written into law. If the formal owner of the land decides to sell, or if newcomers move in, the original occupiers may have little or no redress. As land prices rise, subsistence farmers have little chance of buying it. Conflicts flare up, with

> a rising level of segregation and polarisation … In Nairobi, the collapse of service delivery (water, refuse removal, health, security etc.) has sharpened inequalities, as only those that pay receive decent services. Marginalised inhabitants of informal settlements (for example, Soweto, Mathare, Kibera, Dagoretti and Dandora) resort to their own means of self-provisioning, often including violence … local agents and communities often feel that violent re-distribution of assets is the most effective way to deal with the situation.
>
> (Mbiba, 2002).

As the land available for agriculture shrinks, pressure rises to expand into natural or semi-natural habitat. The land may become over-grazed and degenerate. The peri-urban fringe is where the city's waste is dumped, until that too has to move out further. The newly urbanising region often includes little land for recreation as well as agriculture. The role of green space in hydrological management, absorbing rainwater runoff and flood water, and at the same time re-charging aquifers, may be lost (DFID-UN, 2002).

These changes are often unplanned and unmanaged. Few metropolitan governments extend their jurisdiction out into peri-urban areas. District governments are left to cope with the city's effects, without its resources. Slowly, the rural environment turns into an urban one; those who farmed it become urban people, joining the ranks of urban workers.

New economies on the fringe

As old livelihood opportunities decline, new ones appear. The city's solid waste can be recycled and sold back to it. Its liquid waste offers a cheap alternative to commercial fertilisers, with or without treatment to deal with the health risks (DFID-UN, 2002). The area of agricultural land falls, but demand for its produce rises. Farmers who can shift to high-value, perishable, fresh fruit and vegetables may do well, but need access to land, water and the inputs needed for intensified production. Typically, farming in peri-urban areas becomes increasingly dominated by wealthier commercial producers, often based in the town or city (Tacioli, 2002). Meanwhile, jobs are to be found in the city itself. Cheap transport arrives to bring migrant settlers in, and local people with them. Access to health services and education may improve, but may not. Industrial workshops and factories are set up, providing more opportunities for employment. Gradually, the fields disappear. The interface has moved on.

In principle, the evolution of the peri-urban environment could be greatly improved if the process were better managed. In practice, decision-makers often find it easier to just let it happen. However, it is possible to do better.

4.5 Cities that plan

In the 1960s a group of young architects in the Brazilian city of Curitiba became disillusioned by the failures of city planning. They won the confidence of the mayor, and in 1965 their own master plan was implemented. At the same time they founded the Urban Planning Institute of Curitiba, as an independent agency to provide oversight and continuity for city planning. In 1971 one of the plan's architects, Jaime Lerner, was appointed mayor. The plan and subsequent refinements have steered Curitiba's development ever since (Meadows, 1995; ICLEI, 2002). Lerner served three terms of office, and eventually became state governor. One of the Planning Institute's senior officials, Cassio Taniguchi, became mayor. The continuity Lerner hoped for has been maintained for almost four decades.

The example of Curitiba

Continuity was a vital element of Curitiba's success, but not the only one. 'There is no endeavour more noble' says Lerner 'than the attempt to achieve a collective dream. When a city accepts as its mandate its quality of life; when it respects the people who live in it; when it respects the environment; when it prepares for future generations; the people share responsibility for that mandate. This shared cause is the only way to achieve that collective dream' (ICLEI, 2002, 2). The dream was mainly

Lerner's. It became a collective one because its vision for the future could be shared by a whole city. It came true because Lerner and his colleagues had the skill, as planners, to make it realistic, and as charismatic leaders, to make it happen.

The new mayor, Taniguchi, is as much of a practical visionary as Lerner. 'The value of a city' he says 'is directly proportional to the degree of satisfaction of the people that live in it. So an urban administration can be visibly and critically assessed by the quality of life of the inhabitants. When it makes decisions, it should always keep its citizens in the spotlight. For this reason, successive municipal administrations in Curitiba since the beginning of the 1970s have done their best to answer – in a practical and organized way – this basic question: how can the city welcome an ever increasing number of inhabitants into a static physical space, without losing quality of life?' (Taniguchi, 2001, 1).

Urban form and accessibility

Curitiba's population grew from 150,000 in the 1950s to 800,000 by 1974, and is now 1.6 million. Its growth occurred largely through poor migrants flooding in during times of economic crisis. The practical, organized way in which this was managed revolved largely around transport, fully integrated with land use planning and housing development (Box 4.1). Although the city now has one of highest rates of car ownership in Brazil, people are enabled to leave their cars at home. There are 200 kilometres of cycle paths. The quality of public transport, and its usage, are the highest in the country (locally designed buses carrying about 2.14 million passengers a day). The city has Brazil's lowest rates of air pollution and per capita petrol consumption. A standard fare is charged for all trips, so that shorter ones taken around the city centre subsidise the long commuting journeys made by poorer residents living at the periphery. Much of the city centre is a pedestrian zone.

Urban ecology

Nearly one fifth of the city is parkland, with new lakes designed to absorb runoff and floodwaters. Developments that include green space are given tax incentives. One and a half million trees along the city streets were planted by the residents themselves. Gardens and parks

Box 4.1

Transport planning in Curitiba, Brazil

Measures	*Effects*
Integration of transport and land use policies: Land use legislation enforces higher densities around major transport corridors and roads are geared to land use in each area.	Curitiba has one of the lowest rates of ambient air pollution in Brazil; fuel consumption has been reduced by 25 per cent, with gasoline use per vehicle 30 per cent less than in other Brazilian cities.
Public transport: Main roads have an express bus lane; different lines are integrated for rapid transfers; the system is faster and cheaper than those in other Brazilian cities.	The bus system serves 1.3 million passengers daily, or 75 per cent of all commuters; people spend about 10 per cent of their income on transport, one of the lowest rates in Brazil.
Provisions for pedestrians and cyclists: A network of bike paths links city neighborhoods and parks; pedestrians have priority in downtown areas with high streets closed off to form pedestrian malls.	Once-declining shopping districts are now lively and profitable; there is little traffic congestion although Curitiba has over 500,000 cars (second highest amount per capita in Brazil).
Preservation policies: New development is concentrated in existing urban space; old structures are renovated for new uses.	Historic buildings are preserved and degraded land is reclaimed for such uses as an opera and a botanical garden; green space per capita has expanded from 0.5 to 50 m^3.
Traffic control: Some streets are closed to cars; in others, speed limits and trees slow car traffic.	Curitiba's rate of accidents per vehicle is now the lowest in Brazil.

Source: Bartone *et al.*, 1994

are tended by orphaned or abandoned street children, while others get a small wage and a daily meal in exchange for simple tasks under an official scheme where they are 'adopted' by the city's shops, industries and other organisations. Seventy per cent of the city's solid waste is recycled, largely at source by residents. They learn the habit early, through recycling schemes in schools. Post-collection recycling provides formal employment to the homeless and to recovering alcoholics, and the profits go into social and healthcare services. In those poorer parts of the city where the cost of collection is high, neighbourhood centres have been set up where residents can exchange rubbish bags for bus tickets and groceries.

Education and participation

Curitiba has achieved its goals by winning support from commercial interests as well as the rest of its people. The transport system has kept congestion at bay. Curitiba has established its own Open University, providing low-cost courses in the skills the city needs, and which its people need. Old buses have been converted into mobile training centres. Business parks help small companies establish themselves, and transport is designed to link them to each other and to commercial centres. By these means the city has achieved an economic growth rate of 7.1 per cent over the past 30 years, compared with 4.2 per cent for the country as a whole. Companies have grown and prospered, while average per capita income has risen to 66 per cent above the Brazilian average.

Transformation in modernizing Asia

Similar things have been achieved elsewhere, most notably in Hong Kong and Singapore. Unlike Curitiba, these two cities started with the advantage of being major trading centres, but nonetheless faced similar difficulties in managing their growth. Both have transformed themselves from low income developing countries to high income developed ones, or more accurately in Hong Kong's case, a developed city state within the rapidly developing mega-state of China. Their cultures are different, from each other and from Latin America, but in both cases their success has come from integrated planning, for the economy, environment and people. Their built environments do not teem with biological diversity, but nor do they teem with open sewers and decomposing garbage. Hong

Kong is a vibrant bustle, Singapore an orderly classroom, but neither got through its transformation by haphazard means.

Singapore in particular has been criticised for a lack of public participation (Briffett and Mackie, 2002). This does not mean that it has failed to create an environment that is fit for its people to live in. The 1992 Green Plan (Singapore Ministry of Environment, 1992, from Leitmann, 1999, 6) set out a vision of Singapore

> as a model Green City by the year 2000. It will be a city with high standards of public health and a quality environment. One which is conducive to gracious living with clean air, clean land, clean water and a quiet living environment. A city with people who are concerned for and take personal interest in the care of not just their immediate environment but of the global environment as well. A city which will also be a regional center for environmental technology. The role of the public in achieving this vision is crucial. We will have to educate the people and to instill in every Singaporean a national commitment to protect and preserve the environment at home and globally.

Modernization and public participation

That is not quite what is meant in Western countries by public participation, where dialogue is expected to go both ways. It is not entirely Hong Kong's way of doing it either. It is Singapore's way, or at least it was, before its public's concerns became more sophisticated. Many aspects of it are similar to Curitiba's way (Box 4.2). Any city planner who needs to consult the public in order to discover that they want clean air, clean land, clean water, and the means to earn a decent living, should not be in the job.

Success in managing the city's transformation has little to do with democracy as it is normally understood. When Lerner was appointed mayor of Curitiba, his country was ruled by a military dictatorship. The appropriation of a city's governance by powerful interests, with little regard for its people or its environment, may happen or be avoided under any system of government. There are many factors which determine the way in which a city will develop, of which the ideas and the ideals of a small group of influential people is perhaps the most important. In industrializing Britain, Robert Owen's influence on the country's social

Box 4.2

Lessons from Singapore in addressing environmental issues

Start with the basics Singapore pursued a phased approach to tackling problems, beginning with environmental health issues (sanitation, vector control, food hygiene) and highly visible problems such as river and basin pollution.

Co-ordinate planning in key sectors The integration of land use, public transportation and motorization plans and policies has allowed Singapore to reduce the environmental impact of the private automobile.

Integrate environmental considerations in standard procedures Environmental protection is an integral part of land use planning, industrial siting and building controls, largely negating the need for an environmental impact assessment process.

Get the politicians on board Political will has been an essential force behind successful planning and implementation of environmental measures, including the Prime Minister's support for tree planting, the Garden City campaign and river clean-up, and the Cabinet's endorsement of the Green Plan.

Educate, monitor and enforce Environmental regulation has been so successful in Singapore because public awareness of new environmental measures is followed by monitoring and inspection with strict and consistent enforcement of serious penalties.

Manage through institutions with clout The ENV [Ministry of Environment] provides strong environmental management because it integrates important functions like infrastructure and environmental health, and because it has real enforcement powers.

Try and try again A willingness to experiment, learn and evolve has benefited both institutions such the as the ENV and programmes like the various incarnations of Park-and-Ride.

Combine economic instruments with regulatory measures   Traffic management, one of Singapore's biggest successes, is a good example of how rules can be complemented by economic incentives such as road pricing, the high cost of vehicle ownership and, of course, fines.

Involve the private sector   Singapore has made effective use of the private sector for implementing environmental policies such as partnerships to sponsor recycling centers and licensing for hazardous waste collection and treatment.

Source: Leitmann, 1999

transformation pre-dated the 1832 Reform Act's democratic process by nearly two decades. Owen and Lerner were not politicians but practising professionals, driven by their ideals to develop the political skills to realise them. Such people can emerge in any of the different circumstances of different cities in different countries.

4.6 Pollution havens, pollution heavens and races to top and bottom

If a city sets out to make its fortune from exports, there is a danger of it selling itself cheap, to the detriment of the environment and people. If the consumers to whom it sells can buy cheap, the chances are that they will. If people in developed countries can offload dirty industry to the other side of the ocean, they will. If they can say goodbye to low paying jobs and get better ones themselves, they are liable to do so. As trade becomes increasingly international, opportunities may arise for industry to seek out havens for pollution that will accept effluents without demur, and for poor countries to bid each other down in selling low wage labour. If that happens, there may be a tendency for one side of the world to get dirtier and poorer, while the other gets cleaner and richer.

Trading up or trading down

What the city sells need not be from the city itself, but rather from its hinterland: it may be oil, diamonds, bananas or timber. Countries that are rich in natural resources find it difficult to develop, because the city can prosper by just selling. Oil does not play the same role in the Niger Delta as it did in Texas. When Texans found black gold there was no market for it. The market had to be created, with the help of Henry Ford and the rest of American industry. Now the market exists and the city's global traders can just sell, needing only a small number of high wage employees to extract the product. Statistical analyses of the relationship between a country's endowment of exportable natural resources and its rate of economic growth show an inverse correlation. This is known as the resource curse (Auty, 2000; Sachs and Warner, 1995). In countries that are unencumbered with such easy pickings development can take place more readily, but even there it is difficult.

In many countries the global trading relationships which exist now differ little from those put in place during the imperial era. The dominions and colonies of empire were established as sources of raw materials with low labour costs, traded in exchange for the higher value products of industrialization (Hobsbawm, 1969). Some developing countries have succeeded in industrializing to compete with the West, but many have not. Overseas aid and world trade rules encourage them to stick to mineral resources and agriculture, except for the strategic agricultural products grown on high income countries' subsidised farms (George and Kirkpatrick, 2008). Agriculture earns low wages until it is modernized and sheds labour, which migrates to the cities in search of work. The contemporary neo-liberal philosophy of economic management promoted by the World Bank and others argues that market forces will automatically create new high wage opportunities in the city. Stiglitz and others argue vehemently against this supposition (Stiglitz, 2002). If a city's traders are already wealthy from selling agricultural produce, they may prefer to avoid the complex difficulties of setting up the educational and institutional structures of an industrial economy, and stick to what they know.

Environment versus investment?

When a country does begin to industrialize, its main comparative advantage is the ability to do it cheaply. It will not be able to do it if wages or other costs rise too much, including the costs of environmental protection. There may be an incentive to keep environmental standards low, to encourage foreign companies to invest.

Most studies of the 'pollution haven' thesis show little evidence that multinational companies choose the location of their factories according to how much they can get away with in pollution control. Research conducted by the US government has indicated that environmental protection costs have sometimes been a factor in industry relocation (Grossman, 2002), but studies by the Mexican government of the effects of the North American Free Trade Agreement suggest that foreign corporations have better environmental and safety records than domestic firms. Similar findings have come from a comparison of multinational and local companies' operations in China (Lu, 2001). Other studies have shown that, overall, environmental protection costs are rarely a major factor in international companies' relocation decisions (Grossman, 2002; van Liemt, 2001).

This does not necessarily mean that developing country governments will adopt high environmental standards, nor that wages will rise. Klein has shown how the purchasing policies of transnational corporations keep the labour costs of their subcontractors to a minimum, subject to whatever pressures may be put on them by international standards and consumer protests (Klein, 2001). For the corporations' own facilities however, environmental costs are only one of many factors in their choice of location. In manufacturing, the cost of environmental control is typically 2 to 3 per cent of total costs, although it can be much higher in particular sectors (UNEP/IISD, 2005).

This does not apply so strongly for commodities. Here, economic and environmental efficiency may be less closely aligned, and consumer pressures may have less effect. 'Buyers of copper, for example, want the cheapest copper that meets their technical specifications, and they typically do not care about the pollution created in its manufacture' (UNEP/IISD, 2005, 59). Consumers may be unaware of environmental or social issues until they become crises, such as the Ogoni affair in the Niger Delta, which led to Nigeria's temporary exclusion from the Commonwealth along with financial loss for the international oil companies involved (HRW, 1999).

Small business regulation

In general, it is not through the activities of multinational corporations that the effects of weak environmental regulation in developing countries are mainly felt, but through industries that are owned locally. Large installations, whatever their ownership, are relatively easy to regulate. By contrast, thousands of tiny workshops dyeing textiles or electroplating industrial components are a regulator's nightmare. For example, one of the authors experienced a calamity in which a facility filling cigarette lighters in a multi-storey block of residential dwellings exploded, causing many injuries and deaths as the building collapsed. The solutions to such problems are not straightforward. The obvious answer is to enforce laws that ban all such workshops, but this would put many thousands of people out of work. Meanwhile, other problems arise when large polluting industries are owned by people closely connected to government, including senior ministers themselves.

Close links between government and the ownership of industry can be a problem in any country (UNDP, 2005). So can tracking down the illegal operators who export and import hazardous waste. However, a bigger difficulty faced by developing countries lies in making the transition from thousands of tiny workshops to clean, competitive, high-technology factories, employing far fewer people. Irrespective of the huge investment that is needed, people who are already surplus to the requirements of agriculture are likely to become surplus to industry as well.

Green protectionism

Developing countries have expressed strong concern that their richer trading partners' pressures for stronger environmental controls amount to nothing more than green protectionism (Najam and Robins, 2001). Inefficient as small workshops may be, they are able to compete with high income countries' factories through their low labour costs. The added cost of waste water treatment and environmental management infrastructure may make them uncompetitive. The same problem applies to calls for higher labour standards and wages. Unless the call is accompanied by a demand for higher prices in America and Europe, it may leave surplus agricultural workers with no option but even lowlier service labour, in an environment that is just as heavily polluted. Factory work at least offers hope that people will pick up the skills to start designing, producing, marketing and financing high added-value products themselves, to compete for the relatively pollution-free environment of a high income service economy.

Commodities on the open market

In the manufacturing industry, any advantage multinational corporations may take of weak environmental and labour regulation tends to be indirect, operating through local producers.

As we noted above, this does not apply so strongly to the production of commodities, which is a major component of the economy in many developing countries, and can have major environmental effects. The consumption of commodities meanwhile may be a major component of developed countries' economies, with consequent pressures to keep prices down.

Theoretical analyses of the behaviour of companies have suggested that, in the long term, environmental quality is influenced more by the impacts of international trade than by differing environmental standards and regulatory capability (Liddle, 2001). If international trade strengthens existing economic relationships between developed and developing countries, it may reinforce their differences in social and environmental quality. If on the other hand it helps developing countries make the transition to being developed countries themselves, those differences will be reduced. However, this may not not necessarily occur in such a way as to bring low standards up to match high ones. The external environmental resources on which all the world's cities draw are finite. If convergence is to be achieved by levelling up rather than some degree of levelling down, means have to be found of achieving the desired quality of life in all countries, without potentially catastrophic environmental impacts.

4.7 The global city-region

The sustainable development debate includes an implicit assumption that a developing country's cities will achieve the same level of material affluence as developed ones. Some parts of the debate argue that for this to happen, developed countries will have to cut back on their material affluence. Others believe this to be unnecessary, but still assume that the distinction between developed and developing countries will disappear. Even the United States Treasury has suggested specific targets for developing countries to catch up with developed ones. 'The greater the productivity gap between a country and the United States, the greater should be the productivity growth rate in that country' (Taylor, 2002, 2). For a country whose productivity is a fifth of that of the United States, the suggested target is a 3 per cent higher growth rate than the US, and 9 per cent higher for the poorest countries, whose labour productivity is currently a hundredth of the US level.

Urban development and global impacts

Achieving such goals poses considerable problems for the global environment. If greenhouse gas emissions in developing countries were to rise to the same per capita level as in developed ones, the magnitude of the effect would be magnified by a factor of six to ten, depending on rates of population growth (Brown, 1996). Similar concerns over biodiversity loss derive primarily from the current loss of natural habitat in developing countries, towards the extremely low levels in industrialised countries. Semi-natural habitat remains significant in most industrial countries and is fairly well protected, but fully natural habitat is a rarity (Spellerberg, 1992).

Implicit in the US Treasury's suggested goal for the development of developing countries is the assumption that the necessary growth is attainable without over-burdening the global environment. For biodiversity to be conserved, this means that developing countries' areas of natural habitat must not be allowed to fall below some critical level, which must remain unconverted to agricultural use. For climate change to be contained, the goal implies that per capita greenhouse emissions in developed countries must fall by a factor of six or more.

Economic welfare and the environment

Much of the sustainable development debate revolves around whether these necessary outcomes can be achieved through regulation and planning, or through market forces.

At one extreme, planning would entail detailed codification and enforcement of multilateral environmental agreements such as the Climate Change and Biodiversity Conventions. These would be designed in such a way as to provide equal per capita allocations of, and responsibility for, the available global resource.

At the other extreme, the invisible hand of the market would itself define the allocations, in such a way as to sustain the resource indefinitely. In theory, as the supply of natural habitat falls, the value of its tradeable biodiversity rises, to the point where it exceeds its value as agricultural land; as the climate changes, insurance costs rise, to the point where they exceed the economic gains available from activities that produce greenhouse gases; and as supplies of oil and natural gas decline, their prices rise, increasing the incentive to develop alternatives.

Present international policy is part way between these two extremes, though somewhat favouring market forces over regulation and strategic planning. A series of World Summits, in Stockholm in 1972, in Rio de Janeiro in 1992 and in Johannesburg in 2002, have attempted to push the balance more towards planning, with relatively little success. The Biodiversity Convention contains few requirements that are sufficiently specific to have major effect, while the Kyoto Protocol has yet to be fully implemented and is itself insufficient to contain climate change. The principal difficulty is the absence of an international authority. The World Trade Organisation has the authority to establish global rules and enforce them, but its remit is only to manage trade. The United Nations can organise conferences, but it cannot regulate or plan.

International trade

Meanwhile, international trade has reached the level where cities no longer draw their supplies primarily from their own city-regions or nations. In 1999 international trade amounted to 22 per cent of global GDP, up from 19 per cent in 1990 (World Bank, 2001b). Trade rules encourage traders to capitalize on their countries' existing comparative advantages rather than develop new ones. Those with access to high technology sell high technology goods and services, those with oil sell oil, and those with access to cheap labour sell cheap labour. Market forces dictate that if the average wage in one country exceeds that in another by the cost of transporting goods between them, food will be grown in one to be eaten in the other, irrespective of the wider consequences. High value goods and services will be sent back in return, to be consumed by the affluent sections of society.

At some point in the future, some form of governance may exist under which products are not shipped across the world in response to different wage rates, where the global footprints of the citizens of Stockholm and Johannesburg are the same, where they trade in products whose added-value is similar, where they all live in a built environment that is fit to live in, and where the global rural environment on which they all depend is shared equally between them and the people who live in it. Achieving such a situation presents a major challenge for the democratic process, and also for the management of the shared global environment.

The tragedy of the commons

Biodiversity and the global atmosphere are public goods, in the sense that the public can benefit from them without the intervention of a market. The price mechanism does not operate for public goods, in what is termed market failure. Unlike goods that can be owned, their use is 'non-exclusive', meaning that reduced consumption by one person allows others to consume more. If anyone invests in an improvement, 'free riders' will take the profit. Some public goods, such as fresh water, can be privatized, so that market forces apply. Biodiversity and climate cannot, but an artificial market can be created instead, such as with the carbon emissions trading schemes established under the Kyoto protocol. This creates an incentive to reduce emissions, but the size of the effect depends entirely on the level of credits that are granted by governments, and the means by which the market is regulated. Governments continue to suffer from the problem that any individual contribution they might make to tackling a global issue will result in an economic disadvantage, compared with countries that choose not to follow their example. Garrett Hardin called this The Tragedy of the Commons (Hardin, 1968).

Hardin's original concern was mainly for population growth. Since then it has become fairly clear how that particular problem might be solved, as developing countries pass through their own demographic transition to population stability (Simpson, 1994). The fundamental problem remains however. Whether the commons be local, national or global, they are a free-for-all, until some kind of sovereignty is established over them. Two schools of thought have emerged on managing the global commons (Vogler, 2000); first, voluntary agreements between separate sovereign states and, second, enforceable global law. The second is the more consistent with Hardin's analysis. Following that view, the world needs its planning authority and laws to support it.

The world's cities are becoming parts of a single global entity, each one dependent on every other, drawing their resources from just one global city-region, with no more neighbours left outside it and, ultimately, no more independent sovereignties to compete for it. How the forces which have created this emerging situation will play themselves out in the coming decades remains a matter for extensive debate, but once they have done so the world will be a very different place (George, 2007). Those who would plan for it instead of letting it just happen have much the same task as those who would plan the next Curitiba.

4.8 Conclusions

The cities of the developing 'South' have much in common with those of the developing 'North' of 200 years ago, as centres of wealth from international trade, and magnets for growing populations that are surplus to rural demand. Like the London and New York of the early nineteenth century, many of their built environments boast extravagant affluence alongside abject squalor. The cities of the 'North' have left the worst of that behind, while stamping their consumers' footprints across the entire planet.

In this chapter we have examined the mechanisms that cause cities to grow and the factors that influence the quality of their changing environments. Some suffer from the effects of industrial pollution, some from sprawling slums, and most from the exhausts of ever-denser traffic. Some grow in whatever way the market dictates and accept the consequences. Others have envisioned a better future for their citizens and have planned accordingly. All, in the 'North' as well as the 'South', have to find their own unique place in an interdependent, rapidly changing world. If all of their environments are to be what we would like them to be, the city planner must now be a global planner too.

Further information

United Nations Human Settlements Programme   http://www.un-habitat.org/

United Nations Development Programme   http://www.undp.org/

United Nations Environment Programme   http://www.unep.org/

World Bank Urban Development   http://go.worldbank.org/PQE9TNVDI0

Local Governments for Sustainability   http://www.iclei.org/

ID21 Urban Development, Institute of Development Studies   http://www.id21.org/urban/index.html

Cities Alliance   http://www.citiesalliance.org/index.html

Further reading

DFID-UN (2002) *Sustainable Urbanisation: Achieving Agenda 21*, UK Department for International Development and UN Habitat, London and Nairobi: United Nations.

Jacobs, J. (1986) *Cities and the Wealth of Nations*, Harmondsworth: Penguin.

Newman, P. (2006) The environmental impact of cities, *Environment and Urbanization*,18; 275.

Robinson, J. (2002) Global and world cities: a view from off the map, *International Journal of Urban and Regional Research*, 26.3, pp. 531–4.

Simpson, E.S. (1994) *The Developing World*, Harlow: Longman.

UNCHS (2006) *The State of the World's Cities Report 2006/7*, United Nations Human Settlements Programme, London: Earthscan.

⑤ Towards the eco-city
The physical urban environment

5.1 Introduction

Since their origins, cities have been the sites of intense concentrations of flows of energy and material, alongside the flows of pollution and waste. The concept of the 'urban environmental transition' was explored in Chapter 3 – the bigger picture of urbanization, modernization, and the changing agenda for the urban environment. This showed how the gross pollutants from the early stages of industrialization are being replaced by the more insidious substances of a more complex globalized economy, where even the stress of modern living is another kind of pollutant for many. Overall, it seems that economic development creates new pollutants and hazards as fast as old ones are cleaned up; but meanwhile the traditional air pollutants of 'SO_x, NO_x and VOCs' still hang over most cities (abbreviations in the Appendix).

Clean air, water and land are at the core of the environmental agenda, but this raises the questions – how clean is clean, who causes the pollution, who pays the price, and how much? And if pollution is defined as matter in the wrong place and time, then the definition of what is the right place is all-important, in terms of cultural values and lifestyle habits (Douglas, 1983). So underlying the agenda of this chapter is the question of 'sustainable pollution', as a balance between material flows and human values. This goes far beyond the conventional end-of-pipe approach of cleaning up pollution after the event – it looks towards a new-kind-of-pipe approach, of the restructuring and transformation of technologies, markets, supply chains and consumer choices.

In this chapter

The focus here is the physical dimension of the urban environment. This starts with the 'services' provided by ecosystems and the environment, the 'metabolism' of resource flows, and their impacts both local and global.

The following sections look at the environmental media, solid waste and material flows, water supply and water quality, and air emissions and air quality. At the centre is the crucial role of the energy system and its effects on climate change; and this in turn raises the impacts of climate change, and other environmental hazards and disasters. In order to focus on the urban scale, this book leaves to others the details of environmental science, and other environmental issues such as ozone depletion, radioactive waste, marine pollution and so on.

In each section, the critical perspectives bring insight on the human context, and why air quality policies or waste taxes may or may not work in different cities around the world. Over-arching this again is the linkage between the 'green' agenda of the 'North', and the 'brown' agenda of the 'South' – where local environmental problems come from global pressures, and sometimes vice versa.

5.2 Urban environmental metabolism

Ecosystems and urban systems

The urban trees seen from an average urban window might have many uses – as an ecological habitat; a piece of private property; or a public resource for shading and filtering the air; they might also be a risk or liability if they fall over. In many cases they perform all these functions and more – they are a typical 'multi-functional' resource (Mander, *et al.*, 2007). And yet these trees are not isolated objects – they depend on throughputs of water, sunlight, maintenance and other systems.

The ideas of 'service' and 'metabolism', as outlined in Chapter 2, are
well suited to understanding the complexity of the urban environment.
One way to look at this is as a set of human-environment functions, or
'ecosystem services', in and around human settlements (Douglas, 1983;
de Groot, *et al.*, 2002). A related notion is that of 'metabolism', or the
flows of energy and resources through a city system.

Broadly, the surrounding physical ecosystems can provide services
to urban economies and societies in four main ways: as provisioning,
cultural, regulation and supporting services (Millennium Ecosystems
Assessment, 2005c) (Figure 5.1).

- 'Provisioning services' – tangible goods which ecosystems provide
  directly. This could be fresh water for consumption or production; food
  for consumption; forest and crop plantations for energy and fibre.
- 'Cultural services'– more intangible experiences which are offered or
  enabled by ecosystems. Landscapes, uplands, community forests and
  urban green space are valued for aesthetic and recreational qualities:

Four main types of ecosystem services, together with urban & peri-urban responses
Source: Millenium Ecosystems Assessment, 2005: Ravetz, 2000

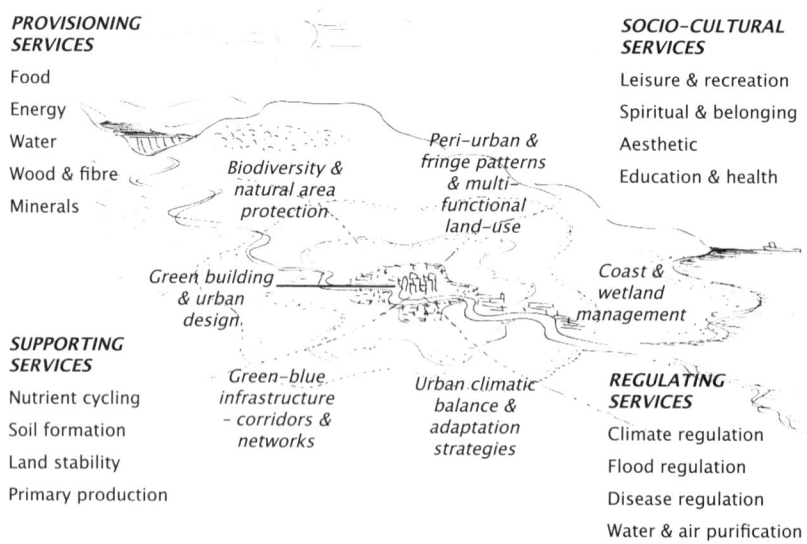

PROVISIONING SERVICES

Food
Energy
Water
Wood & fibre
Minerals

Biodiversity & natural area protection

Peri-urban & fringe patterns & multi-functional land-use

SOCIO-CULTURAL SERVICES

Leisure & recreation
Spiritual & belonging
Aesthetic
Education & health

Green building & urban design

Coast & wetland management

SUPPORTING SERVICES

Nutrient cycling
Soil formation
Land stability
Primary production

Green-blue infrastructure - corridors & networks

Urban climatic balance & adaptation strategies

REGULATING SERVICES

Climate regulation
Flood regulation
Disease regulation
Water & air purification

Figure 5.1 *Urban Ecosystems services*

reservoirs, canals and urban water courses enable social relations and cultural identity.

- 'Regulating services' – benefits from ecosystems concerning regulation of natural processes. Wetlands, dunes, and floodplains for flood and flow regulation; vegetative cover for erosion regulation; peat bogs for carbon sequestration, are all examples of the regulation functions, which urban development ignores at its peril.
- 'Supporting services' – these underpin the provision of other ecosystem services. Soil formation is essential to other services; wetlands, aquifers and riparian habitats for water cycling; soil for nutrient cycling.

This concept of ecosystems services can be extended to include the 'urban metabolism' or flow of energy and materials through a city system, as inputs, throughputs and outputs, from and to the surrounding environment (bearing in mind the question in Chapter 3 of what is a city or a city-region).

'Environmental inputs' describes the physical flows of resources and energy into the urban system, from its hinterland or from global supply chains:

- supply of food for the urban population;
- supply of fuels and/or distributed energy for the built environment;
- supply of materials for constructing the built environment;
- supply of materials for the urban economy.

'Environmental throughput' refers to the physical metabolism itself, which works at many different scales in space and time:

- flows of materials, water and energy, through the *supply side* of industry and the urban economy;
- flows of materials, water and energy through the *demand side* of households and the built environment;
- flows of transport systems and their infrastructure, including by road, rail, water and air.

'Environmental outputs' covers the obvious themes of urban pollution and solid waste. If we extend the concept of outputs to environmental 'outcomes', then we need to consider the final destinations, impacts and indirect effects, which may be distant in space and time:

- climate emissions from energy used in the city, or 'embodied' in products and services imported to the city;
- air pollution and assimilation of pollution on a local or regional scale;
- water effluent and its effect on rivers, seas, groundwater;
- solid waste, its management and disposal, and its environmental 'end-fate' or final destination.

Displacement, risk and responsibility

Each kind of environmental problem can be placed somewhere on a scale between local and global, and between short term and long term (Figure 5.2). This shows that short-term causes, for instance in burning dirty fuels, can lead to long-term outcomes, such as the acidification of ecosystems. Meanwhile, localized problems, such as urban biodiversity, can be driven by globalized pressures, such as climate change; and likewise, that actions now can cause problems for future generations.

Such problems of 'displacement' across space and time are typical of the environmental agenda. These raise the political question of multi-level

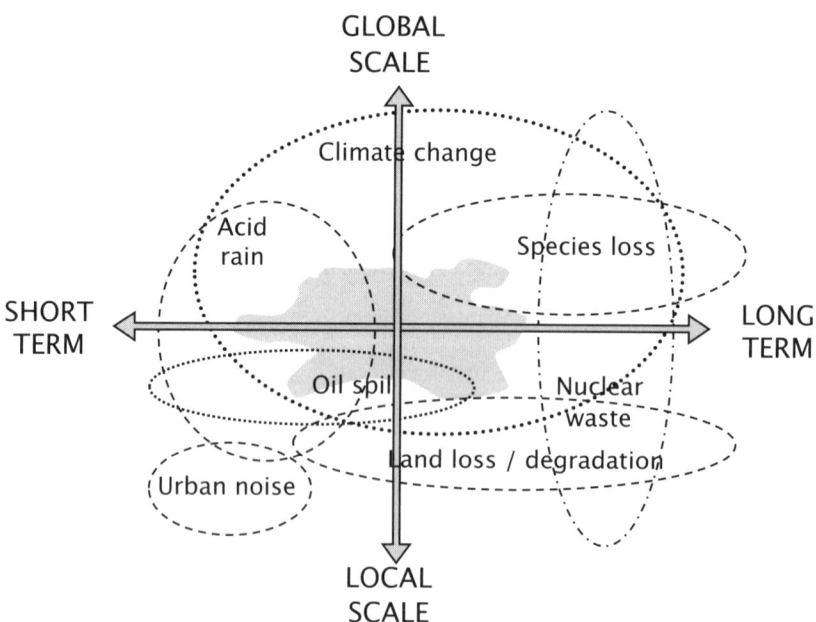

Figure 5.2 *Environmental impacts in space and time*

governance: that is, how to integrate the needs and demands of different groups from local to national to global. It also raises an economic question of costs and benefits, centred around the concept of welfare economics and the principle of the 'polluter pays'. For example, who should pay the cost of cleaning up air emissions from domestic heating? Should it be the fuel suppliers, heating installers, the building landlords, or people downwind who would benefit most directly? And if a tax is put on the polluting fuel, does that then mean that the poor will be unable to heat their homes?

An equally controversial debate is that of risk and uncertainty (see the discussion in Chapter 8). The reality is that most environmental impacts are not in a straight line of cause and effect, but hedged with many layers of risk and uncertainty, on health impacts, technological change, and financial costs or benefits. This means that urban environmental policy has to combine scientific uncertainty with many other kinds of controversy – political, social, economic, cultural and technological. The evidence, for example, on the effect of vehicle emissions on asthma and allergenic conditions, is still a medical controversy and public debate (Vasconcellos, 2001).

Towards sustainable pollution

If pollution is *de facto* wasteful and damaging, can it ever be 'sustainable'? Such a question hinges on the balance of environmental, social and economic values. From a scientific angle, some fundamental goals are shown by the Natural Step 'system conditions', although it is not so simple to relate these to the complex flows of materials through an open city system (Natural Step Foundation, 1999):

● no net reduction in stocks of renewable resources;
● no net accumulation of artificial substances in the biosphere;
● no net transfer of substances from the Earth's crust to the biosphere;
● social and economic development to enable the above conditions to be met – (clearly, this condition could be the most crucial and problematic).

Other measures include the 'environmental burdens' approach: this assesses pollution loads in terms of their eventual impacts, for example on climate, ozone protection or acidification (http://www.scorecard.

org). The indicators of 'environmental space' aim to estimate, for each commodity, an equal distribution of global resources and pollution assimilation capacity (Carley and Spapens, 1997). Each of these focuses on the environmental impacts side, and needs to be accompanied by measures of the management and performance of products and processes. Here for example, the European Integrated Pollution Prevention and Control (IPPC) system hinges on some legal definitions and practical interpretations which are often quite fuzzy. 'Best Practical Environmental Option' (BPEO) and 'Best Available Technology Not Entailing Excessive Cost' (BATNEEC), both raise questions on what is 'practical' and how much cost is really excessive (Gouldson and Murphy, 1998).

In practice such methods either focus on environmental standards, or policies and management processes, and there are many gaps in between. Each method points in the direction of what is meant by sustainable pollution – in the sense of sustaining local and global ecological integrity, and sustaining society and the economy within 'acceptable' levels of risk or safety. One guideline comes in the shape of the 'precautionary principle'; again this is fine in theory, but in practice the risks of uncertain future impacts have to be balanced against the risks of social or economic impacts in the present (O'Riordan 1996). So, if the definition of sustainable pollution is not a single fixed and scientific target, perhaps it should be the outcome of a 'deliberative process' of public debate and evaluation (European Environment Agency, 1995).

Resource flow metabolism

There is a new-found interest in measuring the environmental metabolism of cities and regions, and of industries or products, in terms of Material and Energy Flow Analysis (MEFA). This looks at the resource inputs in terms of raw materials and products, and at the resource outputs in terms of waste and emissions, plus or minus any changes in stocks, and the balance of imports and exports (Eurostat, 2000). The MEFA method divides into two strands of parallel accounting. Production-based analysis includes imports and locally generated pollution and waste: in contrast, consumption-based analysis counts the impacts of the supply chains which reach the final consumer, which could be anywhere in the world.

Such flows can be aggregated up to the total 'direct material inputs' to an urban system. For the average city in the EU, for example, 90 per cent of input by volume is water, and solid inputs are in the region of 15 tonnes per year per person, 10 tonnes in mineral form and 5 tonnes in biomass. One third of the solids go directly into buildings and infrastructure, a fifth is exported, a fifth goes to households, and the remainder goes to waste (European Environment Agency, 1995; Barrett, *et al.*, 2006). Following the calculation through means that the total mass of the city increases every year by about 6.5 tonnes per person. For highly automotive dependent cities such as those in the USA mid-west, estimates are that the built environment with all its infrastructure contains 250 tonnes of minerals and aggregate per resident (Davis, 1998, 80).

Ecological footprint targets

The ecological footprint analysis method then builds on the MEFA approach. As mentioned in Chapter 1, it has gained worldwide attention, as it aims to provide a single index of all environmental and resource impacts, measured by the most finite resource of all – land area (Rees and Wackernagel, 1995; Chambers, *et al.*, 2000). The footprint is an aggregated measure of environmental impact allocated by final consumption, and it aims at a similar role to that of GDP in national economic accounts. It is measured in a standardised area unit, equivalent to a world average bio-productive hectare or 'global hectare' (gha), and

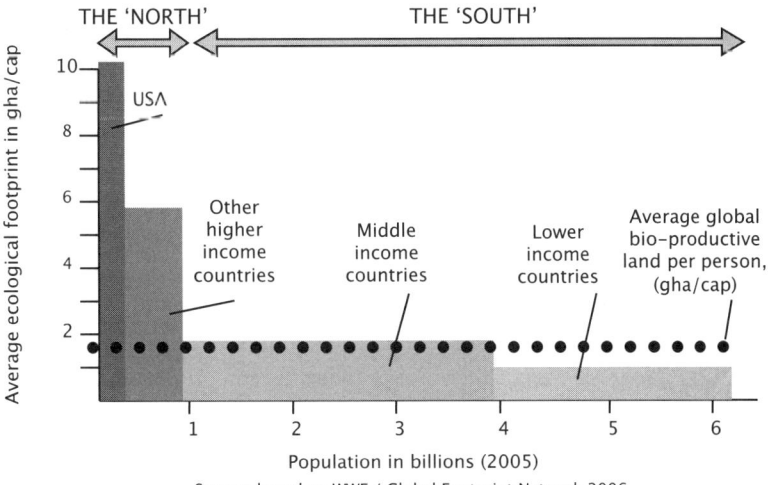

Figure 5.3 *Ecological footprints – a global comparison*

for comparison is expressed as *gha per person* (gha/cap). While the method is very simplistic, and leaves out important questions such as toxicity or bio-diversity, it does have an advantage over climate emissions accounts, in tracking the totality of supply chains, and the global distribution of resources (Figure 5.3).

For instance, if a steel mill relocates from Pennsylvania to Penang, and then ships its products back across the ocean, the climate emissions account would show a big reduction in Pennsylvania. In contrast the footprint account would show a more accurate picture of the total impacts of consumption in the USA, production in Malaysia, and the transport between them.

The key finding of the footprint analysis is that most industrialized nations are exploiting the Earth's bio-productive land area at more than three times their 'earth-share' – in other words, for the world population to live as the affluent 'North' do, would require at least three planets (World Wildlife Fund, 2006). From this very stark conclusion, there follows a set of medium- and long-term targets:

● By 2050 there is likely to be a 50 per cent increase in world population, up to 9 billion people (UNFPA, 2007). There may be additions to bio-productive land area, through reclamation and reforestation, but other damage is also likely, due to climate change, industrialized agriculture and urbanization. So, a mid-range estimate of change in earth-share bio-capacity would be from 1.8 to 1.3 gha/cap in 2050.
● To reduce the current average footprint of OECD nations of 5–6 gha/cap, to the 2050 fair share bio-capacity of 1.3 gha/cap would need reduction by 75–80 per cent, or a 'Factor of Four' (von Weizsacker, *et al.*, 1997). This equates to a year on year reduction in total resource use of about 4 per cent per year (if starting from 2010).
● Setting this against economic growth at an assumed 2.5 per cent (typical of more developed nations), then the required rate of 'de-coupling' or improvement in the resource efficiency (footprint/GDP) would be a reduction of about 6.5 per cent, year on year for the next 40 years. This is over twice the rate of decoupling in the recent past, which held resource use more or less level against continuing economic growth.

This Factor Four target rate – 4 per cent reduction in absolute resource use, and 6.5 per cent in relative decoupling – is the ultimate benchmark for a pathway to environmental sustainability. It can be translated into schemes such as the 'cap and trade' principle of the European Emissions Trading System: using the 'cap' as a ceiling on total emissions, this should in principle be targeted to reduce by 4 per cent per year. This provides a combined sustainability target for both 'North' and 'South' – but the implication is that as the 'South' needs to develop, the 'North' should be more stringent in its footprint targets.

These idealized targets apply not only to national economies but to urban economies and urban development, given that 80–90 per cent of the population of the 'North' is urbanized, and that cities are the locus of growth and modernization in the 'South'. And, as urban systems are increasingly service-based and consumer-based, the footprint measure is in many ways more relevant than measures of direct emissions. In practice, the world is just waking up to the challenge of climate change and the responsibility of cities and nations for direct emissions. Although there is a groundswell of support for the consumption-based approach and the use of the footprint metric, this has yet to translate into firm policy and programme development (Kitzes, *et al.*, 2008).

Global resource flows and urbanization

Applying these measures to the international urban order, we can begin to see how resources flow through cities and urban systems around the world (Figure 5.4) (Ravetz, 2006b). A typical city-region in the 'South' shows large volumes of primary extractions and harvests, a small service sector, large export volumes and small investment in capital equipment and infrastructure. Pollution is often high due to unregulated industry, and the external impacts of this commodity-based economy are also high, in terms of forced evictions, destruction of ecosystems, and chaotic urbanization (see Box 5.1).

By contrast, a typical city of the 'North' shows large import volumes, a slim and capital-intensive manufacturing sector, and a growing service sector. The factors of energy, transport and infrastructure are large, as are capital investments. While emissions in well-regulated industries are small, waste volumes are high, due to the sheer throughput of consumer goods and packaging. While some waste is recycled, growing amounts are exported in return container loads, often creating severe pollution in the destination country.

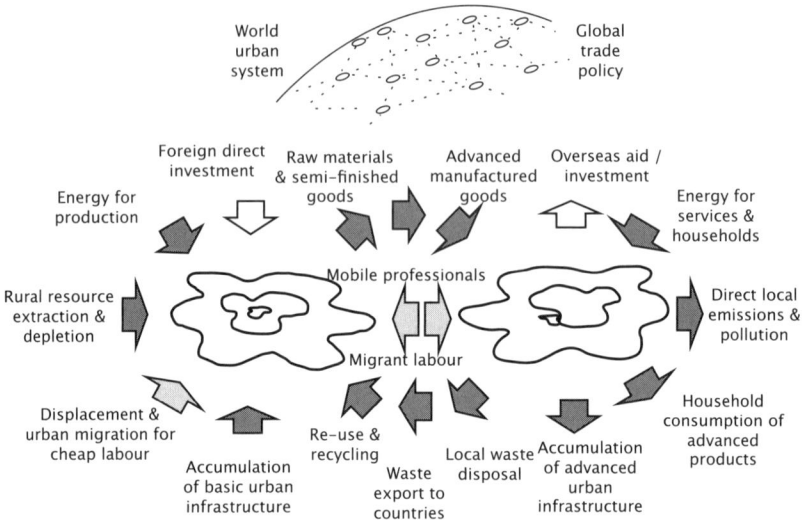

Figure 5.4 *Global resource flows in urban systems*

The picture as a whole shows in a graphic way how the metabolism of urban affluence is inextricably linked with the metabolism of urban poverty in 'Southern' cities. Further investigation shows how such cities themselves are often acting as expropriators of the resources of their hinterlands (Stren, *et al.*, 1992).

Critical perspectives on environmental management

While this chapter is focused on the environment – air, water and waste – each of these is driven by politics, economics and institutional forces. So again, critical perspectives are needed for further insight.

- Globalization and liberalization: the first implication for the environment is in the political economy of utilities and infrastructure such as water and energy. Globalization is also a factor in environmental policy, where raising standards can often risk the migration of industry across the world.
- Governance and public management: the complexities of regulation are shown by air quality or water management, where typically the

impacts are displaced in time and space, from the powers to deal with them.

- Consumption, affluence, identity, locality, cultures of place: again there is displacement, for instance, where the waste management crisis in many cities can be argued as a problem of consumption as well as production.
- Growth and decline, and the pattern of urban hubs and peripheries: this highlights the role of environmental quality and environmental planning in urban competitiveness, 'place marketing', and the rise of the leisure and tourism industries.

There are parallel themes raised by the shift from 'end-of-pipe' pollution control, towards a 'new kind of pipe' which aims towards innovative technologies, markets and supply chains.

First is the question of environmental ethics and environmental justice – the values which underpin the organization of cities and distribution of resources. Such values may be framed as 'economic' values which can be measured and traded, as in the concept of shadow markets (see Chapter 9). They may also be framed as 'moral' values, for the questions of justice and equity between generations, between communities or between other species. In cities as far apart as Mexico City and Beijing, there are NGOs and popular movements based on the theme of environmental justice at local and global levels.

The questions of justice and equity then point towards the deep-rooted divisions of gender and youth and their role in environmental hazards. Basic risks are often borne by women who cook with fuel wood or paraffin in smoky rooms, or by children who play in the street. As the unemployment rate for young adults in some cities is between 50 and 75 per cent, there is pressure on them to engage in unregulated, informal or criminal activities. There is an agenda for environmental justice, equity and empowerment, for all social groups – young, old, ethnic groups, racial groups, differently abled, and so on (Friedman, 1992 and 2005). This also raises the question of property and/or stewardship – both a practical question of tenure of land and housing, and also an ethical question of environmental rights and responsibilities.

5.3 Waste management

Solid waste is the residue at the end of the material chains of an industrial society – a massive resource, devalued to less than zero by its state of entropy or disorder. Until recently the waste industry was mainly concerned with holes in the ground, but in the 'North' such holes are more scarce and costly, and in the 'South' there is an emerging environmental agenda (see Box 5.1 below). In the long term view on sustainable development, the issue is not so much waste management as resource management – where recycling, re-using, re-manufacturing and waste minimization are the only viable options. This involves re-organizing and re-engineering material chains and processes right through the urban economy (Elkington, 1997).

However at present the global trends of solid waste arisings of between 3 and 5 per cent growth per year is set to continue, given the rapid industrialization of developing countries. Responsibility for such waste is often fragmented into many competing agencies and providers. For example in the UK, the Environment Agency regulates, the local Waste Disposal Authorities issue contracts, partnership companies provide processing and transport, other private companies operate the sites, local governments collect municipal waste, and private operators collect and manage industrial and commercial wastes. At each stage there are transaction costs, interlocking contracts, subsidies and taxes, new technologies, changing environmental legislation, and rapidly changing commodity markets – often a recipe for limiting progress to the speed of the slowest.

Waste hierarchy principle

The first principle of good waste management policy is the familiar waste hierarchy; the most desirable option being waste minimization, then material recycling, then recovery of energy or bio-mass, with landfill and uncontrolled burning as the least desirable options. However there is conflicting evidence on the life-cycle impacts of each option, as shown by the arguments over the EU Waste Directive, which requires pre-treatment of all landfill waste and a shift to other methods.

The BPEO principle is invoked, but in practice the definition of the BPEO appears to vary from place to place, making national standards and strategies quite difficult. The 'proximity principle', stating that waste should be recovered or disposed as near as possible to the point of arising, is also complicated by local differences in transport systems, disposal technologies, local sensitivities and so on (Kobus, 2003).

Waste disposal is also subject to ecological taxation in many 'Northern' cities (O'Riordan, 1996). The results are mixed in different countries – there is evidence of increased illegal disposal and shortages of fill materials, but the main benefit is in the strong signals given to waste producers and operators. An improved tax in future could be the central plank of an integrated waste system, extended to other disposal methods, other parts of the material chain, other tax differentials, and incentives for eco-efficient collection.

Recycling, reuse and minimization

Recycling is often seen as the most desirable solution to waste problems; but in the view of industry, the environmental benefits are not always clear. The economic viability depends on volatile markets, and success depends on changing consumer behavior, which is hard to guarantee (Gandy, 2002).

In many cities of the 'South', recycling from waste landfill sites is an important part of the informal economy, but conditions are often extremely hazardous. Re-use of packaging, containers and small items are also major activities for the urban poor. In many cities there are large material flows and cycles, which are outside the formal economy or waste disposal system, and hence unrecorded or unregulated. For instance, many street cafés in India use raw earthenware pots, which after use are compacted and re-manufactured into containers for the next day – while this is apparently very ecological, the risk of contagious disease is high.

Over the past five years, the idea of Zero Waste has been adopted by cities around the world. In 1996 Canberra became the first city to adopt a Zero Waste target for 2010. Some Californian authorities, having achieved their initial targets of 50 per cent waste reduction, are now moving to the next phase of Zero Waste. The approach is to set stretching targets, which then become 'Total Quality Management' challenges, at every level of the organization (Murray, 2002).

Recycling rates of 60 to 80 per cent can be achieved, not just as ends in themselves, but as part of an overall 'market transformation', with both environmental and economic benefits. There is an estimated potential for up to 1 per cent additional employment in most cities, through waste recycling and materials management industries. The eventual costs of full recycling systems can be a fraction of landfill or incineration methods. However there is generally a hump of increased capital and start up costs to be overcome, which could be found by diverting other revenues, or levies on waste streams. Funding can also involve public-private partnership initiatives with potential materials operators and markets – newsprint, metals, plastics and others – and in each case the public sector needs to expand its remit, from waste management, to integrated resource management.

The recycling strategy also has topical implications for a successful post-industrial business – customer focused, ICT responsive, supply-demand integration, just-in-time logistics, and a networked rather than monolithic

Box 5.1

International experience in waste recycling

The *Sydney Morning Herald* reports that rubbish from Europe's recycling bins is being transported half way across the world to blight China's booming coastal cities with plumes of acrid black smoke. Transporting the rubbish between Europe and China is cheap on return journeys of container ships. But its fate is very different from that imagined by conscientious householders when they sort their detritus into piles for 'environmentally friendly' disposal.

The rise in pollution has political repercussions. Outrage at the chemical industry near the south-eastern city of Dongyang last week provoked riots by farmers and elderly women. The streets are piled high with plastic, and resemble a cross between a junkyard and modern installation art. In front rooms, women cut up plastic with scissors, removing tags and zips for 20 yuan (US$3) a day.

Source: Resource Recovery Forum, 2006, http://www.resourcenotwaste.org

structure. Such a business also has to coordinate with existing schemes such as those for packaging. The UK, for instance, brought in one of the first internal trading systems based on the producer responsibility principle, with packaging recovery notes (PRNs) required for all medium and larger businesses. This created a tangled web of rival collection schemes, disputes on who 'owns' the packaging waste, and incentives for trading instead of real environmental improvements. A more pro-active resource management sector would transform systems such as the PRN, by enhanced partnership between packaging producers, consumers and operators. However this is not so simple to translate into practice.

In the longer term cities will need to rethink the concept of waste itself with a further set of R words – re-design, re-use, repair, re-condition and re-manufacture. The alternative is to continue wasting materials, of which the current value is estimated in the UK for one, at 6 per cent of manufacturing industry total profits (Cambridge Econometrics, *et al.*, 2003). Forward-looking companies, such as Interface Carpets, are now leasing services rather than selling products, and aiming at zero waste processes. But while larger process companies are willing to sign up to such schemes, it is still difficult to engage the majority of SMEs, and for these a more comprehensive support network may be needed (Howes, *et al.*, 1998).

Critical perspectives on waste and resources

The waste management agenda is far more than a question of the BPEO and improved technology; rather it cuts right across the political and economic spectrum. There is a very active vision of the emerging thinking on Zero Waste, eco-efficiency and integrated resource management, still mainly centred on the cities of the 'North' (Murray, 2002). In other parts of the world there are strong pressures and trends for rising waste volumes and increasing environmental impacts.

Globalization is again driving the pace of change: major factors are the liberalization of international trade, deregulation of national laws, and falling costs of international transport. This increases the profitability of international transfers of waste for disposal or recycling, as between the EU and China (Box 5.1). This international waste/resource trade can be seen in theory as the outcome of comparative advantage and of transport logistics. In practice it shows the danger of the 'race to the bottom', in terms of corruption, lax regulation, and disregard for social or environmental impacts.

Another facet of globalization is also seen in the privatization of municipal infrastructure such as waste management to integrated utility corporations (Bleischwitz and Proske, 2006). This enables inward investment, access to new technology and expertise, and a market discipline which may be lacking in the public sector. However the first objective inevitably becomes short-term profit, and this will tend to bind municipal authorities to technological end-of-pipe solutions, where waste minimization and recycling are less viable once the infrastructure is there. However, cost recovery models in waste are often different to those in water or energy networks, as the potential for cost avoidance by consumers is greater. The practical effect may be to exclude poorer groups, in order to target municipal waste collection on wealthier areas. Such poorer groups may then practice informal recycling within their communities, or on the landfills of the wealthy as a main livelihood (Rakodi, *et al.*, 2004; Adeyemi, *et al.*, 2001).

Waste management also raises difficult questions of governance and regulation. Waste facilities are generally seen as Local Unwanted Land Uses (LULUs), and by a process of elimination are generally located in poor or derelict industrial areas, often in urban fringe locations. Such 'impact displacement' can be measured in millions of tonnes exported from urban areas.

The production of waste reflects the lifestyles and consumption patterns of social and cultural groups. It can be argued that the dynamic of urban consumption makes increasing waste production almost inevitable (Bell and Jayne, 2006). New products and innovations which are based on recycled materials can be seen in a post-modern context as art, as idealism, or indeed as a new face of capitalism (McDonough and Braungart, 2002).

5.4 Water in the city

The water cycle forms over 90 per cent of the total resource flow in an average city – so water in many ways leads the theory and practice of sustainable development (Newson, 1992b; Hildering, 2004). There is an integrated view of the urban water system, as one in harmony

Showing key stages in the physical & urban water cycle, with alternative urban / hinterland boundaries

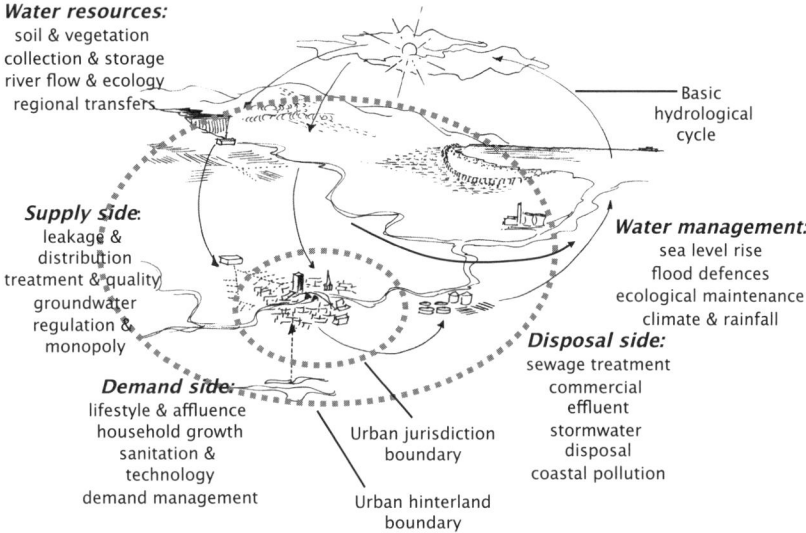

Water resources:
soil & vegetation
collection & storage
river flow & ecology
regional transfers

Basic
hydrological
cycle

Supply side:
leakage &
distribution
treatment & quality
groundwater
regulation &
monopoly

Water management:
sea level rise
flood defences
ecological maintenance
climate & rainfall

Disposal side:
sewage treatment
commercial
effluent
stormwater
disposal
coastal pollution

Demand side:
lifestyle & affluence
household growth
sanitation &
technology
demand management

Urban jurisdiction
boundary

Urban hinterland
boundary

Figure 5.5 *Water cycles and the urban environment*

with its hinterland; from the provision of water infrastructure to water resource management, river quality, groundwater, sewage and effluent (Figure 5.5).

In practice, water has been a source of conflict as long as cities have existed. Water quality problems may originate in urban areas, while their greatest effect may be downstream in other towns or even other countries: likewise, pollution sources from upstream agriculture may impact on urban drinking water. There are also problems from water abstraction leading to lowered river flows, which then exacerbate pollution loads, and make cities more vulnerable to stresses such as storm water peaks and seasonal flooding. Huge investments are made in urban treatment and distribution systems, which are then managed as financial commodities on a globalized utility market. The sustainable environmental management of water systems depends on an integrated approach to investment and governance, involving many stakeholders at different levels (Farmer, 1997; UNEP, 2002b).

Water and urbanization

Water has played a fundamental role in urban development since cities began, and almost all great cities are based on coasts or rivers, from Tokyo to New York. Now, with the continuing growth of larger cities and mega-cities, water is increasingly a medium of political and economic competition and conflict (Niemczynowicz, 1996). Alternative scenarios can be mapped out for ways in which urban development can disrupt and degrade the natural water systems on which it depends (Konig, 1999).

In one scenario, groundwater is available in adequate quantities and quality: but demand for potable water increases with city growth. Meanwhile the recharge of groundwater decreases due to reduced infiltration into sub-soils, and a diminishing groundwater supply is then subject to industrial pollution. In Latin America and Asia, many cities are sited over their own groundwater sources, and this leads to pollution, reducing water tables, and in some cases major ground subsidence (Foster, *et al.*, 1999). Parts of Mexico City, for instance, are subsiding at 6cm per year, with big problems for building subsidence and damaged infrastructure. The city then imports increasing amounts of potable water over longer distances, endangering the rural agricultural economy, and inflaming political conflict (Connolly, 2006).

In another scenario, potable water can be obtained from rivers or lakes, but treatment becomes more costly as growing neighbouring communities also dump their effluent into the same surface waters. The rising cost means that large populations are left without potable supplies and sanitation, and to avoid social unrest the city offers subsidies, which become an increasing burden on the taxpayer. While there are improvements in waste waster treatment and industrial processes, diffuse sources of pollution from agriculture are much more difficult to manage. For instance, it is recognized by the Chinese Ministry of Water Resources that water pollution is perhaps their foremost environmental problem, with over 90 per cent of urban rivers being polluted, and a quarter of the entire population drinking unsafe water (World Bank, 2001c).

Water infrastructure

A supply of clean water is taken for granted in developed cities, yet in many other parts of the world, supplies may be scarce, polluted and costly. Even in cities with treated water supplies, households in informal settlements may have to buy water from street vendors at up to 200 times the tap price, so that much scarce household income is taken up on water (McCully, 2006).

Availability of potable water in urban areas increases rapidly with development. Around 30 per cent of households do not have access to clean water in the least developed cities, and 60 per cent of households in informal settlements, while 98 per cent in cities of the 'North' are supplied. Accordingly, water consumption is much higher in cities with higher incomes, as with most other forms of consumption. Typically people in developed cities use about 220 litres per day, while the average in Africa is 50 litres per day. Households in informal settlements may use 25 litres per day, due to less availability and higher costs, which can be five times the average price in developing cities. Ironically, wealthy

---

Box 5.2

Community water supplies

WaterAid is taking action to save community 'water kiosks' currently up for closure in squatter settlements around the city of Lilongwe, capital of Malawi. It is often inappropriate to sink wells or drill boreholes in urban neighbourhoods where sources may be contaminated. Water kiosks present an effective way of delivering safe drinking water to urban communities where there is a basic distribution network. Water kiosks offer a way to dispense drinking water from a sheltered tap stand, connected to the piped network.

From a technical perspective, water kiosks are easy to implement. However, they often fail unless the community takes into account the social and political problems of flat-rate metering, free-riding and illegal connections, which on the scale of a city can endanger the viability of the water company. NGOs such as WaterAid and the local partner CCODE, have a valuable role to play in facilitation and mediation between all stakeholders.

Source: International Water Council, 2006, http://www.worldwatercouncil. org

consumers in 'Northern' cities may then choose to pay for bottled mineral water, which can be more expensive than either fruit juice or petrol.

The level of household connection to networked infrastructure is a key indicator of the level of city development. The level of connection to each type of infrastructure tends to reflect the relative cost per household of providing the service and the relative importance to lower income households, so that access to potable water (which can be arranged fairly cheaply using communal standpipes) and electricity connections tend to advance most rapidly with development level. Sewerage, as the most expensive service to provide, increases more slowly, as do telephone landline connections which can now be bypassed by mobile networks. There are many creative solutions to be seen in developing cities for low-cost and community-controlled networks for water supplies (Box 5.2).

Sewage and effluent treatment

Sewage and effluent treatment should be an opportunity for completing the nutrient eco-cycle – so that organic wastes return safely to the land from which they came – but this is rarely the case. Worldwide, 40 per cent of the population lack decent sanitation (Black and Fawcett, 2008). There is a 'brown' agenda and the '6-F model' concerning access to sanitation – ('faeces, fluids, fingers, flies, fields, food') – which seeks to explain over 2 million avoidable deaths from diarrhoea, malaria and other sanitation-related diseases every year (WHO and UNICEF, 2000). In contrast to the potable water supply which is easy to disconnect from free-riders, it is more difficult for privatized utilities to operate a strict cost recovery model in drainage and waste water treatment, and the infrastructure often fails to reach lower income and informal settlements.

In the cities of the 'North', the majority of food is imported from overseas, processed, digested, piped and treated at large expense, and then disposed with further impacts and hazards. Pipelines carry wet sludge to the surrounding farmland, although land spreading is increasingly seen as risky and controversial. The goal of a sustainable urban sewage strategy would aim to work at both ends – gradually improving the effluent quality, separating the most hazardous inputs, and re-engineering disposal methods for useful output.

In the least developed cities of Africa, only 8 per cent of waste water is treated; and across the cities of the 'South', less than 35 per cent. As with networked infrastructure, the effectiveness of environmental management increases rapidly with the level of development, but this is increasingly based on a high cost foreign investment model involving the privatization of urban utilities and full cost recovery (Scheumann, *et al.*, 2008).

Critical perspectives on water

Many urban water utilities are now operated by trans-national companies, as urban or regional authorities are unable to access the capital investment needed. The World Bank and related development programmes have until recently encouraged a liberalized cost recovery model for utility supply, which is only now shifting (Bigio and Dahiya, 2004; Hildering, 2004). Meanwhile, in the tightly regulated markets of the 'North', utilities are constrained from excess profits, and so they diversify into other areas such as telecommunications and media. There are continual conflicts between driving costs down for consumers, driving profits up for shareholders, and driving up environmental improvements which are demanded by public regulators. There is evidence that the privatization wave still continues around Asia: neo-liberal governments still promote privatization as the way forward, with some success on some counts. At the same time, there are counter trends, where 'water democracy' is gaining ground, providing an attractive alternative to privatization (Swyngedouw, 2004: Manahan, *et al.*, 2007).

For the governance and public management of water systems there are new models emerging. Possibly the most advanced and holistic approach anywhere is the EU Water Framework Directive, which is being translated into national legislation in each of the member states (see Box 5.3).

On the consumption side, there are obvious parallels between water stress and the culture of affluence. For instance, the Mediterranean coastal zone has at present over 200 proposals for tourist golf courses, each one demanding a water supply enough for a medium-sized town, in the context of an escalating water crisis in Spain and Portugal. In affluent cities, water is seen as a key to local environmental quality, waterfront development and the place marketing which is crucial to urban regeneration.

Box 5.3

EU Water Framework Directive – best practice?

The European Commission has developed what is perhaps the most advanced policy package for water anywhere in the world. The Water Framework Directive (WFD) has the following key aims:

- expanding the scope of water protection to all waters, surface waters and groundwater
- achieving 'good status' for all waters by a set deadline
- water management based on river basins
- 'combined approach' of emission limit values and quality standards
- getting the prices right
- getting the citizen involved more closely
- streamlining legislation

Generally the WFD takes a 'whole systems' approach to the water cycle, from source protection, to river management, supply quality and effluent treatment. However the practice may be more difficult than the principle.

Source: European Commission, DG Environment, 2006, http://europa. eu.int/comm/environment/water/water-framework/

Social exclusion and inclusion enter in the debate on infrastructure and utility investment. Recent studies on urban infrastructure in transition have looked at the changing political economy under liberalization, whereby some consumers are seen as commercially attractive and others are a social burden; this then translates into selective investment and engineering decisions (Guy, *et al.*, 2001). In the cities of the 'South', there is a realization that infrastructure as perceived by centralized engineers is not necessarily serving the interests of consumers and citizens on the periphery of a large system (Bostoen, *et al.*, 2006).

The question of risk assessment and risk minimization underlies larger scale water planning and management – increasingly urgent with the onset of climate change and sea level rise, and the likelihood of water shortages and flood events in urban and coastal areas. The emerging social science of climate adaptation and associated flood prevention looks at risk assessment from different viewpoints – householders, urban authorities, developers, financiers and insurers – and the potential for collaboration between them (Bulkeley, *et al.*, 2003).

5.5 Air quality and pollution

Urban air pollution is linked to up to 1 million premature deaths and 1 million pre-natal deaths each year: the problem is estimated to cost approximately 2 per cent of GDP in developed countries and 5 per cent in developing countries, and overall more than 1 billion people are affected. Rapid urbanization has resulted in severe levels of urban air pollution, especially in developing countries. Over 90 per cent of air pollution in cities in these countries is attributed to emissions from older vehicles coupled with poor maintenance, inadequate infrastructure and low fuel quality (McGranahan and Murray, 2003).

The recorded exceedences or overshoots of official standards are only a starting point for air quality assessment. From the same quantity of emissions, pollution levels vary by season, time of day, location, weather temperature, secondary reactions and synergistic effects, and their effects depend on activities and vulnerability of the receptor population. There are also countless trace elements and organic compounds in urban air and dust – many of them carcinogenic with no thresholds or safe levels, and there is evidence of increased risk of childhood cancer of between 20 per cent and 400 per cent near to industrial sites. Dioxins, for instance, until recently were emitted by waste incinerators in many cities, and now are deemed unsafe at any level. Many ecosystems are also subject to air-borne and rain-borne acid deposition, far in excess of their critical capacity and threshold of irreversible damage, not only in the urban hinterland but at great distances.

Air pollution trends

In 'Southern' cities, both transport and industry have increased rapidly, and pollution levels are often up to eight times the WHO exposure limits (World Resources Institute, 1997). This is generally a function of rapid industrialization, poor control on industrial emissions, and lack of resources for abatement technology (Kaika, 2005). Many 'Southern' cities burn coal for industry and power generation, with additional burdens from hazardous chemical emissions. Asia contains 13 out of the world's 15 most polluted cities: China is particularly dependent on coal technology, and the extremely rapid rate of urbanization means that effective planning is very difficult (WHO, 1999). But the greatest burden may be for women and children indoors, in low-income housing with heating and cooking by dirty fuels, with inadequate ventilation and over-crowded conditions.

In most 'Northern' cities, emissions of NO_x are increasing more slowly if at all, as continuing traffic growth is offset by cleaner vehicles, while most other emissions have declined with the displacement and export of heavy industry. The 'auto-dependency' emissions of each city are then concentrated by geographical factors into photo-chemical smog (see Chapter 6): Tokyo, Los Angeles and Athens are among the worst polluted cities in the 'North'. Ironically, even in more affluent cities, indoor environments can also be hazardous, as trace carcinogens and allergenic dust builds up inside modern buildings which are relatively airtight (Curwell, *et al.*, 1990).

Air quality management

Urban air quality management is a classic example of a localized problem which generally exceeds the capacity of local solutions. Few cities are willing or able to shut down polluting industries or to stop the traffic, and the pollution load is itself an accumulation of small increases from many sources. Many larger cities and agglomerations have split levels of governance, between centers and peripheries, so that responsibility for pollution from transport or industry is often transferred from elsewhere. The urban air quality problem is arguably a function of fundamental economy, technology and infrastructure, and so to change it needs a fundamental shift. However, given the scale of the problem, air quality management is at least in part a success story.

Regulation and planning approaches have been developed since the industrial revolution in many cities in Europe and North America. The UK Clean Air Act of 1956 was a milestone in regulating the use of cleaner coal in households, which in the event was superseded by the shift towards gas for heating and cooking. More recent experience of regulation shows the benefit of a negotiating approach, which seeks the optimum BPEO and BATNEEC.

For economic and market-based approaches, many cost-benefit studies have focused on the common pollutants from fossil fuels in energy or transport. One example is a shadow cost estimate for SO_2, historically the most widespread common air pollutant; this covers damage, adaptation and investment in the international 'Sulphur Protocol', with European estimates of costs of € 2–5 per person, and benefits of € 5–25 per person (Pearce and Barbier, 2000). The first major pollution trading system was set up in the USA in 1995, and with a consensus between environmentalists and industrialists, managed to halve SO_2 emissions within 10 years (Ellerman, *et al.*, 2000).

For this and similar cases, the benefits of pollution control are often valued at many times the external costs of continuing pollution or investment in emissions technologies. If all emissions to air were charged or permits traded, clean technology incentives would increase and could outweigh 'end-of-pipe' abatement. In practice there are market barriers to such apparently rational investment, particularly among SMEs, and this suggests other approaches through technology development and supply-chain initiatives, (as discussed in Chapter 7).

Critical perspectives on air quality management

The uncertainty of hazards and risks in air pollution are especially challenging. For the public in 'Northern' cities, there are improved air quality reporting schemes, but conflicting evidence still creates fear and uncertainty, especially for people with respiratory or allergenic conditions. Again there is a general trend for residents of poorer neighbourhoods to breathe the emissions of the wealthier drivers, but this is not always the case.

Meanwhile most cities in the 'North' and many in the 'South' now have some form of air quality management strategy. In practice their objectives are often beyond their remit, involving national transport strategy, fuel technology, trans-national utilities, and industrial regulation. In other words, environmental governance is often fragmented between sectors and levels, in a classic case of displacement and split incentives.

At the city or community level, air quality strategy has to find ways to engage with transport, housing, ecology, economic development and public health strategies. But there are many examples to point to – cities such as Seoul, Athens, London and New York have each found a workable combination of taxation, regulation, mode shift, demand management and green infrastructure, and have achieved great steps towards more healthy and breathable air.

5.6 Energy and climate change

Cities cover less than one per cent of the earth's surface, but around half the world's population lives in cities, consuming 75 per cent of the world's energy (International Energy Agency [IEA], 2007). Energy is fundamental to the life of cities – but the energy which powers them also causes climate change, the greatest single threat to the global environment. Climate change is a problem for which every nation and every city is beginning to accept some kind of responsibility, in a new kind of global order. Tackling both the causes and effects of climate change needs to combine economics and politics, science and technology, cultures and lifestyles, and urban development. Such integrated energy-climate strategies should also be opportunities for economic growth, environmental improvement and social welfare.

At the same time there is large inertia and time-lag in the world climate system, in that emissions reductions now will take over a century to show results. As the climate begins to descend into turbulence, the expected impacts and hazards for cities in the next half century range from the very serious to the catastrophic. The urban climate adaptation agenda, as in the next section, needs to be combined with the re-engineering of urban systems for climate adaptation and protection.

Energy and climate in cities

Cities represent huge concentrations of energy demand, and the process of modernization and urbanization, as we know it, can only proceed in parallel with energy supply growth. Each city or city-region has a profile of energy demand, in housing, transport, commerce, industry and power generation; and a profile of energy supply, from fossil fuels, nuclear, renewables, or other combinations of technologies. Each city has a

Showing alternative boundaries for urban energy and climate policy

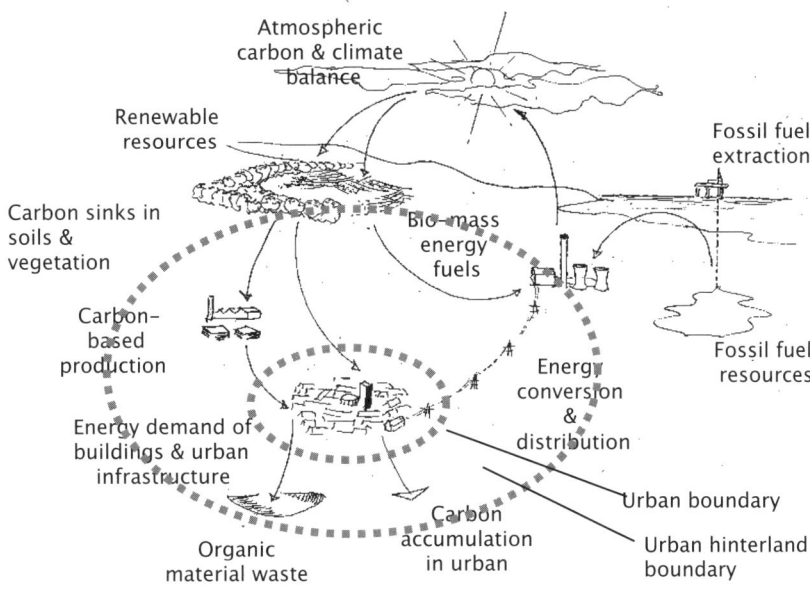

Figure 5.6 *Energy and carbon cycle in the urban environment*

climatic profile with a 'heat island' effect which concentrates the effects of climate change and pollution; and each city also has a climate change profile, with hazards and vulnerabilities to sea level rise, extreme storms, water and other pressures. These profiles are spatialized, so that cities will import energy from greater distances, and alter the environment of their hinterlands; with extremes in the megacities and agglomerations which create their own regional climates (Figure 5.6).

The single largest impact of energy use is through carbon dioxide (CO_2) emissions from burning of fossil fuels, the majority cause of climate change. World CO_2 emissions under BAU (business as usual) projections could rise by up to 25 per cent by 2025, in direct conflict with the scientific advice for a rapid reduction (IPCC, 2007). Alongside climate change there are many other impacts from the energy industry, with the fossil fuel and nuclear plant life-cycles. And climate change also has other human causes, including methane and nitrous oxide emissions from agriculture, and particularly deforestation, generally as a result of industrial and urban development.

Population growth and economic development has led to an average 3 per cent per year rise in energy demand, and the distribution of energy consumption is clearly inequitable. The OECD countries of the 'North' comprise 17 per cent of world population but consume more than half of all its energy, while the 'South', with 75 per cent of the population, consumes only one-third of world energy. Due to rapid urban and economic growth these countries are expected to increase their share of world energy by almost 40 per cent in the next 10 years (IEA, 2007). Urban energy shortage, where demand exceeds supply, is widespread, as is energy poverty, where low income groups spend excessive amounts of income on basic functions (World Energy Council, 2006).

At present the international energy industry is in a period of rapid liberalization, globalization and restructuring, and the main strands of global climate policy, as agreed at the Kyoto meeting, are still being worked through. At present it seems possible that an international carbon emissions trading scheme, based on the European Emissions Trading Scheme (ETS) can be set up post 2012.

Towards sustainable urban energy

The goals of a 'sustainable' urban energy strategy are clear in principle – to manage equitable levels of demand and efficient means of supply within local and global environmental limits (World Bank, 2005). But when we look for such clear limits, we find there are many difficult trade-offs. For instance in most cities, the energy demand sector with highest growth is air travel – while the positive benefits are claimed as the urban hubs and gateways, the responsibility for impacts is disclaimed as belonging to nations, airlines, travelers – anyone, in fact, but the city. Energy taxes to reduce consumption can be seen either as a brake on the economy, or as an incentive for modernization. There are global estimates of the 'free lunch' available through energy efficiency and low carbon technologies. One baseline estimate of global energy demand growth to 2020 is 1.7–2.8 per cent per year, mostly in buildings and transport, and this in principle can be halved using low-cost and high-return efficiency improvements (McKinsey Global Institute, 2007).

In practice there are many barriers, and most businesses and consumers have little real commitment to energy efficiency – progress depends on linking energy-climate strategies with problems and opportunities in housing, transport, economic development and so on, in a diverse and responsive approach (Droege, 2008). An integrated urban energy-climate strategy is likely to include five main strands, each with a range of economic and political incentives (Box 5.4):

- mitigation of emissions through demand efficiency measures in buildings and transport;
- cleaner fuels and supply technologies;
- increase in renewable and low-impact energy sources;
- co-generation (CHP) and networked heat distribution;
- climate change adaptation or defence.

Box 5.4

Integrated urban energy systems, Finland

This urban district heating system relies on a technology that combines the production of electricity and heat. Heat obtained in generating electricity is now used for heating the city, instead of leaking it into the sea. Operating on market terms since its inception, the system currently serves more than 91 per cent of all Helsinki's buildings. The efficiency of energy supply has been raised from 40 per cent up to 80 per cent. The specific heat consumption in buildings connected with the district heating network has also decreased from 65 kWh/m^2/yr, to 44 kWh/m^2/yr, due to the energy saving programmes. The coal-fired power stations have been equipped with desulphurisation plant: by-products of combustion are recycled and used as by-products for cement and for geotechnical engineering.

Source: UN-Habitat Best Practices, 2007, http://www.bestpractices.org

---

Critical perspectives on urban energy

A sustainable energy path is clear in principle and technically viable. But few cities, have the powers and resources to achieve such a transformation, combining 'supply-side' actions on energy fuels and markets, with 'demand-side' actions for housing, transport, industry and agriculture. As with other environmental sectors, there is displacement and fragmentation between causes and effects, and between local and global scales. There is also the challenge of integrating energy-climate strategies with spatial planning, urban development, social policy and economic development.

The context to this is energy as a globalized commodity, and more recently, the attempts to establish carbon also as a global commodity, which can be traded, stored, speculated on, or simply allocated by quota. This continues the approach of the Kyoto Clean Development Mechanism and Joint Implementation programmes, with further opportunities for corporate investment on larger scale carbon offsets and avoided deforestation. But the effects may not always be positive – it is likely to enable large cities and firms of the 'North' to expropriate and colonize the ecological space of the 'South' through funded carbon offset projects which can be as insensitive and exploitative as any other type of cash cropping (Lohman, 2006). The continuing privatization of energy utilities around the world is likely to encourage the stripping out of city governance functions, where the urban energy profile will be commodified on the global market place. Recent evidence shows that clean energy may be short changed, that the incentives are focused on manipulation of fossil fuel and carbon markets, and that pro-poor urban energy policies may be undermined (Redman, 2008).

At the same time it is possible that the emerging 'low carbon economy' can encourage and enable new forms of collaborative governance at the city or city-region level (Bulkeley, *et al.*, 2003). This may work on multiple levels – changing the culture of consumption, adopting a precautionary approach to risk, enabling intermediate labour markets and social enterprise, and driving urban renewal with the energy efficiency agenda. To realize this there is an agenda for transformation – of infrastructure, markets and institutions – which offers great opportunities for city economies and city governance.

5.7 Urban environmental security

The global map shows the areas of emerging environmental stress, overlaid with the cities of over 10 million people (Figure 5.7). It is clear that the majority will be in coastal locations, and can expect a climate-dominated future of sea level rise, tropical storms, tropical diseases, food and water shortages, combined with deforestation and desertification

Showing zones of major environmental and resource pressure and hazard projected to 2030-2050:
with conurbations of over 5 million population.

Adapted from World Urbanization Prospects 2007, UNESA, 2007: Living Planet Report, WWF, 2006

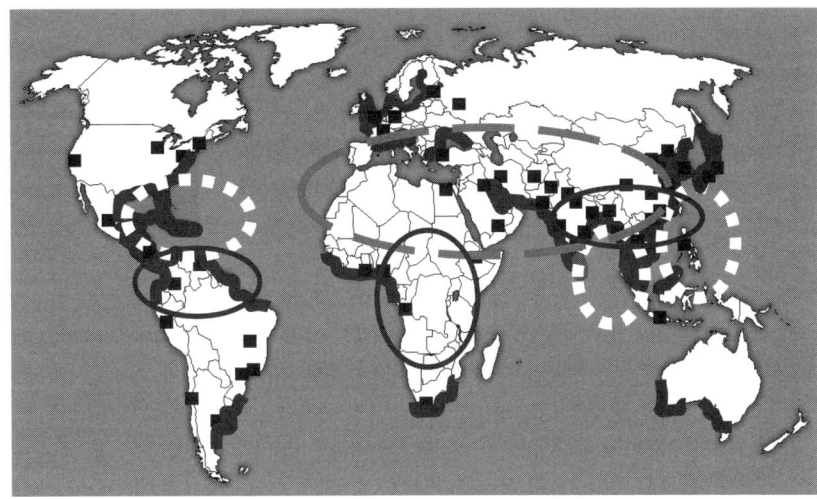

~~~ Severe impacts of sea level rise
Increased water stress
Increased risk of tropical storms
Increased risk of hunger and crop decline
■ Conurbations of over 5 million population

Figure 5.7 *Urban environmental pressures and hazards*

(IPCC, 2007). Out of 33 cities projected to have a population of 8 million or more by 2015, two-thirds are coastal cities. Within the Low Elevation Coastal Zone (less than 10m from sea level), there are 80 million people in China, 30 million each in India and Japan, 20 million each in Indonesia and the USA, and so on.

Meanwhile there is very disturbing new evidence which suggests the possibility of much faster climate feedback mechanisms now in process, which would greatly accelerate the melting of the Greenland and Antarctica ice, producing an eventual sea level rise of up to 25m (Hansen, *et al.*, 2007). If these feedbacks did take place, and their probability appears to be impossible to quantify at the present time, half the world's cities could be made uninhabitable.

Even on conservative estimates of climate change trends, the 'likely range' for global average temperature rise could be as much as

6.4 degrees by 2100, and many cities will suffer much more extreme temperatures than this (IPCC, 2007). There is an urgent agenda for climate adaptation, building design, utilities, urban design, urban green space and landscape design (Satterthwaite *et al.*, 2007).

Such human-induced threats will accumulate in addition to existing natural hazards, such as earthquakes, tsunamis, volcanoes, mudslides, forest fires, pestilence, droughts and unstable land. Cities have an unfortunate habit of locating in vulnerable zones, and their poorest residents are often in the most vulnerable areas. These environmental threats are often compounded by parallel problems of state corruption, corporate exploitation, ethnic or sectarian tension, human pandemics, and political or military conflict. Such political and economic pressures are then exacerbated by the vulnerability of the population – poorly constructed housing, lack of flood defences, lack of basic services and sanitation, malnourishment and poor healthcare all compounded (Handmer and Dovers, 2007).

The destruction of New Orleans by Hurricane Katrina in 2005 was an example of what may lie in store, all the more surprising for exposing the lack of emergency planning, institutional corruption and social divisions

---

Box 5.5

Post-earthquake reconstruction, El Salvador

This housing reconstruction programme in La Paz, El Salvador, demonstrates how successful community development can result from rebuilding after a devastating earthquake. Appropriate earthquake-resistant house designs have been developed and training is provided in construction and maintenance techniques, as well as business and community leadership skills. To date, over 7,460 houses have been completed with households providing the labour, and secure title to their property has been established for those who did not previously have the correct documents. 1,400 small businesses have been established; many of them run by women.

Source: World Habitat Day Awards, 2006, http://www.worldhabitatawards. org

in what is apparently the world's richest nation. Some 70 per cent of major cities in a World Bank survey operate detailed hazard mapping or have accounted for disaster risk in their building codes, with strategic planning for emergencies (World Bank, 2005). A lesser proportion have some form of disaster insurance, though this is not often compulsory.

The building of capacity for urban climate adaptation is now a major issue; unfortunately it is not yet included in the scope of the post-Kyoto international carbon transfer proposals. Capacity building works as part of various pro-poor housing and urban development programmes, focusing on the construction and management of buildings, on the community capacity for governance and participation, and on the resilience of local economic development on overseas donor and investor programmes (Satterthwaite, *et al.*, 2007). In the more affluent cities of the 'North' the climate impacts and adaptation agenda is more concerned with new patterns of building design, urban form and land use, and protection of critical infrastructure (Roaf, 2004).

While the climate response agenda is still mainly in the future, the present day sees a catalogue of earthquakes, floods, hurricanes and landslides. In each of these, there is an agenda for more responsive governance, participatory planning, and appropriate technology (Box 5.5).

5.8 Conclusions

This chapter has looked the physical urban environment in its human context – rarely a straight line from causes to effects to policy responses. Instead there are layers of uncertainty and controversy, a long way from the textbook versions of environmental science or economics.

What stands out clearly is the way in which the physical urban environment is a crucial strand of the dynamics of globalization, liberalization, modernization and urbanization. Environmental quality in air, water and land is segmented along lines of wealth and poverty, order and chaos, both within and between cities. While the cities of the 'North' debate their quality of life, many in the cities of the 'South' struggle for life itself.

Nowhere is this more clear than in the agenda for urban energy, climate emissions policy and climate adaptation policy. While there are clear visions for 'carbon free' cities, a more likely and less aspirational outcome may be that cities will continue to dominate their hinterlands with new forms of energy and carbon economies, while the sea level continues to rise, and the wealthy head for the high ground. Overall, the danger of a simplistic form of urban environmental policy, is that such divisions can easily be amplified. Concepts such as 'water democracy' and 'climate justice' in both 'South' and 'North' are still up for debate.

Further information

C40 Cities for Climate Protection   http://www.c40cities.org/

China Dialogue (China and the world discuss the environment)   http://www.chinadialogue.net and

http://www.chinadialogue.cn

Cities for Climate Protection   http://www.iclei.org/index.php?id=800

Global Footprint Network   http://www.footprintnetwork.org

Inter-governmental Panel for the Scientific Assessment of Climate Change, Working Group II   http://www.ipcc-wg2.org/

International Council for Local Environmental Initiatives   http://www.iclei.com

Millennium Ecosystems Assessment   http://www.millenniumassessment.org

Resource Recovery Forum   http://www.residua.com

Stern Review: The Economics of Climate Change   http://www.occ.gov.uk/activities/stern.htm

World Energy Council   http://www.worldenergy.org

Water Aid   http://www.wateraid.org

World Water Council   http://www.worldwatercouncil.org

World Wildlife Fund   http://www.wwf.org

Further reading

Douglas, I. (1983) *The Urban Environment*, Maryland, USA: Edward Arnold.

Hardoy, J., Mitlin, D. and Satterthwaite, D. (2001) *Environmental Problems in an Urbanizing World*, London: Earthscan.

Hough, M. (1984) *City Form and Natural Processes*, London: Routledge.

Lynas, M. (2006) *Six Degrees: Our Future on a Hotter Planet*, New York: Fourth Estate, Harper Collins.

Marvin, S., Guy, S. and Moss, T. (eds) (2001) *Urban Infrastructure in Transition: Networks, Buildings, Plans*, London: Earthscan.

Millennium Ecosystem Assessment (2005a) *Living Beyond Our Means: Natural Assets & Human Wellbeing: A Statement from the Board*, New York: UNEP (available on www.millenniumassessment.org).

Millennium Ecosystem Assessment (2005b) *Current State & Trend Assessment, Vol 1, Chapter 27: Urban Systems*, Washington DC: Island Press (available on www.millenniumassessment.org).

Murray, Robin (2002) *Zero Waste*, London: Greenpeace Environmental Trust.

OECD (2008) *OECD Environmental Outlook to 2030*, Paris: OECD.

Stern, N. (2006) *Stern Review on the Economics of Climate Change*, Cambridge, UK: Cambridge University Press (available on http://www.occ.gov.uk/activities/stern.htm).

White, R. (1994) *Urban Environmental Management*, London and New York: Routledge.

World Wildlife Fund International (2006) *Living Planet Report*, Geneva: WWF International.

6 City form and fabric
The urban built environment

6.1 Introduction

At the top of the urban agenda is the provision of shelter for homes, jobs and services – the built environment. In this, as elsewhere, it seems that environmental problems cause human impacts, that human problems cause environmental impacts, and that solutions are to be found in many dimensions – political, economic, social and cultural.

The spatial form and pattern of the city conditions the human experience at every level, in blocks, streets, neighbourhoods, districts, city-regions and agglomerations. At the macro level, the pressures and demands of growing populations for spaces and places then drive urbanization and spatial development. At the local level, the character and liveability of urban forms are essential to quality of life. Depending on the quality of the built environment, investors may either come flocking in, or abandon the cities to decline and decay.

In this chapter

At each level of urban development there are implications for the urban environment. On the housing agenda, the energy consumed is the prime cause of climate change, and the everyday environment of people is dependent on the quality of buildings in their economic and social context.

At a wider scale, the pattern of spatial development is highly dependent on environmental qualities and environmental hazards; as ever it displays sharp divisions between wealth and poverty, and growth and decline.

The counterpart to this is transport, the 'maker and breaker' of cities, and a front-line environmental challenge. The transport agenda divides into two strands – the demand for travel or accessibility; and the supply of mobility via various transport modes and infrastructure. Transport growth has up until now been locked into conventional economic growth, and more sustainable alternatives demand a re-think of both spatial form and the meaning of development itself.

As with the previous, this chapter looks at trends and prospects in the cities at either end of a simplified 'South'–'North' axis; recognizing that the reality is always more complex. And again, discussion focuses on the critical perspectives – cross-cutting debates on the real-world tangles of politics, economics, cultures, and their effect on the urban built environment.

6.2 Urban shelter and housing

The first question for any city is how and where it shelters its people. This is an urgent issue for the cities of the 'South', in the throes of rapid migration, urbanization and changing social structures. It is also crucial for most cities in the 'North', experiencing demographic shifts, growth and decline, outward sprawl, and rising expectations for quality of life.

The theme of the human urban environment runs through this debate. The internal and external environments of housing are crucial to public health and wellbeing. Housing itself, in most cities, is the largest single consumer of energy, water and materials. Housing quality is the first indicator of wealth or poverty; and lack of safe, affordable or secure housing is the first link in the chain of ill-health, unemployment and deprivation.

Urban housing in the 'South'

Over the next 30 years, urban dwellers will double in number and account for nearly two-thirds of the global population. Most of these new urban dwellers are likely to be poor, and will suffer from the urbanization of poverty. Slums and slum areas are a physical and spatial manifestation of increasing urban poverty and intra-city inequality. However, slums do not accommodate all of the urban poor, nor are all slum dwellers always poor. The rapid growth in developing cities suggests that all the problems of slum dwelling will worsen in those areas that are already most vulnerable (Box 6.1).

Box 6.1

A world of slum cities

By the middle of this century, two-thirds of the global population will be living in towns and cities. Yet nearly 32 per cent of the world's urban population – roughly 1 billion people – already lives in slums, mostly in, or on the edges of, cities across the developing world. In a process which UN-Habitat, the UN human settlements agency, calls the urbanization of poverty, the locus of global poverty is moving into towns and cities.

Sub-Saharan Africa has the largest proportion of its urban population resident in slums – nearly 72 per cent in 2001. In South-Central Asia it was 58 per cent, while in East and West Asia, Latin America and the Caribbean it was 32–36 per cent. In absolute numbers of slum dwellers, Asia as a whole has by far the largest number at 554 million, or 60 per cent of the world's total slum populations. In the developed 'North' the total slum population is 54 million, just 6 per cent of the total population.

There are five basic criteria for a 'slum' dwelling, as defined by the UNCHS database: insecure tenure; lack of safe water supply; lack of sanitation; unstable land or building structures; and inadequate living space.

Source: Neuwirth, 2005; UN-Habitat, 2006

This problem is partly a matter of rural development and the stability of rural economies; some 70 million people a year migrate from rural areas to cities (about 1.4 million per week, or 130 per minute) (Neuwirth, 2005). Many of these have to set up home in illegal shelters, put together from waste materials, in hazardous or polluted locations, harassed and victimized by the authorities. Currently there are about one billion squatters or informal urban residents; by 2050, on current trends, that figure will reach three billion. A reasonable estimate would be that the majority of city-dwellers in 2050 could be in quasi-temporary, low-tech shacks (Davis, 2005).

While the definition of 'slum' includes a multitude of different types of settlements and populations, two main directions can be seen (UN-Habitat, 2004):

- *'slums of hope'* – settlements showing signs of progress with some kind of a future. These often have new, frequently self-built structures, usually informal or illegal, that are involved in a process of development, consolidation and improvement;
- *'slums of despair'* – declining neighbourhoods, with little future in their present state, where environmental conditions and domestic services are degenerating.

Many slums are now seen more positively by public authorities than in the past, as places for potential improvement, rather than irretrievable decline (UN-Habitat, 2003). National policies on slums are gradually shifting from forced eviction, harassment and involuntary resettlement, to self-help and in situ upgrading, enabling and rights-based policies; however, it is not clear whether the changes are enough to keep up with the growth in the problem. There are opposite pressures, such as the privatization of public utility companies, and the consequent disconnection of water and power supplies, with severe effects on public health. In Mumbai, Jakarta and other cities, slum dwellers have been harassed and forcibly evicted; and in Harare, Gaza and Grozny, slum dwellers have been subjected to military or political oppression (Sharma, 2000).

Housing tenure and finance

Environmental problems and opportunities in housing are underpinned by tenure and finance; without security and resources for investment, problems are likely to multiply, especially at the higher densities and land values in urban areas. Housing tenure can be quite different between otherwise similar countries, depending on the regulatory framework and subsidies, infrastructure provisions, mortgage finance, and the level of affordability and household debt (Mitlin, 2001). The most rapid shift in housing tenure is seen in the transitional countries, where social housing has halved, to be replaced by private ownership.

There are many tenure systems for land and housing, ranging from the 'freehold' or permanent full property rights, to the lack of any formal tenure in many squatter settlements or camps for 'refugees and internally displaced people'. In practice there are overlaps between different forms of tenure:

● Freehold or registered leasehold: the latter is the most common in the 'North', where leasehold periods may be up to 1000 years. One disadvantage may be that maintenance and improvement is more difficult where the title is split. Another common situation is that the property may be more valuable as a commodity for speculation, rather than an actual dwelling or workspace.
● Rental, either public or private, is the most flexible and responsive to changing needs and locations. However in many countries there are few controls on private landlords, various forms of legalized extortion exist, and many rental tenures have the disadvantage that responsibility is split between landlord and tenant, so that energy efficiency and other improvements become difficult.
● Shared equity is a combination of ownership and rental, where the balance may shift over time to be more affordable for lower income families. Co-operative or community tenure, where land or buildings are held in common, in mutual aid or non-profit trust frameworks is perhaps the most desirable but least robust of any of the tenure systems.
● Customary ownership tenures may be part of a native legal system or framework of rights; in more remote areas many residents have little or no written legal title to their dwellings, apart from their kinship histories; there are also religious tenure systems such as those based on Islamic law.
● Non-formal tenure systems – squatters' rights are often marginal and subject to the whims of local landowners, police and armed militias; in many countries there are movements to return the land to the native 'First Peoples', from whom it was taken by extortion or force.

Generally a more secure and long-term tenure will tend to encourage investment in housing improvements and infrastructure. However in more stable cities, such securities can also inhibit the movement of people towards jobs, and the changing needs of the family life-cycle; with the result of mismatch between population and the housing stock. The mortgage finance system is especially crucial in 'Northern' economies, and in many OECD cities there are problems with housing shortages, personal debt and negative equity. In cities of the 'South', access to housing finance is more often informal, and accessed through kinship and ethnic networks; this has a downside, often excluding poorer migrants, and polarizing the divisions of class and caste. In contrast, organized or semi-informal housing investment can have a very positive role on local economic development (Tibaijuika, 2008).

Urban housing in the 'North'

In many cities which are otherwise affluent, up to half of all housing is unfit or substandard, and such physical problems are compounded with the effects of unemployment, ill-health and energy poverty. At the same time, levelling up and the provision of new housing to current standards on the scale which appears to be needed would have large land-use and environmental implications. Underlying this is a structural demographic shift in most developed nations and particularly in cities. While the population is relatively static, social and economic changes are steadily reducing the average size of household, from an average of 2.5–3 people, down towards 2 persons per dwelling. This contrasts with the 'South', where the definition of 'household' can be rather different, often including extended families of three or four generations.

Much of the older housing stock in many cities is also near the end of its life and needing large scale renewal, which puts further pressure on housing land. Much new household demand is for single persons or smaller units, while the residual housing stock may be in larger family dwellings. The demand for space is also changing with new forms of family arrangements, new types of indoor leisure and communications, and flexible home-working (Ravetz, 2008).

Urban space and community

As land is perhaps the most finite resource of all, the question of its distribution is at the core of many environmental issues. In a nutshell, the more private space is used, the less public space is left; by building multi-storey we can increase the internal space per person, but decrease the external space. However there is more to this than simple quantities – space and territory, and the identity of place, are at the foundations of human psychology and culture (Rapaport, 1977; Newman, 1981).

Such factors include:

- 'space to grow', with life-cycle aspirations for territory and identity – people in over-crowded conditions usually aspire to increased space per person;
- 'space to be one-self' – desire for identity through sub-cultures, by sharing spaces with like-minded people, and consequent polarisation of the remainder;
- 'wanting everything', in locations with good access to green fields, jobs and services – the result is seen on the urban fringe and other locations with the highest values.

In many countries there is now an accepted model of the sustainable city – based on neighbourhood clusters of high density, high accessibility, and mixed uses (CEC, 1990; Ravetz, 1999c). In reality, this model can conflict with the perceived free-market consumer psychology, as above, and also the conservatism of the property and construction industry (Rydin, 1995). So the challenge is to turn a problem into an opportunity, and to promote quality of 'place' rather than quantity of 'space' per person, by clustering around positive attractions.

It is also clear that lifestyles, attitudes and community cohesion are essential to gaining added value from proximity. As and where people can share collective space such as gardens, and exchange services such as childcare, the community in principle is better off. This suggests a re-think of modern systems of housing tenure, investment and social welfare: the twentieth-century Fordist model of nuclear family owner-occupation could become outdated, in an age of mobile careers and flexible household types. One forward-looking housing type may be the self-build cooperative, as in the Danish 'co-housing' model, which encourages like-minded people to work together, invest in housing 'sweat equity' and share facilities (Williams, 2005). For mainstream social housing in the 'North' there are newly integrated social housing programmes, such as the UK Housing Plus scheme, which aims to coordinate all facilities and services (URBED and Newbury King, 1998).

Such integrated solutions are also mirrored in the 'South', where the capacity for community self-organization is often greater (Box 6.2).

Box 6.2

Integrated urban housing regeneration, South Africa

The work of the Johannesburg Housing Company (JHC) involves the adaptive re-use of empty city-centre buildings to deliver mixed-tenure, affordable rental housing whilst acting as a trigger for the regeneration of the surrounding area.

JHC was established in 1995 to provide homes for low-income city workers in Johannesburg, transforming dilapidated inner-city neighbourhoods. Abandoned inner-city offices, hotels and apartment blocks are renovated and converted and new build projects developed. To date 2,403 homes have been provided for 8,000 men, women and children in 21 buildings, adding eight per cent to the residential stock of Johannesburg.

All developments are mixed-income, and pioneering participation and management processes have been instituted. Tenants committees are encouraged, and there is a culture of cleanliness and order. Community development workers are employed, and there are training programmes and social support such as crèches for working mothers.

Source: UN World Habitat Day Awards, 2006, http://www.worldhabitatawards.org/

Housing and energy demand

The most direct environmental impact of housing is in energy demand for heating, cooling, cooking, water, lighting and other power. Depending on the location, climate and level of affluence, this can account for over half the total CO_2 emissions of any city.

Housing energy performance can be measured in floorspace efficiency as kWh/m^2 per degree day (i.e. temperature difference between inside and outside over an average year). Average housing in 'Northern' cities has increased efficiency by about 5–10 per cent per decade, but this trend is offset by the increase in space per person, comfort standards and electrical appliances (Lowe, 2007). Such trends will continue, with energy use in buildings projected to rise by about 10 per cent per decade, unless positive action is taken (IEA, 2007). The effect of climate change over several decades may reduce heating demand by 20 per cent, but without changes in design and construction, will be offset by energy-intensive air conditioning for hotter summers. This is crucial for tropical cities in particular, where intelligent bio-climatic design, using natural and vernacular means of shading and ventilation, could avoid most air conditioning altogether (Yeang, 1995).

While energy efficient technology is for the most part viable and proven, there are many financial and institutional obstacles. Although the payback periods for efficiency improvements can be as short as one year, investment is often stopped by inertia, uncertainty and the problem of split responsibilities between landlords and tenants, where neither side has a clear incentive to improve the building.

In practice there are diminishing returns by simply increasing the insulation, and summer cooling is likely to be a greater problem than winter heating. To approach low-energy or 'zero-carbon' standards involves more fundamental re-engineering for heat recovery and integrated element design (Boardman, *et al.*, 2005). The largest growth in energy demand comes from appliances such as freezers and dishwashers; this suggests that city-wide incentives for purchasing and maintenance of low-energy technology could be very effective. A very simple policy would be to replace every light bulb with low-energy fittings, which in an average 'Northern' city could save up to 10 per cent of total CO_2 emissions.

A very topical experiment is now underway in the UK, with a national target for all new housing from 2016 to be zero-carbon, i.e. producing all its energy demand onsite or nearby from renewable sources. It remains to be seen how far the construction industry can rise to this challenge, and what the social and economic effects might be (Communities and Local Government, 2007). On a larger scale are the new zero-carbon city developments at Dongtang in China, and at Masdar in Abu Dhabi (see http://www.arup.com/eastasia and http://www.masdaruae.com).

For housing in the 'South', energy demand is highly dependent on the climate and general level of infrastructure. Electricity supplies can be variable, and fuels for heating and cooking include coal, charcoal and kerosene. Each of these has various kinds of pollution hazards and health risks, and in the case of fuel wood, damage to the urban hinterland. Few dwellings for average incomes have artificial cooling, even in extreme climates. Vernacular designs with built-in natural ventilation, heating and cooling, are often replaced by imported designs which are highly energy intensive. The process of modernization and upgrading of all urban housing to basic standards of adequacy would tend to increase energy demand, unless a large proportion of the multi-storey blocks, which now form the bulk of housing in the 'South', were replaced by other more environmentally sensitive designs. Either way there is a crucial agenda to combine affordable energy services with climate mitigation policies and with climate-responsive design and construction (Zhang, 2000).

Housing design and environment

Many internal environments in urban areas are often seriously deficient. In the 'North', a combination of affluence and ignorance often brings poor ventilation, poor heating and lighting controls, and carcinogenic emissions, and these can be linked to employee stress and sickness (Curwell, *et al.*, 1990). Energy improvements themselves can create health problems, when sealed environments increase condensation and stale air, and such problems are exacerbated by the effects of climate change. A high-impact approach would use more energy intensive air conditioning; while a low-impact approach would use passive ventilation, breathing wall construction, natural cooling, location and aspect planning, and external shading and planting (Roaf, 2004).

Such problems contrast with those in the 'South', where a typical lower income household has to deal with dangerous heating and cooking systems, extreme heat and lack of ventilation, overcrowding, fire hazards, poor sanitation, invasion of pests and so on (Hardoy, *et al.*, 2001). Again, such environmental problems are generally the result of combined political and economic problems, and so any policy responses need to be equally integrated (Payne and Majale, 2004):

- finance for housing, available to lower income groups in a way which minimizes corruption, usury and extortion;

- regulatory systems for environmental and urban planning, with proper legal accountability and transparency;
- policy and governance systems which are democratic, accountable and participative;
- housing management and tenure systems which encourage local economies, vernacular designs, low-impact materials and design for adaptation;
- supply side changes in sectors including construction, engineering, utilities, property, minerals and forestry;
- an underpinning of human rights, active citizenship, gender equality and other hallmarks of an open society.

Critical perspectives on housing and environment

Housing is generally the largest capital investment sector in any city, and is crucial to the performance of national economies, and evidently the global economy. Its supply is dependent on long-term finance for developers, landlords and owners; likewise the demand for housing is driven by employment and income growth. The culture of consumption drives the increasing demand for space, and the outcomes of urban growth and polarization. The location of housing choice in terms of supply and demand represents a complex balance of employment, mobility, migration, lifestyle and community factors. The governance of housing is one of the political defining lines for any city, and the concept of urban 'regime theory' emerged around the provision of public housing (Painter, 1995; Logan and Molotch, 1987).

As many 'Northern' cities have converted most of their housing stock to private ownership, there are problems with price inflation and the exclusion of poorer and younger people from a financial escalator; even while the public housing stock is marginalized and devolved to third sector organizations. In this context, improving the environmental performance of housing is both a technical issue and a socio-technical issue, which involves the complexities of organizational learning, supply chain management, professional skills, and market transactions and incentives (Kats, et al., 2003; Guy and Shove, 2000).

In 'Southern' cities, most dwellings are also in some form of private renting or ownership, often with complex and onerous financial arrangements, while in some areas up to 80 per cent of dwellings are in squatter or informal settlements. So the public provision of housing on a large scale would represent a huge financial investment. One estimate is for an average of $1,800 per person, to meet the MDG of 'improving the lives of 100 million slum dwellers', plus basic facilities for a further 700 million new slum dwellers by 2025 (UN-Habitat, 2007a, 162). The total is similar to the cost of the USA Federal Government rescue of the banking system in 2008. But in practice, large proportions of such costs can be offset by programmes for self-help and mutual aid for housing at the community level, and this a powerful stimulus for local economic development (Hamdi, 2004; Tibaijuka, 2008).

6.3 Urbanization and spatial development

The dynamics of urban development

Urban development has traditionally been analysed as a spatial economic interaction, where jobs, services and housing should be located at an optimum balance of cost and travel time. However, this mechanistic view has never truly reflected the drivers of spatial form and change, which include environmental and social factors. There are now many factors in the mix – global networks, corporate restructuring, lifestyle and cultural shifts, and the flight from urban crime and decline; and there is a realization that urban development is better understood as a complex system (Batty, 1995; Green, 1999; Portugali, 2000).

Meanwhile, the aggregate statistics continue to multiply up the global rate of urbanization. This can be defined by the land area of settlements; many cities in the 'North' have growth rates of 1–2 per cent per year, with a doubling time of 50 years. It can also be defined by the urban population measure; growth rates vary between less than zero in some 'Northern' cities (i.e. shrinkage), to about 50 per cent per decade growth in rapidly developing countries in Asia and Africa.

In a major World Bank study, developing-country cities were found at present to have three times the population density compared to industrialized cities (Angel, *et al.*, 2005). However their current trend for reducing density is 1.7 per cent per year; and if this trend continues, the built-up area of these cities will triple to more than 600,000 km^2 by 2030, while their population doubles. The implication is that cities in both 'South' and 'North' should be making realistic plans for large-scale physical expansion, investing in basic infrastructure, and protecting sensitive or hazardous areas. In other words, strategic spatial planning is not an option, but an essential prerequisite for any city.

Urbanization and sustainable urban form

Behind such mega-trends lies the topical agenda of 'sustainable urban form' (Jenks, *et al.*, 1999). The typical outcome of rapid urbanization is a self-reinforcing spiral of outward sprawl, inner city decline, derelict and vacant land, and increased car-dependency. Meanwhile, spatial policies or land use restrictions may lead to housing shortages, loss of urban open space, and out-migration of wealthier households. So a sustainable urban form agenda for any city is a complex balance of many needs and goals, at larger and smaller scales (Breheny, 1992). At present the 'compact city' approach aims to re-use wasted assets in the urban area, increasing urban densities, while protecting critical assets on the fringe – often called the brownfield agenda. In contrast the suburban approach defends the right of households to occupy a patch of territory, arguing for 'sustainable suburb' models with greater biodiversity and leisure access (Teaford,

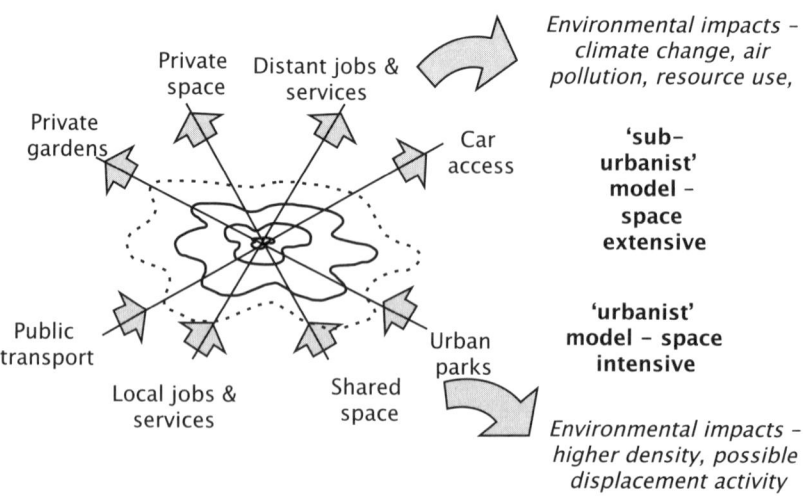

Comparison of the space extensive and space intensive models of urban development. Adapted from Ravetz, 1999a

Figure 6.1 *Spatial development choices and the urban environment*

2007). This can be summed up as a series of choices on the 'urbanist'/ 'ex-urbanist' spectrum, representing not only spatial development alternatives, but economy and society alternatives (Figure 6.1).

There is a policy debate in rapidly urbanizing countries on alternative patterns of development and the optimum size of settlements. One approach looks at the positive factors of larger settlements, such as access to services, competitiveness, urban infrastructure and higher order functions; and sets this against negative factors such as pollution, congestion, and concentration of poverty (Capello and Camagni, 2000). This debate is highly topical for the issue of 'metropol-ization' – where larger cities, agglomerations and megalopolises are not simply growing, but rather spreading urban functions and networks across a much wider area (UN-Habitat, 2004; Cohen, 1998).

Spatial development in the 'South'

Urbanization in the cities of the 'South' is characterized by rapid migration from rural areas, often at a scale which overwhelms any planned infrastructure. The current rate of urbanization across all cities in the 'South' is 64 million per year, or about 175,000 per day. Most of the world's largest cities in the next 50 years will be in the 'South', and the urban agglomerations with over 15 million people by 2030 are expected to include Dhaka, Mumbai, Delhi, Calcutta, Karachi, São Paulo, Mexico City, Jakarta, Shanghai and Lagos. In addition to such mega-cities, a larger urban population will continue to live in so-called 'million cities', of which there are currently over 350 around the world. Even more will inhabit smaller cities of less than 1 million people, often in peripheral locations, and lacking critical mass for investment and competitiveness; such settlements bring particular challenges (Bell and Jayne, 2006).

The shape and structure of many such cities is a more direct result of post-colonial capitalism and globalization than is generally seen in the cities of the 'North', where there are many historic layers of planning and intervention. The dominant trend is that of economic and social polarization as seen in the widening gap between the wealthiest and the poorest, with increasing numbers at the extremes, and increasing variety of social and cultural groups between.

The implication is that the former dual city of native and colonial powers has shifted to a more segmented city structure. There is growing evidence of five increasingly separate residential urban types, each with a parallel city of business and work: the international 'mobile citadels', wealthy suburbs, workers' quarters, slum quarters, and the peri-urban hinterland (UN-Habitat, 2004). The business centres in each quarter are not tied solely to their residential quarters, but bring together a mixture of labour for the range of economic functions. The boundaries between these different quarters are carefully guarded and recognized by people and functions on both sides. While the poor may travel as unskilled service workers into the wealthy citadels, they are walled in by powerful stigmas of address, education and ethnic group.

Environment and urban structure

Each of these segmented urban types then brings an environmental agenda, in both 'South' and 'North'. This can be defined by basic provision of electricity, water and sanitation; and then by levels of pollution and amenity; and then by environmental investments and improvements.

The mobile citadels often wall themselves into gated communities and high-tech security enclaves in the central cities or special zones. While the USA contains some of the largest urban slum ghettos in the 'North', it also has the most gated communities. 'Common Interest Developments' for affluent households now contain over 32 million people, or 10 per cent of the population, in 150,000 projects across the country (Glasze, *et al.*, 2005).

Surrounding these, the gentrified and suburban areas in the 'South' generally aim to emulate the middle-class lifestyles and dwelling forms of the 'North', based on a high-energy and high-mobility model. In contrast, the working-class and ethnic enclaves of the inner city areas and peripheral housing estates may have basic services of energy, water, sanitation, with local services of health and education. Housing may be overcrowded and substandard, but generally within sound building structures with some degree of tenure.

The slum areas, informal settlements and most rapidly urbanizing areas often suffer from gross pollution of air, water and ground, both within and around the dwelling. They are more likely to be on flood plains, contaminated or unstable land, and to be without basic energy, water and sanitation. There may also be an informal environmental economy in 'waste picking', urban food growing, and supply of water and fuel at the micro level.

Beyond the urban fringe are larger hinterlands, where the urban gravitational field can exert rapid and destructive changes on smaller towns and villages. The viability of local subsistence farming changes when the males go to the city as migrant labourers or the females as service workers, as do tradition patterns of kinship and community structures. The environmental effects then follow as cultivated land returns to semi-wild, water systems fall into disuse, and other cash crops are then grown by entrepreneurs for urban markets and so on. The effects of these processes are difficult to measure but could be greater in the hinterlands than in urban areas themselves (McGranahan, *et al.*, 2004).

Spatial development in the 'North'

The pattern of urbanization and spatial development in the 'North' shows how the local agenda for shelter and property intersects with economic and social pressures at the national and global levels. Housing, employment, transport, communications and public services all generate spatial change, which is mediated by the regime of spatial planning and environmental regulation, organized at the national level. For most of the industrial, or post-industrial, city-regions of the 'North', there are at least six spatial trends which can be seen running in parallel (Geyer, 2002; Bontje, 2001; Portugali, 2000; Champion, 1999):

- thinning out – the reducing size of the average household means that the population of most existing areas is gradually dispersing as demand rises for space, privacy, amenity and individuality;
- shrinking cities – in some areas such as the former East Germany or the USA rust-belt, the economic base declines rapidly and large sections of the population leave the city-region altogether.
- conventional urbanization – the traditional spread of metropolitan urban areas at their peripheries, unless constrained by policy;
- counter-urbanization – the wider distribution of urban populations across surrounding rural areas;
- re-urbanization – the return of more affluent professional populations to city centres, inner cities and regeneration areas;

- centralization – inter-urban and regional migration towards the capital cities and economic growth poles.

While these are broad geographic trends, what happens on the ground is due to many contingent factors, and there is evidence of a 'fractal city', with endless levels of self-organizing complexity (Batty and Longley, 1994; Soja, 2000). At the root of such trends are the needs and desires of individuals and households – not only economic as in the models, but also social, cultural and behavioural. Many social surveys show general public aspirations for sustainable communities, with high levels of social justice, environmental efficiency and economic prosperity (ODPM, 2006; ODPM, 2000).

Whether this ideal combination can be realized for the whole population, not only the wealthier sections, is a question of huge importance. At the same time there is much evidence on the 'de-territorialization' of urban communities, to say that the former urban order is rapidly splintering and transforming into a kind of 'meta-city', composed of individuals organized in global networks with little interest in the local (Graham and Marvin, 2001). The sustainable community agenda in future may need to focus on networks as much as neighbourhoods, and this applies both in cities which are expanding and those which are shrinking (Box 6.3).

Box 6.3

Turning around a shrinking city, Germany

The municipality of Leinefelde-Worbis in former East Germany has effectively engaged with the challenges of depopulation, a failing economy and a lot of empty and deteriorating housing stock in a shrinking city. This has been achieved by the Zukunfts WerkStadt, through an innovative and integrated, participatory approach. New job opportunities have been created, the urban infrastructure and living environment have been significantly upgraded, and over 2,500 apartments have been refurbished to high environmental standards.

Source: World Habitat Day Awards, http://www.worldhabitatawards.org

---

Environment and spatial structure in the 'North'

Each of these trends generates problems and opportunities for different social groups. It also raises a policy agenda for *integrated planning*, bringing together social, economic and environmental goals, based on the underlying dynamics for each area type, across a typical city-region (Breheny and Rookwood 1993; Ravetz 2000):

- opportunity areas – areas of rapid change and restructuring with a legacy of negative assets such as contamination and dereliction; investment should capitalize on potential assets such as location and large scale sites;
- regeneration areas – with a legacy of social, economic and environmental problems; investment needs to be generated indigenously as far as possible, through a linked programme of social, economic and environmental improvements (Roberts and Sykes, 2000);
- consolidation areas – general stability of infrastructure and socio-economic resources, as in the affluent suburbs; indirect environmental impacts have to be tackled through the gradual and strategic development of urban strucures, form and fabric;
- protection areas – there may be positive environmental assets under pressure, where change and development is directly constrained by critical capacities and environmental assets.

In each of these area types, the environmental assets or stocks vary from negative to positive quality, the pressures or flows vary from internal to external, and the impacts from direct to indirect. Somewhere in the middle are large areas of suburban housing typical of 'Northern' cities, where apparent stability conceals large indirect environmental impacts (Gwilliam, *et al.*, 1999; Teaford, 2007). Overall, the sustainable development agenda for urban planning and management envisages a balanced development, in which the elements of urban change are encouraged to contribute to common objectives, to be coordinated and integrated within a city-region framework (Ravetz, 2000; Roberts, *et al.*, 1999b).

Urban density, transport, energy

Much of the debate on urban form focuses on the relationship of urban size and density to transport and energy demand. An international analysis of urban density and transport demand showed a 500-fold difference between Hong Kong at one extreme (10,000 persons per hectare, or pph) and Phoenix, Arizona at the other (10 pph) (Newman and Kenworthy, 1999). Such density-energy analyses can be extended to include the balance of energy use in buildings, infrastructure such as combined heat and power (CHP), and urban food cultivation.

In principle there may be an optimum for urban size and density, at the balance point of the density-transport-energy functions, and the urban social and economic functions (Alonso, 1971; Capello and Camagni, 2000). However it is clear that cities are more like nodes on a network than island objects; this is described by the concept of 'polycentric' city-region or agglomeration (Hall and Pain, 2006). The implication is that the question of the optimum size and density concerns the network as a whole, which will contain a variety of sizes, locations and functions.

At present there a worldwide reaction to the problems of uncontrolled urban sprawl, and the 'compact city' offers an alternative model (Jenks, *et al.*, 1999). This partly depends on the level of 'urban capacity', i.e. space for absorbing new development within existing urban areas. This is both a technical and social issue – development viability depends on the social acceptability of densities, mixed uses and other factors, and European cities such as Paris or Vienna show that much higher densities are highly viable, given the right combination of lifestyles, housing tenures, public transport and public services. While density studies are useful, raising urban capacity depends also on a creative and shared vision of the future city; as noted in the previous section, quality of place-making can be a good substitute for quantity of space (Ravetz, 1999a).

Urban fringe, peri-urbanization and the edge city

The urban fringe of many metropolitan areas is a tangle of raw materials and residues from the urbanization process – economic innovation with landscape change, and opportunity alongside dereliction. The growth of 'edge cities' is indicative of a new kind of networked 'metropol-ization', sweeping away the previous pattern of cities surrounded by countryside, and producing new and more diffused urban forms (Garreau, 1991). For instance, in Tyson's Corner near Washington DC, a former rural crossroads is now the centre of a major spread of business, retail and housing development covering 100 km². In such areas, reducing pollution and re-using vacant land are worthwhile goals, but little is achieved without looking at the wider dynamics of the peri-urban fringe and hinterland (Wood and Ravetz, 2000).

One of the primary driving forces is often the city or regional airport – a new determinant of location and value in the globalized economy, just as highways were in the twentieth century and railways in the nineteenth century (although as Hall and Pain (2006) and others argue, railways are again competing for dominance in a post-peak oil world). The concept of the 'aerotropolis' represents a powerful trend which is often *ad hoc* and unplanned at present. But in size, throughput and value added, many such airport-centred development zones have started to overtake the former central business district to form a new 'technopole' hub and gateway to the region (Kasarda, 2004; Arend *et al.*, 2004; Hall and Castells, 1996). The environmental management of such sites poses special challenges, both locally and globally, but recent trends in oil prices may mean that airport-centred development is a short-lived phenomenon.

Box 6.4

Urban food production, Argentina

In 2001, Argentina was in turmoil as public anger over a deepening recession and widespread poverty sparked riots and angry protests. Rosario City in the Santa Fe province, with a population of 900,000, was no exception. The Urban Agriculture Programme (UAP) was initiated after the economic crisis, which manifested itself in Rosario with poverty levels rising to 60 per cent of the population.

The programme aimed to enable sustainable means of food production in urban centres for a population whose poverty line is US$90. It has helped to make low-income families feel valued, especially women. So far 791 community gardens have been established, and more than 10,000 families are directly linked to the production of organic vegetables, which are consumed by 40,000 people. This has been possible through the creation of an economy of solidarity network that includes 342 productive groups.

Source: UN-Habitat Best Practices, 2006, http://www.bestpractices.org

Overall the dynamic of metropol-ization is a combination of economic, technological, and social change, which is (sometimes) mediated by various forms of spatial planning. The environment then becomes a crucial factor in shaping the result in spatial development; high-quality and high-value environments are favoured for business parks and leisure parks; low-quality and polluted residual environments are preferred for urban infrastructure such as landfill sites, sewage plants and power stations.

There are many negative effects of such economic and spatial growth. For farming and land resources, these dynamics can often produce instability and lack of investment; the productivity of conventional farming drops near urban areas, small producers are over-ridden by commercial interests, and there is much land vacancy and fragmentation (McGranahan, *et al.*, 2004). In response to such problems there are moves to reclaim peri-urban landscapes for urban community use (Nicholson-Lord, 1987). There is a movement in both 'South' and 'North' to reconnect the urban food market with the potential of the urban fringe to provide food and livelihoods (Box 6.4). For instance, in regions such as Shanghai or Greater Manila, various planning frameworks have aimed to retain a patchwork of intensive cultivation, as part of a mixed use and diverse economy across the wider conurbation. Even here, however, there are problems with rising land values, labour markets, and water resource demands between rural and urban communities (Junde and Zaide, 1996).

Green belts and protection zones

A crucial theme in spatial development is that public policy is the essential counterpart to market forces; not so much that planners are popular, rather that most private sector development is dependent on public investment in roads, drainage, utilities and so on. One of the foundations of twentieth-century spatial planning, the UK green belt, is practiced around the world under different guises and regimes. However in all peri-urban areas, there is increasing economic pressure for motorways, airports, business parks and other infrastructure, and much of the green belt areas are damaged and polluted. There are similar policies such as landscape areas, green wedges and river valleys, but it is clear that green belt policies and functions and boundaries may need a rethink (Elson, *et al.*, 1993). Some of the problems and opportunities include:

- degraded or derelict land where development would promote enhancement and after-use;
- rural areas where diversification needs leisure or ecological development;
- smaller settlements where development would enhance viability of local services;
- green wedges in the urban area which would benefit from green belt extensions.

At the same time there is a need for long-term stability of protected land boundaries and policies, as any uncertainty tends to inflate hope values and undermine investment in existing development. Meanwhile on the ground, large areas of such zones are often degraded and under-used, and there is an agenda for working towards a more integrated and sustainable 'eco-belt', within or between cities (Ravetz, 2000).

Planning for sustainable urban form

In the wider view, the question of urban form is much more than simple density and brown-green choices, it is about the spatial structure and form

"cities in evolution"

showing outward 'push' of urban area along radial routes, with inward 'pull' of open space needs

"social city–region"

showing planned clustering or 'nucleation' of central & satellite cities, suggested size 30–60000 population, with radial & orbital routes

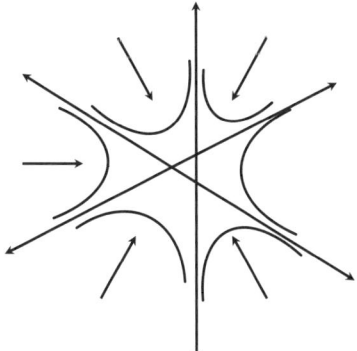
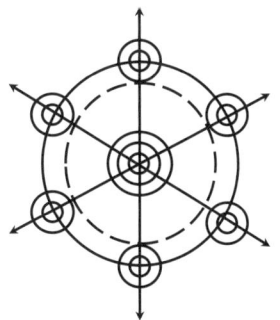

General dynamics and patterns for conurbation forms:
Source: based on Geddes 1915: Howard 1898.

Figure 6.2 *Planning for sustainable city–regions*

of human activities. This is not a new theme – over a century ago, the Garden City or Social City concepts aimed at planned communities as a response to the chaotic overcrowding and pollution of English industrial cities, coupled with the deprivation of rural dwellers (see Figure 6.2). The agenda promoted by Ebenezer Howard is very similar to the modern interpretation of the sustainable community model (Howard, 1898; Hall and Ward, 1998; Roberts, 2008).

The agenda now facing developing cities of the 'South' is how such an idealistic model can be achieved in the face of chaotic urbanization, economic growth, social conflicts and fragmented government. The agenda for the post-industrial cities of the 'North' is not only about new settlement forms, but about restructuring existing cities and city-regions for social and ecological goals (Ravetz, 2000). At each end of the 'South'–'North' spectrum the theme of strategic clustering for viability and cohesion, to achieve organization out of chaos, is equally valid.

Neighbourhoods and the human scale

At another spatial scale, such a creative vision might be more practical at the neighbourhood level – but in reality most neighbourhoods contain many communities, and most communities are spread over many neighbourhoods. While networked communities and sub-cultures are part of the 'richness' of the city, the physical quality and human scale of local neighbourhoods are equally important in public perceptions (Jacobs, 1961). So the principles of human-scale neighbourhoods are nothing new, but it seems they have to be re-interpreted for the post-industrial city (Rudlin and Falk, 1999). The starting point is the linkage between homes, jobs and services – a viable level of education, retail and public services requires a 10,000–20,000 population range, and a viable walking distance for most people is about 4m–1,000m, depending on the quality of the environment.

Both these can be achieved by clustering higher density housing and mixed employment around local centres, in pedestrian access units of up to 1km radius. Clusters of Transit Oriented Developments can be arranged around public transport loops and district centres with specialized services, with an interlocking green infrastructure or matrix of green spaces (Calthorpe, 1993). Highways should be in a tree pattern with restricted through routes – walking or cycling within the neighbourhood, and public transport between neighbourhoods, should be the first choice. Some versions stress the importance of design controls and social controls in order to foster community cohesion (Duany, *et al.*, 2003).

Again, the environmental agenda works several ways. The neighbourhood concept is the first way to reduce transport and building-energy demand: it builds in environmental quality, biodiversity and green space, and it helps to foster social cohesion and diverse local economies. Against this, there are many neighbourhoods living with poverty and social stress, and the neighbourhood unit is only one level of an increasingly networked city-region.

One underlying concept for the peri-urban areas, green space policies, and the neighbourhood agenda, is that of 'multi-functionality' – that land use is more effective with social, economic and environmental activities in combination (Forman, 1995; James, *et al.*, 2000). This can be seen working at different scales, such as the peri-urban farming in Manila; neighbourhood vitality through mixed uses; and the community management of urban green space. Where residential streets can be converted to Home Zones, for example, then the available open space for leisure and childrens' play can be doubled (Ward, 1978).

Urban environmental planning

This brief review has shown how spatial development is inextricably linked with the urban environment agenda. For the policy response – urban environmental planning – there is a diverse range of approaches over the last century, as summed up by Peter Hall (1996). Each of these urban planning models can be linked to an environmental planning model:

- mass-transit suburban development, as in the world cities of London, New York or Tokyo; the suburban dwelling in its suburb with its carefully managed environment emerged as the most enduring of twentieth-century forms;
- garden city models, and the more easily achieved garden suburb, as first advanced by Ebenezer Howard in 1898; this agenda arose partly out of the gross hardship and industrial pollution of the nineteenth-century industrial city;
- cities as hubs for regional planning, focused on urban–rural

interactions and ecological resources; initiatives such as the Regional Planning Association of America paved the way for twenty-first-century concepts of the sustainable city-region;

- cities as monumental creations, often in colonial or totalitarian states, and often in capitals such as Brasilia, Abuja or Chandigarh; many of these were built with little consideration of the local environment or climatic conditions;
- cities of 'sweat equity', focused on more autonomous self-help communities, some of which were informal forms of tenure, promoting low-impact lifestyles and local economies;
- cities of highways; the postwar Fordist model of decentralized production and consumption – the classic American suburban dream, now associated with the worst problems of traffic congestion and climate emissions;
- cities of late capitalism, anti-planning and entrepreneurial action, leading towards a more post-modern city of global networks and transient communities; the environmental challenges in this model are perhaps even greater, as supply chains are globalized, and policy is fragmented.

In reality most mature cities show influences and layers from different periods, with different planning approaches, which may or may not succeed before another generation comes in. Around the EU there are various experiments in spatial planning and environmental planning, with the aim of integrating economic growth with environmental protection through more flexible and responsive means (Salet, *et al.*, 2003). Overarching all areas of public policy, the EU system of Strategic Environmental Assessment is extended in some countries to 'sustainability appraisal' (see Chapter 9). In North America and elsewhere there is strong pressure for citizen participation, combined with corporate and business lobbies, which assumes the co-ordination role of an otherwise fragmented city and regional government. In one of the birthplaces of modern urban planning, for instance, the Chicago Metropolis 2020 programme brings together the six counties together with major businesses and a wide range of civic organizations (details on http://www.chicagometropolis2020.org).

A counterpart to this new spatial planning approach has also emerged at community level. In the EU the sustainable community model was adopted by the member states in 2005 as the Bristol Accord (ODPM, 2005). This sets out the new spatial planning and management challenges at the community level, and provides guidance on the comprehensive and integrated planning and management of places for people. This model brings together the eight acknowledged components of a sustainable community, with a ninth – 'place-making' (Roberts, 2007). The eight components include: active, inclusive and safe; well-run; environmentally sensitive; well-designed and built; well-connected; well-served; thriving; fair for everyone. These can be applied to all communities, from the poorest slums to the wealthiest suburbs, although of course the policies to achieve them will vary.

Critical perspectives on spatial development

The spatial development of cities is a battleground for many discourses, from economic, social, political and cultural directions; and now the agenda of the human urban environment.

Globalization and global economic forces clearly dominate the growth and development of cities, and their main effects on the human urban environment can be seen in the brief history presented in Chapter 4. An economic Fordist approach sees the spatial development of cities through the lens of capitalist accumulation, and the environmental segmentation of high and low value locations across cities and regions. A more post-Fordist perspective is also current, which focuses on citizens as consumers, subject to behavioural fashions, risk and insecurity (Bell and Jayne, 2006).

The political economy of urban infrastructure is closely involved in spatial development. Across both 'South' and 'North' there are various degrees of liberalization, privatization and cost recovery models, in the provision of transport, telecommunications, water, sanitation and waste collection, many of which were formerly public and universal. The effect is often to reinforce the segmentation of supply and demand; i.e. so that private suppliers will favour profitable consumers at the expense of public provision (Guy, *et al.*, 2001).

The segmentation of urban structures then follows from this. At the global level the international hierarchy of cities shows polarization between centres and peripheries across continents (Sassen, 2006). At the city level there is continuing segmentation, and for new development in peri-urban areas it is an urgent policy issue. For urban regeneration and re-urbanization, the effects of gentrification are well known for displacing local communities in order to reshape the urban structure for the dominant order (Harvey, 1995). The transformation of many 'Northern' cities to knowledge-based and cultural activities, and the competition for global investment, highlights the culture of consumption – a city to be enjoyed by *flâneurs*, with clean waterfront environments for affluent shopping and leisure. This tends to polarize urban environments as desirable or undesirable, to be walled off in so-called 'landscapes of fear' (Tuan, 1979). The soft or hard defensive measures which are produced tend to reinforce public choices on housing, transport, education, health and in some cases ethnicity, and so the spatial segmentation of the city and city-region continues. Meanwhile, many cities in the 'South' have inherited spatial structures from their colonial history, and recent development paths have tended to exacerbate these divisions.

6.4 Travel and transport

Mobility and the motorized city

Mobility has been fundamental to the conventional view of social and economic development. But urban transport systems around the world are pushing at environmental limits, as well as strangling under their own congestion. Many now argue that the long-standing link between economic growth and transport growth has somehow to be 'de-coupled'. In the more affluent cities of the 'North' the philosophy of 'predict and provide' has shifted, at least in the rhetoric, to one of containment and reducing impacts. But on a global scale, there is little rescue yet in sight from a seemingly inevitable demand and desire for mobility. At present there are over 500 million cars on the road, and this is projected to double by 2020; and by 2030 there may be more cars in China than in the rest of the world together (Lee, 2007). However such projections are based on the assumption of continued supplies of cheap oil, and this is now challenged both by environmentalists and many in the business world (Financial Times, 2008).

In terms of the human urban environment and the urban environmental transition, the over-riding trend can be seen as 'auto-dependency' or the 'motorized city' (Newman and Kenworthy, 1999). This represents the combined effect of technology, economy, lifestyles, and the spatial development of the city-region, to lock in continuous growth in private car traffic. At an international scale, a similar trend applies to air travel; in a globalizing economy, no city can afford to be without an expanding airport, and the effect is to redraw the world map of cities around the major airline schedules. The culture of 'hyper-mobility' is highly seductive in both 'North' and 'South', but on current trends is helping to tip the global climate into unpredictable and catastrophic changes (Adams, 1996).

Having said that, the discussion here does not focus on international air travel and marine transport, although both are hugely topical, with new concepts such as 'sustainable aviation' to grapple with (Anderson, *et al.*, 2008). Freight transport is part of the economic agenda in Chapter 8. So here the focus is on urban passenger travel, from both demand side and supply sides, and its effect on the urban built environment.

Transport demand and spatial development

For industrialized cities, a producer and consumer market of several million people within a one hour travel time has been the past norm and current standard, with unprecedented levels of choice and specialization. Economic competitiveness had hinged on such specialization, and the long-term growth of transport has reflected long-term economic growth, subject to variations in oil prices. The social agenda in transport focuses on choice and accessibility: despite the anti-car lobby in many cities in the 'North', recent car traffic growth has been due to increased driving by female, younger, older and poorer people. In the 'South', the rapidly growing middle classes show their new positions by owning and driving cars; for instance, the new primary motorway network in India is seen as a major step in development, but a potential disaster for the cities (Vasconcellos, 2001; Pucher, *et al.*, 2005). For most households car ownership has been a major threshold – once the investment and fixed costs are paid, every mile travelled apparently becomes cheaper – hence its lock-in effects.

On the physical level, transport systems have shaped the interaction of people and places long before mass car travel; the first wave of suburban development in Europe and North America was enabled in the late nineteenth century by railways and rapid transit, and in many ways they were co-financed. The commercial property sector has relied on transport infrastructure development for competitiveness and the added value of location. In theory, although not always in practice, technical innovation and infrastructure increases speed and accessible distance, travel demand expands to social time thresholds, and the economic viability of land-uses follows accordingly (Ortúzar and Willumsen, 2001).

If cheap transport fuels can be assumed, the implication is that demand will tend to exceed supply, until a balance is reached by shifts or constraints on either side. However as of 2008, such twentieth-century assumptions appear to be outdated, with the debate now more about how far the world is past 'peak oil', rather than when peak oil might take place (Deffeyes, 2005). The unfortunate implication is that transport bio-fuels, already taking 20 per cent of the USA wheat crop, will take even more bio-productive land from food production, risking the hunger of millions in order to fill the fuel tanks of the affluent (Gilbertson, *et al.*, 2008; Worldwatch Institute, 2007).

Within cities and city-regions, the constraints on transport growth include:

- increases in supply – these are expensive, and tend to increase environmental impacts;
- physical saturation and congestion – economically inefficient, socially costly and environmentally damaging;
- direct regulation – this can be very unpopular with users;
- direct road or fuel charges – unpopular and potentially socially regressive, unless matched by provision of affordable public transport.

The question of the transport mode centres on the divide between private and public. Some environmental lobbies ignore the social and cultural roles of cars – as symbols of status and identity, and as providers of mobile living rooms, offices and store-rooms (Freund and Martin, 1993); these multiple roles enable and encourage new patterns – not only in journeys, but networked lifestyles based on flexible and continuous mobility. Meanwhile the car lobby tends to ignore the pollution, hazard and spatial exclusion consequences of cars. Economic agendas revolve around specialization of labour, competitiveness through access to markets, dual career households, and access to larger consumer markets and facilities. These are each strong dynamic forces which increase the range and diversity of travel destinations. Such patterns have enabled new land uses and activities, with the result that public transport would now be unsuited and inefficient for many journey patterns, even if it was available. The result is that reducing automobile dependency is not only about transport, but the restructuring of urban spatial development and urban economies.

Urban transport in the 'South'

The above shows the aspiration of most cities around the world, even while it is set to cause environmental and potentially economic disaster. At present, the urban population in the 'South' is growing by 6 per cent per year, while motorization is growing by 10–15 per cent per year (World Bank, 1996). Car ownership is spreading through lower income groups, with the result of more polluting and unsafe vehicles. Many cities have a fraction of the roadspace which is available in the 'North', and congestion and pollution build up to a point where vehicles are abandoned by the roadside, in the 'Bangkok effect' (Vasconcellos, 2001). For commuting, about 17.5 per cent in the 'South' use cars, compared with over 50 per cent in the 'North'. Trains are more common in transitional countries, and bicycles and motorcycles in Asia.

Long travel time is an obvious sign of urban dysfunction, a result of traffic congestion, lack of public transport, lack of traffic management, and many accidents. Reducing travel time is a challenge for transport planners in fast-growing mega-cities; average commuting times in Tokyo by subway and moped are over 90 minutes, similar to those by walking, for example in Kinshasa, one of the least wealthy cities in the world.

Road traffic accidents kill nearly 1.2 million people annually, of which about 90 per cent are in the 'South', at an estimated cost of over $500 billion (WHO, 2004). As motorization increases, road traffic injuries are predicted to rise to become the eighth leading cause of death by 2030 (WHO, 2004). Scandinavia has a fatality rate of one per 30,000 vehicles, while in parts of Africa it is more than one per 200. For children between five and 14, traffic injuries are the biggest single source of death, and roads are second only to HIV/AIDS in killing people aged between 15 and 29. In the 'South', accidents are dominated by pedestrians, cyclists and public transport passengers. This is one of the great front lines of global environmental justice – at present, pedestrian or cyclist victims are generally assumed as guilty, with little redress against drivers (WHO, 2004).

Urban transport in the 'North'

While transport in the 'South' might appear to be out of control, 70 per
cent of the world's cars are as yet in the 'North'. If current trends
continue, total vehicle numbers and urban traffic in the European Union
could increase at 1.4 per cent per year up to 2030; freight traffic is
on a faster growth trend at 2.1 per cent per year (CEC, 2003). Across
the OECD, the car population in cities is approaching half the human
population, and in many cities the road space is greater than the human
space, as more female, younger, older and poorer people continue to
acquire cars. The take-up of ICT is crucial to future travel patterns –
its effect might reduce routine urban journeys more suited to public
transport, while increasing more irregular and longer-distance journeys,
more suited to cars, high-speed rail, and air travel (Banister, 2005). On

---

Box 6.5

Integrated transport systems in Singapore

Singapore has a population of 4 million people on an area of 400 sq. km.
The Singapore initiative in urban land transport development and
management seeks to solve worsening traffic congestion and its many
associated problems by building a world-class land transport system.

The initiative is premised on four key principles of: (i) integrating transport
and land use planning; (ii) expanding the road network and maximising its
capacity; (iii) managing demand for road usage; and (iv) providing quality
public transport choices. The average peak-hour travelling speed within its
city centre is about 30 kilometres an hour which compares favourably with
peak-hour speeds of 10 to 12 kilometres per hour in London, New York,
Manila, Calcutta and Lagos.

In addition, the city was the first to attempt to use road pricing to limit the
growth of urban traffic. From April 1998, Singapore replaced its system of
central area access charges based on paper licenses (first introduced in 1975)
by electronic tolls that vary according to time of day.

Source: UN-Habitat Best Practices, 2006, http://www.bestpractices.org

the supply side, ICT may have great potential to improve the capacity and safety of roads, and the flexibility and performance of public transport, and examples of 'intelligent infrastructure' can be seen in cities such as Helsinki and Singapore, as in Box 6.5 (Sharpe and Hodgson, 2006).

Transport and the urban environment

The environmental impacts of transport are familiar – it is the fastest growing source of climate emissions and urban air pollution, with the oil and automobile industries together causing half of global ecological damage (WWF, 2006). The social impacts of noise, pollution, congestion, disruption and accidents, fall heaviest on the poor, while massive investment goes into road infrastructure.

The impact of travel depends on how it is done – the energy consumption of different transport modes per passenger mile varies by a factor of twenty, from large petrol cars to fully laden buses; see Figure 6.3 (Banister, 2005; RCEP, 1994). But the emergence of automobile-dependent cities and city-regions is due to more than the efficiency of single trips – rather it is based on diffused, networked, single-occupant,

Comparison of energy demand per passenger mile, in MJ / pass. mile: for average and fully laden operation: also showing adjustment for aviation climate forcing effect.

Source: European Environment Agency 2006: Royal Commission on Environmental Pollution, 1995

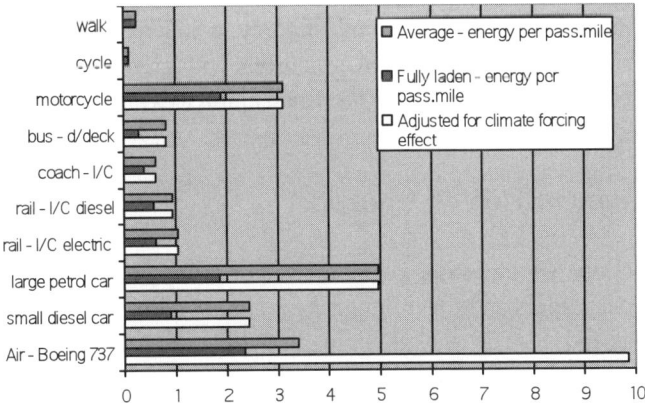

Energy in MJ per passenger mile

Figure 6.3 *Transport modes and the environment*

flexible patterns of mobility, in parallel with the general pattern of a post-Fordist globalized economy. There is an alternative in the 'transit metropolis', based on a European urban model, that is thriving in parts of many cities around the world (Cervero, 1998).

So what can the average city or city-region do? First, it can reduce the use of cars and trucks through access restriction and road pricing. This is a solution that can be introduced locally, and which can be implemented in advance of other more long term changes. The merit of this approach is that it prepares the city or city-region for a future based on public rather than private transport.

Second, it can pave the way for a full supply chain and market transformation of the vehicle fleet through a combination of public procurement, fuel infrastructure, petrol-free zones, employer subsidies and tax breaks, public service fleet management, and subsidized car clubs (Ravetz, 2008). A 50 per cent efficiency improvement is viable with current technology across the European fleet by 2030, while prototype hyper-cars are operating at over 300 miles per gallon (King, 2007).

Third, in the longer term, cities and city-regions can redirect their infrastructure investment towards a fully integrated public transport system, accessible and affordable to all communities. Rising oil prices may have a significant influence in shifting the balance of costs and benefits of public investment away from freeways and flyovers towards mass rapid transport systems. Fourth, the city can evolve a spatial development and economic strategy, which brings together improved social and environmental conditions with greater mobility and enhanced economic performance (Newman and Kenworthy, 1999).

Critical perspectives on transport

Globalization is one of the drivers of growth in transport, both for passengers and for freight. The potential of an integrated transport system as above, is often challenging to public authorities, many of whom themselves are locked into automobile dependency, and who currently lack the legal powers and corporate commitment to provide alternatives to the growth of road transport. Many urban transport networks and infrastructure systems have been franchised or privatized; the results can show environmental improvements, but often lead to higher costs and lower equity of access (von Weizsacker, 2005). However, many city authorities are now realizing the benefit of shifting from the failed car-based paradigm towards a revitalized transit metropolis.

The culture of consumption has also emerged in parallel with private transport growth. Here, private cars become totemic objects, personal privacy becomes an end in itself, and local cultural identity is shaped by its accessibility by road. Clearly, private transport has strong effects on social exclusion, and affordable public transport can often promote social inclusion; however there is a common perception of public transport as risky and dirty which has been a strong incentive for further shifts to private transport. Overall the transport infrastructure is instrumental to the segmentation of the city and polarization of areas of growth and decline. Such issues are now being addressed, and this is shown in the resurgence of public transport particularly in larger cities.

Each of the possible measures for transport improvements also raises controversial arguments. Financing of new roads through tolls or public-private partnerships, often with overseas finance, can be socially regressive and divisive. The taxation of fuel, vehicles, road space, inner cordons, or congestion in general, is an emotive issue which rewards or penalizes different groups. Surveillance by automatic cameras for fines or congestion charges may seem to infringe civil liberties; however the absence of surveillance can also exacerbate safety and equity problems. Meanwhile there is a strong environmental justice agenda in the reclamation of road space by local residential communities, who are disrupted and disempowered by passing traffic. Gender and generational issues are also highlighted by access to or exclusion from transport – both the motorized and the public transport lobbies argue for the opportunities for the female, disabled, elderly or young.

Over-arching all this is the political economy of oil, the automobile-aviation culture, climate change, and now access to bio-fuels – the crucial dividing question being, whether or how far the automobile-aviation complex will survive the forthcoming post peak oil era (Kunstler, 2006).

6.5 Conclusions

This brief review has explored some of the dynamics of the urban built environment – focusing on the overlapping agendas of housing, spatial development, and transport. Although each of these can be measured in hectares and kilometres, the dynamics are local and global economic and social forces, and the outcomes are often environmental on a local and global scale.

To draw these together, there are alternative urban 'paradigms' or mental models, which represent aspirations and fears for the future of the built environment. 'Smart growth' aims to balance economic development with lower impact, better planned spatial development. 'Transit metropolis' takes this further and sees a pedestrian and public transport based organization of economic and social life. In contrast, the 'Aerotropolis' takes a counter view, focusing on the global hub role of the city, with the expectation of technological solutions for environmental problems. The 'zero-carbon city' also pursues an engineering dream of autonomous development, where social and economic life is expected to fall into place. In the background is the looming nightmare of the Planet of Slums. With these and other competing visions, it is clear that the urban agenda is at some kind of crossroads, without the luxury of time to decide.

Further information

American Planning Association   http://www.planning.org

Cities Alliance: cities without slums   http://www.citiesalliance.org

Commonwealth Association of Planners   http://www.commonwealth-planners. org

Environmental Transport Association   http://www.eta.co.uk

Global Urban Observatory   http://ww2.unhabitat.org/programmes/guo/

International Council for Local Environmental Initiatives   http://www.iclei.org

International Federation for Housing and Planning   http://www.ifhp.org/

Next American City   http://americancity.org/

Town and Country Planning Association   http://www.tcpa.org.uk

United Cities and Local Governments   http://www.cities-localgovernments.org

United Nations Centre for Human Settlements   http://www.unchs.org

Victoria Transport Policy Institute   http://www.vtpi.org

World Carfree Network   http://www.worldcarfree.net

World Urban Forum   http://www.unhabitat.org/categories.asp?catid=535

Further reading

Banister, D. (2005) *Unsustainable Transport: City Transport in the New Century*, London and NY: Routledge.

Cervero, R. (1998) *The Transit Metropolis: A Global Inquiry*, Washington DC: Island Press.

Davis, Mike (2005) *Planet of Slums*, London: Verso.

Jacobs, J. (1961) *The Death and Life of Great American Cities*, New York: Vintage Books.

Neuwirth, Robert (2005) *Shadow Cities: A Billion Squatters, a New Urban World*, London & New York: Routledge.

Newman, P. and Kenworthy, J. (1999) *Sustainability and Cities: Overcoming Automobile Dependence*, Washington DC: Island Press.

Ravetz, J. (2000) *City-Region 2020: Integrated Planning for a Sustainable Environment*, London: Earthscan.

Roberts, P. and Sykes, H. (eds) (2000) *Urban Regeneration: A Handbook*, London: Sage.

Rudlin, D. and Falk, N. (1999) *Building the 21st Century Home: The Sustainable Urban Neighbourhood*, Oxford: Architectural Press.

UN-Habitat, (2003) *The Challenge of Slums: Global Report on Human Settlements 2003*, Nairobi: UN Habitat (available as of Dec 2008 on www. unhabitat.org/categories.asp?catid=555).

UN-Habitat, (2006) *The State of the World's Cities Report 2006/2007: The Millennium Development Goals and Urban Sustainability: 30 Years of Shaping the Habitat Agenda*, Nairobi: UN Human Settlements Programme, London: Earthscan.

Vasconcellos, E. (2001) *Urban Transport, Environment and Equity: The Case for Developing Countries*, London: Earthscan.

 **Cities in the global market**

The economic urban environment

- 7.1 Introduction
- 7.2 Changing economic environments
- 7.3 Business and industry
- 7.4 Employment and livelihood
- 7.5 Finance and investment
- 7.6 Urban economy-environment strategy
- 7.7 Conclusions

7.1 Introduction

From effluence to affluence

The growth and development of cities is largely driven by their economic role and activity. It is an interesting question as to whether this revolves more around production or consumption, or simply the dynamic of 'capital accumulation' – in simple terms, making money. Local pollution from industry, the impacts from services and infrastructure, and the impacts of global supply chains, are all economic issues as much as environmental ones.

However, most economics textbooks don't mention global environmental impacts and limits. Neither do they mention the sharp contrasts between poverty and wealth in the 'South' and 'North', in some ways as shocking as those of the industrial revolution two hundred years ago. In a nutshell, the poor suffer the 'effluence' – the environmental impacts of industrial pollution; while the wealthy enjoy the affluence – rising material consumption in apparently cleaner and greener environments. What has now changed is the global scale of operations, the unprecedented power of international finance and trans-national corporations, and the extreme and dangerous pressures on global environments and resources.

The context and direction for this is summed up by the idea of 'ec
modernization' – a structural transformation of systems of production,
distribution and consumption (Hajer, 2003; Hawken, *et al.*, 2005).
Such a transformation can be seen in every section of the community
– trade unions, women's movements, chemical engineers, trans-
national utility firms, ethical consumers, neighbourhood activists, and
investment managers are all involved. Equally, the process of ecological
modernization in cities and regions has profound consequences for
their spatial development and environmental performance (Roberts and
Gouldson, 2000). What is up for debate is how much this is a technical
and economic question, and how far it has to challenge existing lifestyles
and structures of power and wealth.

In this chapter

This chapter brings together some of the many angles. First, the context
for the changing economic urban environment is the current pattern of
globalization, which determines the economic role of each city system,
and thereby its environmental flows and impacts.

The 'sustainable business' agenda looks directly at the material demands
for energy and resources, and the pollution and waste from industrial
processes. In response there is an emerging framework for environmental
management and eco-efficiency in products, firms and sectors.

The 'sustainable livelihoods' agenda looks at employment and the
prospects for work, workers and workplaces. It also looks beyond the
mainstream money-based economy at other kinds of activities and
trading systems in the social economy, which may be essential to local
environmental improvements.

According to the 'sustainable investment' agenda, the first question on
any policy intervention in production or consumption is – how much
will it cost, who pays and who benefits? There are parallel strands in the
greening of finance and environmental taxation, from the public or private
sector.

Bringing all these together is the theme of 'urban economy-environment
strategy' – the process of ecological modernization across a whole
urban or regional economy. There are many examples which bring

together industrial ecology, supply chain innovation, social enterprise, re-manufacturing and ethical finance.

In each of these areas, as in previous chapters, there is a wide range of experience from cities of the 'South' and 'North', bearing in mind that these are simplified types on a complex spectrum. In each of these areas the application of critical perspectives then throws light on the tangle of debate and controversy.

7.2 Changing economic environments

Multiple layers

The globalization of city economies, and the changing pattern of wealth and income, drives the urban environmental agenda. This can be seen in each of the many layers and dimensions of urban development.

One layer is the global economy itself which centres on the world cities and international hubs, such as Tokyo, London and New York, with global-level activities in financial services, media, education, producer services and other sectors (Sassen, 1994). At this level these cities are not so much trading as separate entities, rather as parts of a wholly integrated system; and at this level the environmental agenda concerns the indirect impacts of financial and professional services.

On another layer, each city is situated in a national context and position in the global trading system, as divided into a set of blocks according to relative wealth, although the precise boundaries are always debatable. This book refers simply to the 'North' as the OECD countries (the EU, North America and Australasia, Japan and South Korea); and the 'South' as the Rest of the World, including China, South and South East Asia, the Newly Independent States (former Soviet Union), Latin and Central America, Oceania and Africa. One alternative grouping is to separate out the BRIC countries – Brazil, Russia, India and China – as the most populous, with nearly half the world's population between them. While this book has to summarize the extremes of cities in terms of 'North' and 'South', the reality is of course much more complex.

The role of the city in its national or regional economy is the key to its physical environmental profile. 'Primate' capital cities dominate their countries, while secondary and provincial cities are more or less dependent on the economic periphery. Smaller settlements may have quite distinct roles and profiles, such as dormitory or retirement towns, market or frontier towns, factory or military towns, and so on (Badcock, 2002). Depending on its position in the global hierarchy, the environmental flows and impacts for such a city may be focused on primary commodities, industrial production, distribution and services, or on consumption and waste.

A further layer is the economic role of a city in relation to its regional or rural hinterland, or 'functional territory' (Friedmann and Weaver, 1979). This is most obvious in rapidly urbanizing cities of the 'South', where chaotic urbanization is often the result of unbalanced development and insecure rural livelihoods. It is also a factor for cities of the 'North', which may show the opposite trends of counter-urbanization due to business decentralization, retirement migration, leisure and tourism. Looking at the city as part of a functionally inter-connected city-region, is a key to understanding the environmental flows, and hence the environmental policy agenda (Ravetz, 1999b).

Within the economy of the city itself, there are internal divisions which determine the patterns of production and consumption. Through the processes of capital accumulation and social stratification, urban economic structures show sharp contrasts – between cores and peripheries, growth and decline, wealth and poverty – and in turn there are environmental impacts and inequalities which reflect each of these. The concept of 'displacement' shows how high-income groups construct high-quality environments, by shifting supply chain impacts to low-income areas; and this applies at all the levels from local to global (McGranahan, 2006).

Each of these layers is then subject to events and contingent factors, such as political instability, military conflict, economic collapse, or environmental hazards. For instance, the closed cities of the former Soviet Union were built in isolated locations as part of a military-industrial complex, and in the rapid transition to a market economy they face a very uncertain future.

In practice many of these layers will overlap. Just as physical development tends to merge together former freestanding settlements, the economic dynamic of cities is also tending towards that of the 'extended metropolitan region' (Hall and Pfeiffer, 2000; UN-Habitat, 2005b; Roberts, 2000). Such regions can then develop more specialized economic spaces: from the central business district, to zones for leisure, retail, logistics, science parks, and other features of the edge city. They will contain elements of a global cosmopolis, with fractal-like patterns of local segmentation (Soja, 2000; Kundu, 2006). As most metropolitan regions do not fit political boundaries, and their economic linkages are increasingly trans-national, this can only increase the challenge for the urban environment.

Political economy of the urban environment

By far the most pressing environmental problems are found in the rapidly urbanizing cities of the developing nations – anyone who has set foot in Nairobi, Mumbai or São Paulo will know vividly the heat and dust, the poverty and pollution. A structural or political economy perspective sees not only a simple problem of local disorganization, rather the current phase of a long history of first colonial, and now globalized, capitalism. So, just as urbanization is an integral part of the global development process, so are urban environmental problems – and the implication is that the chaos and pollution in developing cities are also problems for the affluent 'North'. A typical urban development scenario can contain overlapping stages (Gugler, 1988):

- rural agriculture is displaced by industrialized plantations, and population growth results in migration towards the cities; rural environments are heavily damaged by extraction, urban environments by chaotic growth;
- profit is extracted from rural industries by foreign firms, and either repatriated, or invested in the urban property market;
- industrial manufacturing then concentrates in the urban centres, and a new service sector and bureaucracy begins to emerge to support it; industrial pollution accelerates but only in certain locations, as the city form begins to segment, in order to protect the new middle classes;
- labour is then attracted en masse to industrializing cities, and the state provides a bare minimum of infrastructure; many live in illegal or informal slums in hazardous locations;
- as the city over-reaches its optimum size, and suffers from congestion and pollution, investment looks for growth areas further afield; urban

areas are left in chaotic conditions with only the informal economy for support;

- in some areas this has produced vast areas of industrial 'exo-urbanization' in extended emergent metropolitan regions, such as São Paulo or Greater Manila.

Overlaid on this fairly grim picture is the international financial and development system. Since the 1970s the International Monetary Fund and the World Bank have made lending to developing countries conditional on structural adjustment, with extensive privatization of public services and trade liberalization programmes. While this has aimed to accelerate entry into a globalized economy, there are many critics who see it as another phase of post-colonial expropriation (Stiglitz, 2002; Reed, 1996).

The implications for urban environments are both positive and negative, but in many cases, disastrous for the poorest and most vulnerable. Public water systems have been privatized under the name of 'cost recovery', and prices raised so that many of the world's poorest people pay more for their water than the richest (International Rivers Network, 2006). The combination of international loan conditions, unrestricted speculation and capital flows has undermined the entire financial balance of many lower income countries, as in the East Asia crisis of 1997. The result is that often urban housing and infrastructure programmes are stalled, while environmental pollution goes unchecked (von Weizsacker, 2005; Hertz, 2005).

By contrast in the cities of the 'North', local environmental improvements are seen as a selling point or image bonus in the race for competitiveness and inward investment, and the 'discourse' of local sustainability is adapted to this global agenda (Brand and Thomas, 2005). Such improvements can also be a means to recover economic self-confidence and social cohesion in damaged communities (Box 7.1).

7.3 Business and industry

On the production side of the economy, the unit of analysis is the sector, the firm, the process, or the product. The question then arises, what is the connection between these firms and the city where they happen to be located? Even in a highly globalized economy there is an explicit spatial

---

Box 7.1

Revitalizing post-industrial port cities, UK

Orchard Park and North Hull were established as resettlement areas
for residents following the decline of the fishing industry in the 1960s
and the subsequent slum clearance of the Hull docks. Since then, these
two neighbourhoods have suffered from high levels of unemployment,
exclusion, dependency, drug abuse and a high crime rate.

The Orchard Park and North Hill Enterprises (OPNHE) was established
in 1989 as a community–business non-profit partnership, aiming at job
creation, self-belief and dignity. Since its inception, over 4,000 residents
have been assisted in getting employed, 339 new businesses have been
formed and 735 new jobs have been created. Every year about 34,000
people visit OPNHE and seek advice on employment, training and business
support issues. The UPBEAT project is a successful intermediate labour
market model that provides a year of full-time, waged employment in a local
small-scale micro enterprise, with related vocational training.

Source: UN-Habitat Best Practices, http://www.bestpractices.org

---

and urban dimension to the business-environment agenda (Roberts, 1995;
Ravetz, 1999b).

The emerging business-environment agenda contains several layers –
management of the local impacts of production: global supply chains,
product life-cycles, and wider trading and investment linkages. In each
case there are physical questions on materials and processes, and wider
questions on management techniques and market structures. There are
also wider questions on goals and values, inside and outside the firm,
and the firm's reason for being (Welford, 1996). In response, at least in
the 'North', the environmental regulation of business has moved from
a 'smokestack' approach to a more responsive style of negotiation and
advocacy. A ladder of business sustainability can be seen, from the 'push'
factors of regulation and cost reduction, to the 'pull' factors of new
markets, shareholder value and employee commitment (Figure 7.1)

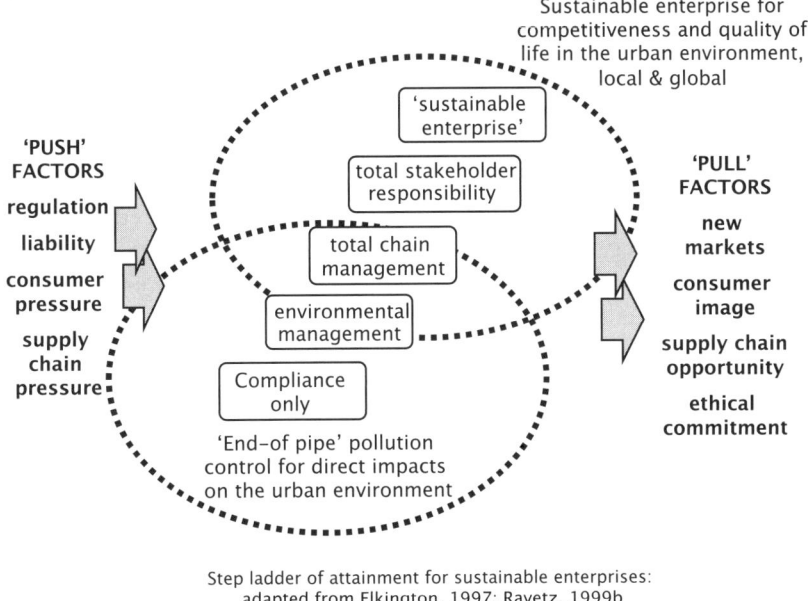

Step ladder of attainment for sustainable enterprises:
adapted from Elkington, 1997; Ravetz, 1999b

Figure 7.1 *Sustainable business and the urban environment*

Industrial impacts in the 'South'

In many cities in the 'South' there is a kind of 'dual economy' with two distinct circuits of capital and production (Cardoso, 2001). The upper circuit is generally capital intensive, export oriented, government regulated and often subsidized. In contrast the lower circuit is more labour intensive, domestic oriented, unregulated and informal, and paying little or no taxation. The upper circuit is then in a strong position to exploit and expropriate from the lower one. One regional study showed that 80 per cent of foreign-financed firms in the Asia-Pacific region retained illegally the payment of wages for up to three months (Yeung, 2004).

The effects of a dual economy are not always consistent with the goals of growth and integration with the global economy. The dynamic of culturally-driven household consumption tends to generate low-cost indigenous production, rather than higher value production for export. This may be seen as a sink for local savings and investment, or alternatively a positive step towards indigenous wealth creation

(Jacobs, 1986). Again the environmental implications can be severe for unregulated and informal employment within corrupt legal practices, using obsolete and hazardous processes, producing low-cost consumer products.

In many cities of the 'South' the informal economy is the majority part of total production; child and female labour is generally cheaper and less resistant to exploitation (International Labour Organization, 2002). There is much informal trading at the margins, from street hawking and protection rackets to begging. The average urban household economy becomes dependent on low-waged informal or illegal incomes, increasing the environmental burden in workplaces and in dwellings used for home-working (Hardoy, et al., 2001).

Industrial impacts in the North

The cities of the North also experience rapid differential change, international pressures and internal divisions. The pressures on the economic urban environment again result from economic globalization, but this also brings new opportunities. Such effects include:

- the rapid growth of trans-national corporations (TNCs) and mobile finance which determine the organization of supply chains – the pressures on urban environments tend to be from advanced stages of production, distribution and consumption, including transport emissions, trace toxics and carcinogens, packaging and post consumer waste;
- de-industrialization and 'tertiar-ization' – the hollowing out of former heavy industry from urban areas is a key driver of environmental change; gross industrial pollution of air and water may rapidly reduce, and former industrial land is then reclaimed and greened over time, but there may be more widespread effects on derelict land and falling property values;
- fluctuations in the market value of natural resources and commodities, where basic prices in agriculture, forestry and minerals have dropped in the last 20 years – in many developing countries their economic base has collapsed; in the urban environment of the North, there has been a rapid increase in freight transport, and solid waste arising;
- the growth of leisure and tourism – in many urban fringes and hinterlands of the North, the visitor economy is now larger than

agriculture or forestry; this is then a driver of environmental pressures such as land use, water, waste treatment and transport;

- finally, the growth of environmental technology as a sector – the emergence of pollution control, renewable energy sources, energy efficiency, advanced transport and similar products, has led to very rapid growth, now estimated at 10 per cent per year (OECD, 2008); there is a wide recognition of environmental technologies as prime opportunities, with added value and competitiveness through 'first mover' advantages, employee commitment and brand recognition.

Innovate + grow tech

Business-environment agendas

In parallel with these urban-economy-environment trends are structural shifts which concern not so much what business produces, but how it produces – how firms and supply chains are organized, and how business as a 'sector' interacts with other public and civic organizations. This is another facet of ecological modernization, as a structural transition with technical, economic, political and cultural dimensions (Hajer, 2003). Its ultimate goal is the de-materialization or 'de-coupling' of economic growth from environmental impacts: one way to frame this is as the Factor Four principle of 'halving resource use, doubling resource efficiency' (von Weizsacker, *et al.*, 1997). Such changes are now emerging in practice in the agenda for business sustainability, or Corporate Social Responsibility, depending on how it is framed (Hawken, *et al.*, 2005; Elkington, 1997; Worldwatch Institute, 2008):

- values – greater emphasis on the trust and commitment of shareholders, consumers, employees and other stakeholders; this is a key factor in the social/environmental responsibility shift;
- transparency – increasing stakeholder interest and consumer awareness through NGO campaigns, often using new forms of media and ICT;
- integrated supply chain management, with life-cycle technology for cradle-to-grave responsibility – many incentives for environmental improvement have come from larger firms in the automotive or aerospace sectors, who are in a strong position to organize larger trans-national supply chains;
- extensions in time, in both shortening of horizons to just-in-time logistics movements, and in lengthening of horizons to inter-generational responsibility, for instance with the climate change agenda;

- corporate governance – from shareholding to stakeholding, and then to *stake-owning*, with new configurations of equity and accountability; this has many positive effects on the environmental agenda, but it also makes it more difficult to tell the message from the medium. Companies such as Interface have shown the way, but such best practices often depend on rare qualities of leadership (Box 7.2).

Box 7.2

A corporate journey to sustainability, USA

Interface is a resource-intensive company whose largest divisions are petroleum dependent. With sales in approximately 110 countries and manufacturing facilities at 28 sites on four continents, the company makes a significant impact on the planet's commerce and ecology. Employees were brought together in 1994 for an intercompany network called QUEST (Quality Utilising Employee Suggestions and Teamwork), which engages all 7,400 associates in every plant or facility, in a search for creative solutions.

Interface is constantly striving to reduce its eco-footprint: it examines every facet of its operations, from the shop floor to the boardroom; reinventing commerce in the process, completely changing relationships with employees, customers and suppliers. The company also introduced a product based on recycled materials, a leasing programme for its carpets and plants trees all around the city. There have been over 400 sustainability initiatives undertaken, with internal metrics of waste, material and energy flows for the Ecometrics score card.

Source: UN-Habitat Best Practices, http://www.bestpractices.org

---

Environmental regulation and management

The beginnings of an integrated approach to business-environmental management emerged in the 1970s with the US Environmental Protection Agency, in the UK with the Integrated Pollution Control system, and later by the International Standards Organization. These became standard practice for site-based environmental regulation, based on the far-reaching but fuzzy concepts of 'best practical environmental option' (BPEO) and the 'best available technology not entailing excessive cost' (BATNEEC) (Howes, *et al.*, 1998). The standard is now set by the European IPPC Directive on Integrated Pollution Prevention and Control (Directive 2008/1/EC), which goes beyond former environmental regulation to cover wider issues, including:

- freight, business travel and commuting;
- energy use and climate emissions;
- integrated supply chain and life-cycle analysis;
- human and ecological risk assessment;
- ethical criteria for suppliers, contractors, distributors and investors.

The IPPC framework also raises the possibility of integration within an urban environmental strategy. This highlights the link which is often overlooked between individual firms, sites and processes, and the city or region where they are located. In future, with increasingly complex industrial supply chains, and rising sensitivities in the local environment, there will be a number of ways to frame urban environmental policy – by local pollution 'bubbles', by ecosystem service trading, or by local consumer responsibility. It may be in future that all these can be combined in an integrated sustainable business framework (Ravetz, 2000).

Industrial ecology

While most environmental management systems focus on firms, sites and processes, a wider view looks at the combined impact of whole sectors, supply chains and industrial clusters, at the city or regional scale. The sectoral profiles and prospects for high-impact industries such as chemicals, textiles and light manufacture, should be coordinated with the urban or regional environment strategy. These profiles are also the basis for developing 'industrial ecology' systems which cascade energy, material and waste flows between a network of industries (Ayres and Simonis, 1997). Such co-ordinated networks of material flows are suited to areas of heavy industry, and are common in large chemicals or automotive complexes. At the urban or regional level, established materials industries are starting to extend their markets and trading networks, anchored on a cluster of materials reclamation, disassembly and remanufacturing plants. Such a green growth pole can help to stimulate an entire urban economy, in the way that chemicals or car plants did in previous decades; equally, waste processing located at such hubs can provide the basis for a cluster of waste material based industries (Roberts, 1995 and 2006b).

In the 'South', there are industrial ecology clusters driven by necessity as much as good intentions. In India, for instance, there are examples such as the textile and leather industries in Tamil Nadu, and the Damodar Valley near Kolkata; a large watershed where mining, industry, agriculture and urban development are all in competition for water resources and environmental capacity (Dwivedi, 1995). As such economies globalize, there is a parallel growth in activity on Corporate Social Responsibility, which begins to make the links between trans-national firms and local suppliers (see http://www.csr-asia.com).

In the North, some of the largest and most globalized industries are setting the pace; BASF and Dow Chemicals have pioneered eco-efficiency benchmarking systems, while BMW and VW-Audi have developed an advanced supply chain management system linked to a sophisticated waste management system (Bourg and Erkman, 2003). While there are clear benefits for larger firms, there are issues for the urban business-environment policy, dealing with large numbers of SMEs, who are less equipped and less motivated to change practices. A wider range of solutions is required, from supply side capacity building, to demand side market stimulation and public procurement.

Innovation, competitiveness and urban development

Much reliance is placed on technological innovation for new processes and products which will reduce impacts and increase competitiveness – whole cities and regions now face the choice of innovation or decline. Environmental technology is one of the key industrial sectors on the agenda. However it is clear that innovation is more than technology, and that it involves more actors than only the firm and its suppliers; investors, regulators, planners, agencies, market developers, higher education, are all involved in the wider 'regional innovation system' (Braczyck, et al., 1998). Industrial ecology systems, for instance, depend on management, markets, infrastructure, and a financial/legal framework in order to take shape (Green and Randles, 2006; Ravetz, 2006b). The concept of the 'learning region' is one where people and organizations are mobilized, pro-active and ready to evolve – hence the importance of the cultural sectors in the urban environment, as in the next section (Cook and Morgan, 1998). One of the largest public investment programmes in the world, the EU Structural Funds, has aimed to integrate environmental objectives with regional development strategy (Roberts and Colwell, 2007).

Critical perspectives on business and environment

The rise of environmental regulation in cities around the world raises topical issues on new modes of governance and public management. At the start are such practical questions such as whether emissions targets are best met by flue gas regulation, by local environmental limits, sector strategies, emissions trading, national taxes or by market development for supply chain innovation. There are wider issues of corporate responsibility, the ethical role of private finance, and the balance of power

---

Box 7.3

Environmental justice and trans-national corporations

The world's worst ever industrial disaster took place in the city of Bhopal, India, on 3rd December 1984. Shortly after midnight, poison gas leaked from a factory in Bhopal owned by Union Carbide Corporation. There was no warning, as none of the plant's safety systems were working … within hours, thousands of dead bodies lay in the streets.

Bhopal isn't only about charred lungs, poisoned kidneys and deformed foetuses. It's also about corporate crime, multinational skullduggery, injustice, dirty deals, medical malpractice, corruption, callousness and contempt for the poor. Nothing else explains why the victims' average compensation was just $500 – for a lifetime of misery … Yet the victims haven't given up. Their struggle for justice and dignity is one of the most valiant anywhere. They have unbelievable energy and hope … the fight has not ended. It won't, so long as our collective conscience stirs.

Source: http://www.bhopal.net

and responsibility between market and state. A common response is that businesses will follow the money, and innovate as and when the market is right, in which case an environmental economic price range should suffice. But there are many examples of 'market failures' which do not produce innovation, environmental justice, risk management or public accountability (see Box 7.3). In such situations the public and the public sector is left to deal with the resulting tragic deaths, health impacts, pollution, dereliction and contamination.

Much of the environmental management agenda revolves around the concept of risk management and hazard protection. In 'Southern' cities the sharp disparities in risk exposure show in direct poverty, disease and the divisions of a dual economy. In most 'Northern' cities, environmental risk levels are now less obvious, and more embedded in occupation and public health issues, so that the poor still die younger (WHO, 2004). Overall there are encouraging signs that the trans-national firms, major producers and international investors have realized the benefits of more sustainable business practice, but there is still a long way to go in translating the aspirations to cities and regions around the world (WBCSD, 2007; Worldwatch Institute, 2008).

7.4 Employment and livelihood

On the other side of the business-environment agenda is the human resource question. Improving business-environment performance depends on the employees, their skills and motivations; improving local environments depends on local communities, and how far activity can be mobilized in the informal economy or social enterprise. Far-reaching transitions are underway – the boundaries between work and leisure are shifting, there are multiple stakeholders to any enterprise, and there is a new kind of networked economic geography (Rifkin, 1995; Fujita, *et al.*, 1999). At the same time, the global economy expands its reach in both formal and informal activity, bringing material development at great cost to workers, communities and cultures (Klein, 2004).

Livelihoods and the urban environment

In the cities of the 'South' there is often a rapid and chaotic transition from rural livelihoods to urban employment, with an emerging dual economy which affects the most vulnerable the hardest – women, children, the unskilled, migrant workers, and minority groups. Employment relationships can be seen in a spectrum, from fixed-wage work to short-term wage work, to insecure self-employment, to career self-employment, and in each case there is a workplace regime which is likely to be polluted and hazardous (Lawn, 2008; Gugler, 1988).

The shifting boundaries between work and leisure, and between formal and informal activity, are also crucial for environmental policy in both 'North' and 'South'. Many environmental campaigns and local improvements come from the third sector – otherwise called the voluntary, non-profit, civic, social or community sector – which combines elements of work, social enterprise and leisure. For urban food issues, for instance, some communities are keen to grow food for leisure, thus helping to reclaim derelict land; while for others it is counted as work, with a different balance of costs and benefits.

Meanwhile, the traditional divisions between public and private sectors are shifting, and the urban environment is a kind of front line for change. In the 'South', there is a huge spread of pro-poor and anti-poverty programmes which have emerged as a counterpart to the structural adjustment approach. Micro-finance works to extend credit and working capital to potential enterprises (Yunus, 2008). Social assets and social capital programmes work on the boundaries of the informal economy and

---

Box 7.4

Self-built and self-managed vertical housing, Brazil

Working to support community-led, high-density urban-housing initiatives, this project has worked with nearly 5,000 families from more than 15 urban popular movements across São Paulo. Appropriate technologies have been developed to allow for the construction of complex multi-storey buildings by the residents themselves. Following construction, community facilities and income-generating activities are developed, including community bakeries, childcare facilities and professional training courses.

Source: World Habitat Day Awards, http://www.worldhabitatawards.org

social enterprise (World Bank, 2002). Participation and capacity building focus on the building of institutions and networks which then enable collective action (Kumar and Chambers, 2002). Each of these is crucial in slums and informal settlements, where environmental improvements are most effective when motivated from within the community (Box 7.4).

Many 'Northern' cities are also racked with unemployment, social exclusion, ethnic conflict and environmental degradation (Moulaert, *et al.*, 2003). The 'intermediate labour market' and similar initiatives are one way to overcome barriers, build capacity and open up hidden resources for public services; although it also risks the devaluation of employment and skills. Such intermediate labour is often used for environmental improvement such as landscape planting and reclamation which are not viable in public or private sectors.

A further variation is in local currencies and social exchange systems, recently re-invented in the form of Local Exchange Trading Systems (LETS); credits are based on time or task values, with accounts, overdrafts, and zero interest rates. On a wider scale, this idea can extend to public services for social trading, in order to mobilize the energy and resources of communities. In the Brazilian city of Curitiba, for instance, recycled materials collected locally are exchanged by the authorities for free public transport passes – increasing the viability of both recycling and public transport, and benefiting low-income people (see Chapter 4) (ICLEI, 2002).

Employment and the urban environment

Improving the urban environment has the potential to change the *employment intensity* of many sectors. For instance, substituting public for private transport in most cities could increase jobs in the transport sector by 10 per cent; shifting from waste disposal to waste recycling increases employment by up to five times (Murray, 2002). The shift of business-environment activity towards service-based leasing, re-manufacturing and re-use, also changes the mix of occupations and activities (Lawn, 2008). However the process of ecological modernization could also be damaging to the local labour market if polluting or energy intensive industries are under pressure from environmental regulation, and migrate or restructure with large job losses. An estimate of the jobs at risk in vulnerable industries in a typical European industrial city could be up to 5 per cent of the total if environmental controls were applied too rigidly and rapidly (ECOTEC, 1994).

The challenge is to balance environmental goals with the needs of human resources, labour skills and business competitiveness. Clearly the business-environment agenda is a learning curve, where innovation and networking capacity is seen as crucial for competitiveness (Randles and Green, 2006). This suggests different approaches to an environmental employment policy, which reflects the needs of the regional economy.

One is to re-skill the labour market for environmentally led growth industries, where knowledge-based skills and human resources are needed to maximize future opportunities (Gallie and White, 1997). Another is to prioritize environmental actions for their employment benefits, combined with new forms of service sector activity, social economy and human resource development. For instance, an integrated public transport programme should be combined with spatial planning for neighbourhood centres, with growing demand for retail and construction activity in the local economy; this helps to increase formal employment and informal enterprise, and also to reduce social exclusion (Newman and Kenworthy, 1999).

Cultural industries and the environment

In the transition to a post-industrial economy, creative and cultural industries can help to bridge the gap between voluntary sectors, community identity, and the cultural *cachet* favoured by investors, entrepreneurs and professionals (Landry, 2006). For instance, Manchester, UK has managed to re-invent its image and economic role as a regional capital with world-class music, sport and fashion (Peck and Ward, 2002). It now contains an archetypal mix of a cultural quarter, urban heritage park, gay village, waterfront cultural complex, international sporting complex, and a growing 'aerotropolis', as seen in Chapter 6. São Paulo has managed to work with powerful youth sub-cultures and channel some of that energy into creative industries (Goldenstein, 2007).

Such creative and cultural industries can be a powerful catalyst for environmental and social improvements. City centre regeneration and redevelopment schemes now aim at zones of urban liveability, where cultural industries overlap with leisure, retail, and convention facilities; this then provides the logic and the funding for water quality improvements, derelict land reclamation, pedestrian zones, and urban green spaces (Girard and Nijkamp, 2009). Environmental improvement is inevitably tangled with the gentrification process, a kind of cultural and economic colonization, often at the cost of local communities and small businesses.

7.5 Finance and investment

The modern economy contains many contradictions and paradoxes. For instance, the energy consumed in cities is known to be causing dangerous and irreversible climate change, and yet there are few economic incentives for saving it. This section looks at the economic and investment case for urban environmental policy. There are guidelines on environmental 'market based instruments' (MBI) in Chapter 9, but here the questions arise – how do these relate to the urban and local level, and what are the prospects for investment in the re-engineering of urban housing, construction and transport in the city? Urban environment strategies need to provide economic levers and incentives for the physical changes needed, and these can only succeed if they bring opportunities rather than costs to the urban economy (Baue, 2007).

Economics for the urban environment

Most environmental or 'ecological tax reform' (ETR) policies and market-based instruments are highly controversial and debated between local, regional and national levels. In federal nations such as the USA, environmental initiatives often move further at state level, such as the emissions trading schemes and the climate programmes declared by 56 major cities at the time of writing (US Climate Action Partnership, 2009).

Some of the most powerful urban authorities are the social providers of health or housing, or the utility providers of energy or railways, and many of these are privatized, franchised or marketized in some way. But the application of fiscal and market measures is often most practical at the city or city-region level – for instance parking charges, road pricing or waste disposal levies. A spectrum from national to local shows the range of possibilities of economic measures relevant at the local or urban level (Ravetz, 2000):

● national economic measures or taxes with local effects but little or no local control;
● local taxes, levies, charges or differentials, controlled and managed locally, with revenues to be invested locally.

There is also a range of options for economic or fiscal intervention; from direct tax or subsidy to trading schemes to public procurement and supply chain partnerships. A coordinated urban environment strategy would combine these for a strategic market transformation, rather than one-off gains. An integrated energy efficiency programme, for instance as seen in Amsterdam or Copenhagen, combines national direct energy taxes with city-region level rating bands, progressive utility tariffs, infrastructure investment, area-based improvements, supply chain innovation partnerships, and public procurement with incentives for accredited businesses.

The question of which measure works at which level, and how the tax revenue is used, is often controversial – many urban environment strategies originate from a more interventionist approach to urban governance. There is also a fine line between positive discrimination via local levies, and the raising of anti-competitive tariffs, which could conflict with the general trend of lowering trade barriers, reducing state aid, deregulation and liberalization (World Bank, 2001a). It is also essential that fiscal and market measures are designed to be socially progressive, on the principle that all citizens should have equitable access to energy and basic resources (Roberts, *et al.*, 1999a).

Institutional economics in the urban environment

Urban environmental economies are in continuous development, with changing economic and institutional relationships between public, private and community sectors. This is nothing new – the reality behind the textbook 'perfect market' of firms and consumers has always been a rich mixture of cooperation, exploitation and dependency between interlocking sets of institutions (Ormerod, 1994; Paavola and Adger, 2005). Such economic alliances and linkages include three basic

General linkages and alliances between public, private, and civic /
non–profit / third / community sectors. Adapted from Ravetz, 2000

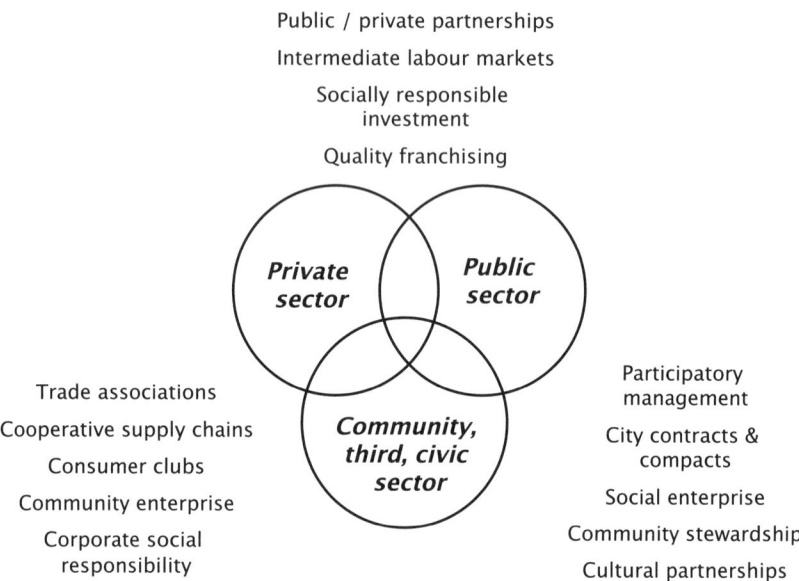

Figure 7.2 *New institutional economies in the urban environment*

combinations – public-private, private-community and public-community
– where each sector has particular strengths and roles to play (see Figure
7.2). This three-way picture is a start in mapping the complex mutual
interactions which may occur at every level of any urban or regional
economy, and it provides a context to the more focused worldview of
eco-taxation.

Public-private linkages around the world include many variations on
the 'private finance initiative' model (PFI), various types of partnership
and consortium, ethical investment and purchasing, and many forms of
local economic development (LED). Historically, public intervention
has aimed to close market gaps and lower barriers; in contrast, private
finance is now seeking to close a public gap in access to capital and
entrepreneurial skill, and many public services are franchised or sold off,
particularly environment-related services such as energy, water, waste
and transport. There are often conflicts of interest – private finance has
to maximize profit and shareholder value, which can undermine the
public goals of universal welfare. In turn, shareholders such as the large

insurance and pension funds, are required to maximize value, rather than re-investing in the cities and communities that generate their assets and use their services. In the USA there is a movement for such community re-investment in environmental, social and economic programmes (Institute for Local Self-Reliance, 2005); while in some EU countries there is national legislation to require pension funds to be invested regional or nationally.

Activities on the private-community axis include local business or regeneration partnerships, social investment funds, what remains of the 'mutual' financial institutions, corporate trusts and guarantee companies, producer or consumer clubs and networks, cooperatives and community development trusts. In some ways the consumer market for any product or service can be a form of alliance or 'community of interest'. Such potential synergy and influence can be mobilized either for commercial gain, as with advertising, or for social benefit through consumer lobbies, NGOs and other networks.

For the community-public axis, the mandate of city government itself depends on alliances or social contracts, such as between municipalities and their electorates. Linkages and alliances include voluntary sector compacts, neighbourhood partnerships, resident or customer charters, social trading schemes and many others. Newly emerging groups and networks in local and wider communities demand new responses from city governments and, where hard currency is scarce, there are alternatives in LETS schemes as above.

Urban environmental investment

Urban environmental improvements need capital investment, and this can take a variety of forms – equity stake in infrastructure and utilities; direct investment in environmental technologies (often pollution control); or new forms of environmental leasehold, such as that of the Energy Services Companies (ESCOs). There is a very topical question as to how far this potentially huge investment market could or should involve city or regional authorities, as partners or enablers, and how much this depends on a privatization agenda (Davis, *et al.*, 2006; von Weizsacker, 2005). The size of this in the 'South' is an order of magnitude greater, and nowhere more so than in China with 1 billion new urban dwellers anticipated by 2025 (World Bank, 2001a; McKinsey Global Institute, 2008).

Socially Responsible Investment (SRI) took a step forward in 2003, with a UNEP report on the performance of sustainability-indexed stocks. The leading investment bank Goldman Sachs reports that SRI leaders outperformed the general stock market by 25 per cent over two years (Baue, 2007; Ling, 2007). Managed funds targeted at community development and regeneration, and aiming to build up a local equity base, are now active both in developing countries and in North America (Grogan and Proscio, 2000). Insurance companies are now taking a keen interest in the risks of climate change and other environmental liabilities, and venture capital firms are hunting for growth prospects through environmental regulation.

The world's largest environmental market, the EU Emissions Trading System is at present launching its second phase, following many problems of over-allocation of permits. There are various proposals for eco-systems credits also to be traded (Daily and Ellison, 2003). In many areas of financial services the market transformations already in progress are focusing on the shareholder value factor and its 'triple bottom line'. (Holliday, *et al.*, 2002). In the USA and elsewhere, there are faith-based community development corporations which are taking on the role of an urban-regional managed fund, venture capital and equity scheme, targeted at urban regeneration and infrastructure, combining public backing and social responsibility with financial sector expertise (Davis, *et al.*, 2006).

For the public sector, the 'new public management' ethos is an opportunity to re-examine all activities against objectives, performance standards and cost-effectiveness (Leach and Barnett, 1998). However, this relies on the boundaries and the criteria set. At present only a fraction of the money saved in urban or lower income areas is re-invested locally – the poor are unwittingly subsidizing the affluent. So there is a vital role for the third sector, in mobilizing local consumer approval and customer loyalty for new types of stakeholding alliances, which can support and re-invest in the local social economy and trading networks (Social Investment Forum, 2006).

7.6 Urban economy-environment strategy

The simple reality is that the economic dimension of urban environment improvements will be achieved only as far as it fits into an urban economic strategy or local economic development (LED) programme – very few cities will invest funds without some clear return. Environmental policy may have positive or negative impacts on economic activity, and for each policy or strategy, there will be economic impacts either in shorter term adjustment or longer term adaptation (Ravetz, 1999b). For a typical older industrial region, there are areas of employment growth, as in the example from the USA (Box 7.5):

- environmental industries, and environmental activity within other regulated industries;
- other sectors with increased competitiveness through price increases in regulated sectors;
- sectors dependent on environmental quality, such as leisure and tourism;
- public sector investment environmental infrastructure.

In contrast to this, sectors at risk of economic or employment decline as a result of stronger environmental policy would include:

- producers of defensive products and older end-of pipe technology;
- sectors which are economically vulnerable and more sensitive to increased costs of environmental compliance; combined with the effect of market shifts, redundancy of plant, and willingness to migrate to less regulated locations.

For these vulnerable industries the style of regulation is crucial, as seen in the negotiation process of the BPEO philosophy of regulation, and more recently with climate change agreements, as a substitute for direct taxation (Howes, *et al.*, 1998). The vulnerability of a sector can be seen in terms of its energy or transport costs, length of the investment cycle, and the proportion of SMEs who are less able to adapt and invest; it concentrates in sectors such as paper, textiles, food, chemicals and distribution (McKinsey Global Institute, 2007). There is also the case for counter vulnerability – in the sense that weak environmental regulation can undermine incentives and competitiveness in global markets where higher standards are already enforced elsewhere and where first-mover status is a crucial advantage (Porter, 1990). European experience suggests

---

Box 7.5

From derelict city to eco-city, USA

In 1969, Chattanooga was the most polluted city in the USA; by 1990 it
was recognized by the US Environmental Protection Agency as the nation's
best turn-around story. Chattanooga's vision is to become a city where
ecological initiatives generate a strong economic base, nurture social
institutions and enhance the natural and made environment. Numerous
collaborative efforts have generated the capital resources, the political
commitment and the civic momentum to tackle complex problems such
as affordable housing, public education, transportation alternatives, urban
design, conservation of natural areas, parks and greenways, air and water
pollution, recycling and job training, downtown river-front development and
neighbourhood vitality.

Source: UN-Habitat Best Practices, http://www.bestpractices.org

---

that the overall impacts of environmental regulation so far tend to be
positive, although there may be some accelerated obsolescence in some
older and more vulnerable industries (Hitchens, 1997).

In the longer term there may be wider effects of an active environmental
policy on the economy, from urban and regional to national levels.
One effect is the share of manufacturing to services – in a low-impact
de-materialized closed-loop economy, service shares may increase, as
products are leased and re-used within the service sector. There is also a
changing balance of investment to consumption, from internal/financial
costs to external/societal costs, where internalization requires longer
term investment. There may be effects on the balance of capital and
labour, where social goals may tend towards low-skill job creation, while
economic competitiveness goals may favour the creation of skilled and
capital-intensive jobs. However, these are not fixed relations, and in areas
such as waste recycling and re-manufacturing there are opportunities to
create employment at all skill levels (Roberts, 2006b).

Finally there are effects on the balance of the private, public and third
sectors – and longer term market transformation may see new kinds of

institutions take the place of conventional share-equity firms (Davis, *et al.*, 2006). Many environmental activities raise questions on these roles. For instance waste recycling is generally labour intensive, but depends on unpaid inputs by householders in pre-sorting their waste, suggesting a new kind of enterprise where householders are also stakeholders.

Green economic strategy

As urban and economic development documents around the world now invoke sustainability on every page, it is crucial to distinguish between rhetoric and reality. Each theme in a typical urban-regional economic strategy can be linked to environmental policy:

- knowledge base: environmental expertise, R&D and technology diffusion should be linked to resources in further and higher education;
- workforce skills: environmental awareness and expertise should be an essential component in human resources development;
- restructuring and redeployment: vulnerable sectors need assistance in adaptation and promoting environmental market opportunities;
- small firms support: the SME sector is more vulnerable to regulation, and needs support, incentives and access to markets;
- inward investment: should focus on low-impact, high value added, clean technologies;
- strategic sites for business and industry: these should be based on low-impact development and transport modes;
- telematics and ICT development: to enable clean technology diffusion, skills development and low-impact consumerism;
- arts, leisure, culture and image: to promote quality of life investment and the spread of low-impact lifestyles.

This spread of activity depends on a social partnership, which extends from municipal government to health and education, business and trade associations, civic bodies and media, voluntary and community organizations, major employers and utilities and so on. There is no simple formula to bringing the green economic agenda to such a mixed community – it may be carried out within such organizations, as a free-standing initiative, or a mixture of both. In almost every city there is some kind of partnership agency to provide lobbying, research, advice, assistance and stimulation. In larger cities or regions there are different models suitable for each business band, from large firms with overseas investors, to local SMEs with regional supply chains.

Green development fund

The above suggest the possibility of an urban environmental or 'green development fund'; this would bridge the gap between dwindling public resources, narrow business programmes, and short-termist bank finance. In Germany for instance, banks, innovation agencies and municipalities have established partnerships for long-term investment and equity holding (Gudgin, 1996). In other OECD countries there are emerging models for urban or regional investment funds targeted on energy productivity (McKinsey, 2008b). Such funds can be run by a partnership of commercial banks and venture capital funds, large employers, and public authorities, with the aim of overcoming market barriers and enabling long-term investment. Their remit aims to combine a number of strands:

- clean technology growth pole, with technology transfer network, linking innovation, higher education and industrial bodies;
- market development programme, linking green investment and purchasing policies to venture capital and potential suppliers;
- central database of green and ethical goods, services, recyclable materials and business opportunities;
- preferential finance for environmental and SRI accredited businesses, with community and regional re-investment;
- infrastructure development; long-term equity or collateral for environment-led schemes such as energy and public transport;
- employment development programme, tackling unemployment, training and skills, supply chain initiatives, and local business development;
- social enterprise development programme, financing and promoting local and community activity in the third/non-profit sectors.

Such green development funds are a natural extension of existing local or regional banking systems, along the lines of the North American Community Banks, or the regional development banks of Europe, which combine commercial success with a social and economic role in the region (Mayo, 1997).

Critical perspectives on green development

The debate on a sustainable urban economy is situated within a larger frame – how far can the current economic development paradigm be adapted or adjusted in the context of liberalization, or is more radical transformation needed? What is the most effective role of a local or urban economy in the context of globalization? And in practical terms, will the green economy improve the urban environment by widening the wealth-poverty gap? There are few clear answers on these questions, but it is clear that many boundaries are shifting at a time of great transition.

It is also clear that cities should contain the solutions to climate change problems, in terms of energy demand, infrastructure, consumer markets and access to finance. But it is not clear yet how these can be mobilized rapidly when the development machine is locked into conventional financial models, when the institutions can only shift their positions at the pace of the slowest and when the majority of the public, and their leaders, still appear to prefer consuming today at the expense of tomorrow.

7.7 Conclusions

The transformation of economic development to a more environmentally viable path is a huge challenge, and this applies as much to cities and regions as to other levels. It is clear that there are different models, which lead to different scenarios, and there is an ongoing debate on how far these are in opposition or counterpoint. A 'globalized growth' modcl sccs solutions through the lens of ethical finance, corporate responsibility, technological innovation, urban-regional competitiveness, and infrastructure development. In contrast, a 'community development' model sees more scope in the social enterprise and third sector, community-based firms, local trading schemes and local re-investment.

Finally, the questions turn up again – how much does the average city have any stake or responsibility for the production of its businesses, and the consumption of its citizens? Or is the city simply a set of factory locations and transient workforces to enable production and consumption which is financed and managed from elsewhere? While globalized supply chains and producer services apparently reduce their stake in the city, there are opposing forces which confirm the role of the local. In terms of urban environmental agendas, these forces work on different layers:

- direct environmental policy – regulation and fiscal incentives, applied to air quality, water, ground and waste;
- local influences on global supply chains and corporate responsibility, through media, public awareness, consumer initiatives;
- local supply side factors – clusters and industrial ecology systems, skilled labour, environmental infrastructure, industrial and commercial sites;
- local demand side incentives – including public procurement, market transformation, product labelling and public awareness;
- local investment strategy; ethical and environmental financial vehicles for green development funds and community re-investment.

In this way, any progress towards more sustainable consumption and production is likely to be a combination of globalizing and localizing forces. It is up to the cities to mobilize these for positive results.

Further information

Corporate responsibility in Asian business   http://www.csr-asia.com

Institute for Local Self Reliance   http://www.ilsr.org

International Labour Organization   http://www.ilo.org

New Economics Foundation   http://www.neweconomics.org

Social Investment Forum   http://www.socialinvest.org

SustainAbility   http://www.sustainability.com

Sustainable Business Institute   http://www.sustainablebusiness.com

UN Industrial Development Organization   http://www.unido.org

UNEP Sustainable Consumption and Production programme   http://www.unep.org/themes/consumption

World Council for Business Sustainable Development   http://www.wbcsd.org

World Social Forum   http://www.forumsocialmundial.org.br/

Further reading

Gardner, G. and Prugh, T. (eds) (2008) *State of the World 2008: Ideas and Opportunities for Sustainable Economies*, Washington DC: Worldwatch Institute.

Hawken, P., Lovins A. and Lovins, L.H. (2005) *Natural Capitalism: The Next Industrial Revolution*, London: Earthscan.

Holliday, C.O., Schmidheiny, S. and Watts, P. (2002) *Walking the Talk: The Business Case for Sustainable Development*, Geneva: World Business Council for Sustainable Development.

Jacobs, J. (1986) *Cities and the Wealth of Nations*, Harmondsworth: Penguin.

Klein, N. (2004) *Fences and Windows: Dispatches From the Front Lines of the Globalization Debate*, London: Picador.

McDonough, W. and Braungart, M. (2002) *Cradle to Cradle: Remaking the Way We Make Things*, New York: North Point Press.

Randles, S. and Green, K. (eds) (2006) *Industrial Ecology and Spaces of Innovation*, Cheltenham: Edward Elgar.

Ravetz, J. (2009) *Pathways to a One Planet Economy*, Godalming, Surrey: WWF-UK (available on http://www.ecologicalbudget.org.uk).

Roberts, P. (1995) *Environmentally Sustainable Business: A Local and Regional Perspective*, London: Paul Chapman Publishing.

Sassen, S. (1994) *Cities in a World Economy*, Thousand Oaks, CA: Pine Forge Press.

Stiglitz, J. (2006) *Making Globalization Work*, New York: W.W.Norton and Co.

8 Community and lifestyle

The social urban environment

8.1 Introduction

The economic perspective in the last chapter might seem to take for granted a textbook 'economic person' – apparently quite selfish, materialistic and disconnected from their community. But it is clear that this is not usually the reality – that citizens, families, communities and public services are the foundations of a civilized society. And yet, in many if not most cities, they are fragmented or under attack.

The context for this is the changing nature of cities and their communities in a globalizing world. More people are connected than ever before, via multiple communications channels – although many people are still excluded from such an idealized virtual lifestyle – and more international migration has led to more displaced communities and diasporas. With many gaps and exceptions, health and education are gradually improving, as are security and human rights. There is a burgeoning middle class in countries such as India and China, and workers everywhere are joining the ranks of global consumers, subject to influences of fashion and media.

There are many implications for the human urban environment. The shift of much of the economy from production to consumption activity, means a re-thinking of conventional environmental management. The emergence of human rights for women, children, ethnic groups and other minorities raises expectations which are often unfulfilled. The aspiration for local quality of life is in tension with the forces of globalization, and the question of how cities can respond is likewise up in the air.

In this chapter

This chapter explores the social and community dimensions of the critical perspectives raised in Chapters 1 and 2. It traces some of the key linkages between these and the practical issues of lifestyles, public services, social policy and urban governance.

First is the perspective on 'globalization and liberalization', as the overall dynamic of urban societies and economies. This points to the agenda of 'consumption and affluence', with direct and indirect effects on the urban environment.

The counterpart to this is the challenge of 'poverty and exclusion', and the segmentation and polarization of communities and social groups. A top-down view on this is through the international development consensus; a bottom-up view is to look at the 'livelihoods' and 'coping strategies' for individuals and communities.

Human health impacts are the first sign of a dysfunctioning urban environment. There are two huge differences in the 'disease burdens' between South and North, and between people in affluence and poverty. Coupled with rising standards and extended lifespans, these bring huge challenges for public services in an age of privatization.

Urban crime is an key indicator of a dysfunctional society, and its environmental dimension includes the geography of crimes and victims, the effect of crime on the shape of the city, and strategies for containment and prevention.

Education likewise has a strong environmental dimension – as physical schools and colleges in their communities; in the educational curriculum; and in the wider agenda of human resources, skills, innovation and creativity for urban environmental management.

Finally each of these themes come together on the agenda for 'urban

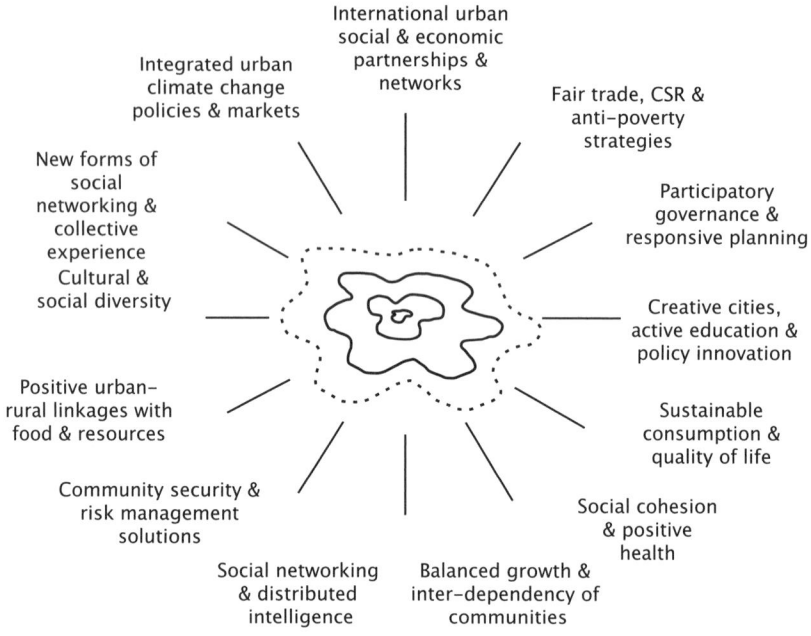

Figure 8.1 *Human urban environment – critical pathways and opportunities*

governance' and managing change – in other words, how to run the city, provide public services, and provide for more sustainable and inclusive communities. In the face of the tide of globalization and liberalization, this is a huge challenge, but there are many inspiring examples to follow.

These themes are not so simple to visualize – the graphic here follows on the critical perspectives shown in Chapter 2, with a snapshot of 'critical solutions' which points towards a more integrated and sustainable human urban environment (Figure 8.1).

8.2 Globalization and neo-liberalism

Globalization is a dominant force in cities around the world. It takes economic forms in the structure of business and finance, political forms in the liberalization of trade flows and public services, and cultural forms through media and ICT. There is also a countervailing strand, sometimes called 'localization', where the cultural identities of people and places are re-interpreted and re-constructed in a local context.

Globalization of capital and financial markets is the foremost driver of change in the economy of cities around the world. The goal of macro-economic growth stability was driven by the World Bank and International Monetary Fund, acting with the US Treasury and development agencies, in the so-called 'Washington consensus', as in Chapter 4 (Stiglitz, 2002 and 2006). Broadly speaking, international capital investment flowed into export-focused commodity industries in the 'South', which in many cases helped to destabilize traditional rural economies and communities. Meanwhile, development aid funds flowed towards national governments, but tied up with strict conditions on liberalization, modernization, privatization, and opening of local markets to world competition.

The result can be seen in typical cities of the 'South' – rapid and chaotic in-migration and urbanization in polluted and hazardous locations where the burgeoning need for housing and infrastructure is far beyond the capacity of fragmented and under-funded city governments (Hardoy, *et al.*, 2001). Meanwhile the cities of the 'North' also suffered the effects of globalization, through rapid restructuring, closure of heavy industries, and in many cases a crisis in local or municipal government.

Global networks

The globalization trend seems to run parallel with a network effect, which is enabled and accelerated by ICT, international media, and international travel. The concept of 'spaces of flows' observes how the urban structure – conventional industrial cities and regions – is shifting to a globally networked and poly-centric pattern, in which the levers of power and wealth are concentrated in the hands of a highly mobile and often clandestine elite (Castells, 1996). The physical and social structures of cities then begin to reflect this new distribution of power and wealth, which is segmented and splintered from the rest of the urban community (Graham and Marvin, 2001). Human migration and labour mobility is also increasing, although still much more constrained than the international movements of capital and production, and many large cities are multi-cultural hubs for international diaspora and migrant communities (Geyer, *et al.*, 2002; UN-Habitat, 2004). The 'international division of labour' analysis reflects this globalized hierarchy and segmentation of occupations and livelihoods (Froebel, *et al.*, 1980).

Global cities and regions

Global cities, or 'world cities', are now the subject of intense debate and competition between cities and city-regions (Sassen, 1994; Knox and Taylor, 1995; Scott, 2000). World cities are the gateways for investment and information flows for their continents. They contain intense concentrations of advanced producer services such as finance, law and the professions. They are also world centres of consumption – through cultural creation, as trend setters and 'style capitals', as the platform for property-based consumption investment, and as major hubs with international airports and other links. There are new terms – mega-cities, megalopolis, hyper-cities, or meta-cities – to describe conurbations and agglomerations such as Tokyo-Yokohama with 35 million people, the USA Eastern Seaboard with 49 million, or Shanghai-Nanjing with around 80 million people (Douglass, 2000; UN-Habitat, 2004).

Globalization and urban environments

Globalization and the environmental agenda are closely linked in both 'North' and 'South'. On the production side, the influx of global capital has a rapid restructuring effect on the urban industrial base, and the demands of new enterprises for space and infrastructure. This accelerates the modernization transition towards oil-based 'automotive' cities and agglomerations. In the cities of the 'South' there are globalized cost pressures pushing for 'informalization' of the economy and workforce – cutting the local tax base, and exacerbating environmental pollution and risks to employees and communities (Klein, 2004). In the cities of the 'North', globalization has driven rapid restructuring or actual shrinkage of cities, with large areas of derelict or contaminated land, failed housing markets and dis-investment in local services.

Finally, there is also a globalization of environmental impacts and environmental technologies. Household waste from Europe is now sent to East Asia for recovery; acid rain produced by urban transport emissions can travel across continents. Climate change emissions are for the most part produced by wealthy cities in the 'North', and threaten the livelihoods of mainly the poorer cities of the 'South'. This over-arching global challenge of our time revolves around cities, the global urban hierarchy and the global divisions of wealth and poverty.

Neo-liberalism and the urban environment

Capitalism as an organizing principle is often seen as infinitely resourceful and flexible; and one way to describe its most recent phase is through an underlying ideology of 'neo-liberalism' (Harvey, 2005; Brenner and Theodore, 2002). From a critical perspective this can be seen as another arena for ideological hegemony and capital accumulation, with a powerful restructuring effect on the urban environment:

● restructuring of economic power and competitiveness away from the level of nation-states towards cities and city-regions, even while political power might be shifting the other way, towards national and international bodies;
● restructuring of local and urban governance, with a focus on enterprise, privatization, and other partnerships with the private sector;
● restructuring of labour markets as a result of industrial change, and restructuring of social welfare systems to reduce government welfare spending and encourage individualism;
● restructuring of urban infrastructure towards deregulated and privatized operations, in order to access foreign investment and development finance, with the exclusion of large parts of the urban poor;
● restructuring of the institutional economy towards service sector dominance, with an emphasis on the middle class focused 'creative city' and 'cultural economy', together with major cultural and sporting events and initiatives (Florida, 2005; Landry, 2006).

Each of these has clear implications for the neo-liberalized urban environmental agenda. In almost every city there is now a cultivation of a local green image which is mainly designed to encourage inward investors, creative industries, high-value professionals and global tourists. Behind this is an often fuzzy aspiration for 'sustainable cities', which is generally flexible enough to be adapted to the agenda of the urban elite. A critical perspective would observe this as an agenda for burden shifting, from the local to the global level, from the state to the market, and from providers to consumers and their lifestyles (Brand and Thomas, 2005).

Meanwhile, broader critical perspectives on globalization already fill many libraries and campaign websites. In terms of the human urban environment, several challenges and dilemmas stand out:

● the urban challenge – as practiced up to now, globalization has left

behind many of the poor cities of the world, in terms of infrastructure,
shelter, services and employment;

● the enviromental challenge – current patterns in globalization have
generally increased the 'externalization' of environmental impacts in
the 'race to the bottom', again with peripheral communities and cities
at the bottom of the pile;

● the political challenge – while economic integration has proceeded apace,
it is arguable that political integration has fallen behind, so that there is
a profound democratic deficit in controlling global financial and trade
flows, and their effects on urban environments (Monbiot, 2005).

8.3 Consumption and lifestyle

Consumption and affluence in both wealthier and poorer cities is a major
driver of individual and cultural identity and community. This then drives
new perceptions of local places, particularly through leisure and tourism.

While globalization is often framed as a supply side production agenda
– finance, trade, industry and services – there is a counterpart agenda
on the demand side in consumption by households and individuals.
Rising incomes and the spread of domestic technology frees up money,
time, mobility and space in the home and garden: global images and
ideals spread rapidly, from the burger chain stores to the nuclear family
suburban household models, complete with domestic technology and
green lawns. In the majority of cities of the 'North', and in the more
affluent areas of cities of the 'South', the volatile and media-driven
patterns of affluence then determine which environments are improved,
which social groups have access, and what balance of market and state is
involved.

Consumption and urban development

Consumption is not only a matter of products in the shops, but very
much part of the economic cycle at all levels. Investment in the built
environment (housing and commercial property) is perhaps the single
largest sector of consumption, with a key role in balancing out economic
swings (Harvey, 1985). This then has profound effects on the shape and
function of the city and city-region. Giant shopping malls spring up on
the periphery, traditional urban markets are redeveloped for high value

offices, and former industrial areas are gentrified. Suburban areas burgeon outwards, to suit the space and amenity demands of affluent consumers. Overall, the reshaping of cities in both 'North' and 'South' can be seen as driven by the agenda of high-value producers and conspicuous consumers (Zukin, 1998).

Consumption and leisure

Tourism and related leisure activity is the most dominant and conspicuous face of the consumer society (Lash and Urry, 1993). The spread of low-cost air travel, broadband internet access and disposable income is transforming many cities into 'emerging destinations': urban property development is accelerating, not only on the coastline of Europe, Australasia and Latin America, but in historic cities in the NIS, North Africa and the Middle East. The concentrations of hyper-rich elites in 'Southern' cities, such as Mumbai or São Paulo, adds to the pressure for restructuring cities into zones of production and consumption. For instance in many parts of the Mediterranean coastal regions, the largest users of scarce water resources are not industry, or agriculture, or households – they are the golf courses which, in defiance of local climates, project an image of endless green grass and leisure time (Giaoutzi and Nijkamp, 2006).

In each of these, the urban environment has a key role to play. High-quality environments are one of the keys to attracting foreign tourists and international elites, and massive funds go towards 'image enhancement' and 'place-making'. Meanwhile, environmental 'bad neighbours' such as heavy industry, waste disposal and water treatment, are displaced towards lower income communities. Informal housing and local businesses are often evicted to make way for high value sporting events such as the Olympic games. In response there is an agenda for cultural tourism as an enabler of local sustainable development (Girard and Nijkamp, 2009).

Consumption and environment

In each of these stages, the urban environment plays a key role. Local environmental quality is a key factor in added value – for instance in some areas the presence of street trees can add up to 10 per cent to property prices (Willis and Garrod, 1993). Local environmental externalities, such as traffic noise and emissions, are pushed away from high-value areas towards low-value ones. Environmental infrastructures such as sewage and waste disposal are unwittingly situated in low-value areas, encouraged by Cost-Benefit Analysis (CBA) based on local property values and social time values. Informal housing and businesses are much more likely to be located in areas prone to industrial pollution, flooding, contamination, unstable land and other hazards. In the zones of decay and withdrawal, derelict and contaminated land is likely to be widespread (Agyeman, *et al.*, 2003).

In this sense, environmental quality can be seen as a kind of commodity to be consumed, alongside other more material goods. If there were effective shadow markets in environmental pollution, as suggested in the textbooks, then the finance for cleaner technology and environmental remediation would be more readily available. The fact that such funding is generally scarce is again a political question of economic markets versus social responsibility. Meanwhile there are examples of 'sustainable

---

Box 8.1

Sustainable urban living quarter, Germany

The city of Freiburg is developing a 42 hectare site for a population of 5,000. Forum Vauban, an NGO created in 1994, has become the official body for citizen participation, with services and support to children, youth, families and sub-cultures. Carbon dioxide emissions from buildings have been reduced by 60 per cent, and public heat supply systems utilise solar power and a co-generation plant. Traffic congestion is minimal, with 35 per cent of households shifting from car ownership to car sharing.

Building owners have formed an organization to enable low-income earners to become homeowners, with a very high level of citizen participation. The ecological housing scheme has made solid contributions to the European sustainable development debate on energy conservation, reduction in car use and the creation of a green living environment for children and community activities.

Source: UN-Habitat Best Practices, http://www.bestpractices.org

living' which show that where high standards are designed into development, and combined with social and economic measures, then a quite different balance is possible (Box 8.1).

Sustainable consumption and production

The agenda for 'sustainable consumption and production' links local with global issues. Consumer behaviour is key to the impact that society has on the environment. The actions that people take and choices they make – to consume certain products and services or to live in certain ways rather than others – all have direct and indirect impacts on the environment, as well as on personal (and collective) well-being. This is why the topic of 'sustainable consumption' has become a central focus for national and international policy, with many aspirational policy documents produced (Jackson and Michaelis, 2003; Sustainable Development Commission, 2005). The key theme is 'internalization' – while the external impacts of production on an industrial site are more easily tracked, the external impacts of consumption are generally spread around the world in a myriad of supply chains and environmental effects. Some would argue that the 'externalization' of impacts is on a par with most firms' profit margins, that internalization is in contradiction to basic profit motives, and that most rhetoric on CSR is public relations 'greenwash' (Athanasiou, 1997). The European initiative aims to change the rules of the game as a whole, with a co-ordinated raft of policies covering eco-design, eco-labelling, procurement, life-cycle analysis, and capacity building with producers, retailers and consumer organizations (European Commission, 2008).

8.4 Poverty and exclusion

Poverty and urban environments

Over 10 per cent of all city dwellers are destitute, living on less than US$1 per person per day, and a further 20 per cent live on less than US$2 per day (UNDP, 2004). But is poverty an urban problem? In absolute terms, urban dwellers even in the poorest slum areas are better off than their rural counterparts in basic food security, and this is the main driving force behind in-migration. But they are likely to suffer many other types of relative poverty, ethnic or racial oppression, overcrowding, vulnerability to crime and exploitation – a wider definition of poverty and exclusion. Extreme poverty as in famine, is seen as a political question of access and distribution, rather a material lack of food (Dreze and Sen, 1989). This is a huge agenda, of which this section focuses on the human urban environment strand – the many linkages between urban environmental conditions, and the dynamics of poverty and exclusion:

- Poverty as a material condition compounds with exclusion, which is a more socio-political set of barriers; a combination of education and skills, class, location, expectations and attitudes.
- Poverty tends to concentrate excluded communities in locations of low value, with high environmental risk and pollution.
- Poverty also forces the poorest social groups into dangerous workplaces, and unsafe or unhealthy residential environments.
- The poorest social groups are then segregated and segmented further by gender, age, religion, race and ethnic divisions.
- Rapid and chaotic urbanization of cities in the 'South' means that urban infrastructures are lacking or inadequate. Basic services such as water are available at inflated costs from private vendors – so 'the poor pay more for water than the rich', both in proportions of income, and in absolute money (Manahan, *et al.*, 2007).
- In the 'North', absolute destitution is less apparent, but there are many shades of homelessness, malnutrition, and environmental conditions which approach those of slums.

Social exclusion

It is clear that poverty is not so much a simple lack of income, but a 'cumulative causation' of problems coming from unemployment, low paid work, poor housing, lack of education and healthcare, as well as direct environmental factors, all adding up to what is now termed 'social exclusion'.

Exclusion is not only a class issue, but a dividing and segmenting force, which can apply to genders, generations, religions, ethnic and racial groups. For instance, it is well known that women and children suffer the greatest environmental burdens of all in the slum areas of substandard housing lacking safe water, sanitation or energy supplies (Hardoy, *et al.*, 2001). Another type of exclusion is generational – half the world's population is under the age of 25, partly due to improved healthcare in urban areas – but in many cities such young people suffer unemployment rates of 50 per cent and more. In other cities, there are large populations of migrant, informal or illegal workers, who have no formal rights or securities, and often live in appalling conditions.

In response, there is a myriad of solutions in cities across the 'North' and 'South', from community action, to micro-finance schemes (Box 8.1). The 'third sector' – combining the enterprise of the private sector with the responsibility of the public sector – is seen as an enabler of partnerships for rebuilding social capital and community enterprise (Lipietz, 1995). Communitarianism is a philosophy and set of practices which encourage mutual rights and responsibilities (Etzioni, 1996): political ecology aims to interpret this in context of anarchist and co-operative philosophy (Bookchin, 1997).

The urban environment again is both a fault line and a front line, which demonstrates both the challenge of social exclusion, and the potential

---

Box 8.2

Slum Improvement Project, Bangladesh

Dhaka is one of the fastest growing mega-cities in the world, and over 2 million people live in slum areas, at densities up to 750 people per hectare. These areas have little or no water, sanitation, drainage or communications. Through the Slum Improvement Project (SIP) participatory approach, the local authority in partnership with urban communities, has made a breakthrough in providing basic physical, social and economic infrastructure services to the urban poor.

Of all the SIP components, the micro-credit programme has been found to be particularly successful and most attractive. The SIP has particularly empowered women through community involvement and the savings and credit programme, thereby raising the status of women in families and communities.

Source: UN-Habitat Best Practices, http://www.bestpractices.org

for social cohesion. Environmental justice movements and slum housing initiatives in the 'South' depend on enhancing citizenship and community cohesion, and combining this with economic development (Box 8.2). Community open space and regeneration projects in the 'North' also depend on finding new kinds of partnerships which unlock the potential of diverse and fragmented communities in 'mongrel cities' (Sandercock, 2003).

The capabilities approach to development

These various forms of support for community self-help have become one of the main approaches to poverty reduction adopted by development aid agencies. The development theory on which they are based comes partly from the 'capabilities' approach to development advocated in the late 1980s by Amartya Sen, which contributed strongly to the design of the United Nations Human Development Index (UNDP, 2004). The 'capabilities' approach stresses the importance of non-economic aspects of human well-being, particularly personal freedoms, and the two-way inter-relationships between such freedoms and economic development (Dreze and Sen, 1989). Although Sen emphasized the positive role of the state in undertaking public actions and in co-ordinating public and private enterprise, the approach fits neatly with the macro-economic development theory advocated by the World Bank, the International Monetary Fund (IMF) and the United States Treasury (Stiglitz, 2002). This 'Washington consensus' argues that economic development, and hence poverty reduction, requires free markets rather than planning, both within a country's internal economy and in its international trade. To comply with IMF and World Bank requirements, developing countries' poverty reduction strategies have to include privatisation of state enterprises and the removal of trade barriers. These policies are readily complemented by the micro-level interventions of Sen's capabilities approach (World Bank, 2002). The 'micro-finance' initiative, which started in Bangladesh by combining micro-business loans with social capacity building, is a demonstration of the huge potential for mobilizing the latent talent of the poor in both urban and rural areas, as in Box 8.2 (Yunus, 2008).

Meanwhile there are critiques of the Washington consensus, some from direct experience of liberalization and privatization of urban public utilities. Cities were generally not the focus of the development aid and finance machine, which tended to roll out standard solutions focused on national economies (Satterthwaite, 2006). In many cities of the 'South', the combination of privatization and deregulation, loan finance conditions, and dependency on costly imported technology, meant that water or electricity became more costly and less available for the majority of urban dwellers (Arend, *et al.*, 2004; Swyngedouw, 2004).

Urban and rural poverty

A further factor in the mix is the relationship between urban poverty and rural poverty. Trade incentives and development aid for agricultural support both encourage the modernisation of agriculture, which reduces rural labour requirements, so accelerating the in-migration to the cities (see the discussion in Chapter 4). During the agricultural revolution associated with Europe's industrial revolution, urban poverty became severe, despite much of the excess labour emigrating to North America and other colonies (Hill, 1985). In South East Asia the severity of such problems was reduced through integrated urban–rural strategic planning, which is less readily applicable under market-led approaches (McGranahan, *et al.*, 2004). There is also a new awareness of the problems of 'refugees and internally displaced persons' (R/IDP), and new solutions for such transient communities (Box 8.3).

Box 8.3

Integrated community shelters, Azerbaijan

Working with communities to create 2,105 permanent, expandable homes in 55 culturally appropriate villages across 13 districts of Azerbaijan, this project has provided security and stability for some of the one million refugees and internally displaced persons (R/IDPs) created by the conflict with Armenia. The project provides much needed infrastructure and has helped to establish over 350 small enterprises.

Source: World Habitat Day Awards, http://www.worldhabitatawards.org

---

Whatever approach is adopted, the goal of a poverty reduction strategy is in principle to achieve a comprehensive transformation of a country's

social, economic and institutional structures, in a manner which is
environmentally sustainable for the country itself, and globally. This
is certainly a challenge. At the city or country level, increased energy,
transport and resource use is clearly needed to modernize the economy,
deliver services, and provide urban infrastructure. But there is an over-
arching contradiction between national and global sustainability goals
– broadly speaking, if global poverty reduction programmes succeed in
bringing the population of the 'South' towards the material affluence of
the 'North', then within a few decades there will be simply not enough
bio-productive capacity and minerals resources to go round (WWF,
2006). This suggests that new and sustainable forms of poverty-focused
development are not only desirable but essential, involving new forms of
governance and participation (Rakodi, *et al.*, 2004).

8.5 Health and well-being

Healthy cities agenda

Public health was one of the main forces in the development of local
government in the nineteenth century, and the supply of public water and
sanitation dates back to the ancient cities of 8000 BC (Chapters 1 and 3).
Now the 'healthy city' agenda is a world-wide movement centered on the
human urban environment (Edwards and Tsouros, 2006). One starting
point is the World Health Organization (WHO) framework of 'health
fields' (WHO, 1992); here the focus is on the last two themes:

- healthcare – community, primary, specialist services: resources, access,
  participation;
- biological factors – genetics, nutrition, demographics;
- lifestyle – work, leisure, risk exposure, food, cultural effects;
- urban environment, both physical and social – housing, pollution,
  accessibility, employment, security.

The health agenda is closely linked to the agenda of poverty and
exclusion, as above, and particularly to the grossly unhealthy conditions
in the poorer cities and slum neighbourhoods of the 'South'. But there
are particular features and links to the environmental agenda across both
'North' and 'South':

- affluent cities generate a different set of health agendas, such as obesity and carcinogenic risks, and these are increasingly found in both 'North' and 'South';
- demographic trends and pressures create new health challenges and opportunities. This is most obvious in the ageing of affluent societies, or the burgeoning of youth populations, and the migration of workers and sub-cultures;
- there are new lifestyle and cultural pressures for risk-taking activity, most obviously in drug addiction, sexually transmitted diseases, gang cultures and crime of all kinds;
- there is an agenda for healthy city policy, in both 'North' and 'South', which aims to mobilize social and institutional capacity, and focuses on 'positive health' through lifestyle and workplace changes (Davis and Kelly, 1993).

The City Development Index (CDI) shows the range of conditions across the 'North' and 'South' (UNCHS, 2001: details in the Appendix). As a combination of average life expectancy and child mortality, the CDI 'health' index ranges from 94 to 44. The 'infrastructure' index of connections to water, energy, sanitation and communications, ranges even more widely from 99 to 22.

What follows is an overview of 'South' and 'North', with some critical perspectives on urban environmental health.

Health in the 'South'

Many cities in the 'South' are at a similar stage of industrialization and urbanization as in nineteenth century Europe, but the process is often more rapid and chaotic. In India, for example, there is no nation-wide urban sanitation programme due to a lack of funding, lack of political priority for the poor and a culture of indifference (Hardoy, *et al.*, 2001). HIV/AIDS is an increasing threat particularly in African cities, where sexual and social attitudes are different from those in rural areas. Infant mortality in poor areas of less developed countries (LDCs) can be up to four times that of wealthier areas: there are increases in depression, alcohol abuse, drug abuse, suicide, violence and murder, and as the rich do more to defend themselves, the poor suffer more as a consequence.

The urban poor in the 'South' are more susceptible to the effects of increasingly frequent heat waves, and infectious diseases, usually associated with rural areas, are becoming more commonplace in urban areas due to crowded, insanitary conditions. In Africa for instance, lead is still added to petrol, affecting the development of children in urban areas. Mexico City and São Paulo are extreme examples of the many developing cities where the public transport infrastructure is too poor to cope with the expansion of the city. As a consequence, car travel is the option of choice for the affluent, further increasing congestion, air pollution and accidents for the poor (McMichael, 2000). At the extreme end of the spectrum, the world's worst industrial accident was a direct result of corporate negligence, in Bhopal, India (see Box 7.3).

Health in the 'North'

Clearly public health is related to environmental quality, but there are many links and indirect effects. There is a persistent 'health gap' between rich and poor communities in most 'Northern' cities, due to the combined factors of lifestyle, diet, risk-taking, alcohol and smoking, poor housing, and workplace hazards. The gap then shows up in standard mortality ratios (SMR) of over 10 years difference (Townsend *et al.*, 1988).

Now that the large-scale epidemics of the early industrial cities are more or less contained, there is a growth in the 'diseases of affluence' – cancers, cardiac conditions, allergies, food additives, drug and food abuse. Many of these are not only physical conditions, but linked to stress, insecurity and depression. Ironically, the continuing advances in public healthcare are mainly responsible for different kinds of 'demographic time-bombs', putting pressure on public health provision. In the 'North', there is an increase in average life expectancy beyond the age of retirement; while in the 'South', the reduction in child mortality causes a demographic bulge of young people.

Meanwhile there is growing awareness of health as a positive quality, more than the absence of disease – a holistic fulfilment of human needs at every level. For instance, linkages can be traced between income, housing and health conditions, in the context of needs and services (Byrne, *et al.*, 1986). This shows that definitions of health and illness vary with expectations, family, religion, employment, leisure and lifestyle, and that a wide range of activities and institutions surrounds the statistics for official health services (Box 8.4).

Urban environmental planning and design then becomes a crucial factor – the quality of secure spaces, convivial spaces and green spaces is seen as instrumental to social cohesion and quality of life, which in turn encourage positive health.

Box 8.4

Healthy living for the homeless, Austria

Initiated by a self-help association, the Lighthouse Wien project involves the refurbishment of existing buildings on environmentally sustainable principles to provide high-quality accommodation for 60 vulnerable persons with special needs, many of whom have been homeless for years. With a focus on the preservation of health through substitution and HIV therapy, the project provides residents with a wide range of medical and social support services as well as training opportunities.

Source: World Habitat Day Awards, http://www.worldhabitatawards.org

---

Critical perspectives on urban public health

The globalized 'Washington consensus' tends to put the liberalization of health services at the centre of the political debate. There is continuing tension around the world between the state provision of health services, and the privatized 'cost recovery' model of health insurance and private healthcare. The privatization initiatives of the 1990s for public water and electricity supplies have slowed down, due to operational problems, but unfortunately have not yet been replaced with equivalents from the public sector, which often suffers a crisis of governance (von Weizsacker, 2005).

In more affluent societies the consumption agenda is a key issue, with burgeoning costs of healthcare, and controversial choices around the responsibility of individuals for their lifestyle. Issues such as smoking, drinking, diet, exercise and sexual behaviour are emerging dividing lines, and these are entangled with ethnic or religious practices. The 'risk society' agenda is also instrumental – with multiplying risks, real or perceived, from food chain contamination, workplace equipment, pandemic infections and so on. And at the root of all health issues are the divisions of poverty and exclusion – where environmental hazards are compounded with physical, social and economic hazards, under healthcare services which are often selective and inadequate.

8.6 Risk and security

The risk society

The hedonistic agenda of consumption is also increasingly vulnerable to a host of risks, which may be actual or perceived, physical or social. As a response, consumption is shifting focus from material possessions towards a wider concept of security – security against environmental pollution, crime and social disorder, illegal migrants, or more recently, political terrorism (Graham, 2004). This cross-cutting theme of risk and security was famously titled as the 'risk society' – a combination of technological, environmental, political or economic hazards and vulnerabilities (Beck, 1995; Lupton 1999).

In parallel there is a new awareness of the importance of trust in governance and society – arrangements of mutual aid and reciprocity between citizens, community networks, governments and other stakeholders. This is often framed as 'social capital' and measured by indicators of civic involvement, but also extends further to informal social and cultural networks (Putnam, 2000). This is realized in the aspirational agenda for the 'liveable city' and the 'sustainable community' (Chapter 10). This combines physical, social and economic aspirations: high-density mixed-use urban quarters, high-quality public realm and open space, safe public transport, vibrant local economies, and multi-cultural communities with low crime rates and high social capital. In practice there is mixed evidence – affluent inner urban neighbourhoods tend to score highly, while dense mixed-use slum areas have other inbuilt problems. Some physical factors such as local transport seem to be more clear – surveys in the US showed that social interaction goes down as car traffic goes up (Appleyard and Lintell, 1972).

The urban social environment can then be seen as a geography of more risky or safer zones, which then determine the daily patterns and experiences of city dwellers – so called 'landscapes of fear' (Tuan,

1979; Taylor, *et al.*, 1996). The affluent can reduce their risk, either by locating housing and work in high-value areas, or by target hardening with physical security and guard services, or more recently by video surveillance and electronic recognition technology. The extreme case is a defensive partitioning of cities and suburbs into gated communities with advanced security technology, which is becoming the norm for urban elites (Chapter 5). São Paulo for instance, contains over 1,000 private helicopters, seen as the best way for the wealthy elite to avoid both congestion and violent crime on the roads (Leite, 2007).

Crime and urban environments

Crime has wider social effects of insecurity and stress for the affluent and poor alike – the city-scape becomes divided into zones of security and danger, with rising tension between genders, classes, age-groups, ethnic groups and sub-cultures. Crime may be perhaps the largest single factor in the trend of migration to the suburbs, and policies for re-urbanization may achieve little until public security is increased. Crime is a perennial theme for social theorists, and there are several approaches to understanding its causes and effects:

- 'social disorganization' focuses on the loss of social 'norms', particularly in situations of rapid social and economic change;
- a 'sub-culture' approach looks at pressures and incentives on individuals and groups in defining their identity;
- the 'new criminology' or 'radical criminology' looks at crime in its larger social and economic context (Taylor, *et al.*, 1973).

Here we focus particularly on 'environmental criminology': this focuses on the physical built environment, the urban geography of crimes and victims, and the effects of crime and security on the changing urban environment (Brantingham and Brantingham, 1991; Bartol and Bartol, 2006).

For 'rational' crime such as burglary, with direct economic incentives – there are obvious links with unemployment and drug abuse, and the environmental implications point towards increased physical security (Field, 1990). For 'impulsive' crime such as vandalism, there are linkages to failures in education, housing and cultural exclusion, and likewise decline in the urban environment. A wider trend is anti-social behaviour,

---

Box 8.5

Crime and security trends in Brazil

São Paulo, Brazil has six times the world average of homicides.

- In 2007, at least 25,000 homes in São Paulo had security cameras to monitor entrances and exits from their grounds.
- In 2007, there were 35,000 armed cars in Brazil, with a 33 per cent increase since the attacks on police in 2006.
- In 2007, the Brazilian security market had reached US$49 billion, about 10 percent of GDP.
- In 2006, private security companies employed 1.5 million individuals in Brazil.
- Over 1000 private helicopters are the default travel choice for the wealthy elite.

Source: UN-Habitat (2007b)

---

harassment or 'disorder', often a result of conflict of different lifestyles and cultures in public places. A general picture emerges that neither law enforcement or social norms are containing social tensions and conflicts in many cities, and that more integrated solutions are needed.

This is a problem for cities worldwide, but it is concentrated particularly in areas of high unemployment and slum dwellings; it affects particularly women, children and young males (Moser, 2005). One survey showed that 60 per cent of people felt unsafe walking home in the dark in Brazil, South Africa and Bolivia – around three times the number in most countries in the 'North' (UN-Habitat, 2007a). Such pervasive insecurity concerns not only individual crime incidents on the streets, but the whole range of inter-ethnic conflict, organized gangs, protection rackets, trafficking of people or drugs, political terrorism, tax avoidance and other white collar crime (Box 8.5).

Security and urban environment

The links between urban design and security can be controversial –
one approach tends towards secure private enclaves, while another
aims at natural surveillance in public streets and spaces. The concept
of 'defensible space' was coined to describe the layering of territory,
overlooking and public security. The policy of a 'community of interest'
aims at clustering of homogeneous populations where social norms
and controls can be reinforced (Newman, 1981). The 'communitarian'
philosophy goes even further towards the 'gated city', which encloses
community enclaves in order to deter or contain crime and disorder
(Etzioni, 1996). This has gone furthest in the USA, where up to 10 per
cent of the urban population lives behind some form of security fence,
and is increasingly the model in high-crime cities such as São Paulo,
Moscow or Cape Town (Davis, 1998).

For existing residents the result is found in advice on 'crime-free housing'
– avoiding dark corners, undefined spaces and escape routes without
overlooking (Poyner and Webb, 1991). This is a challenge for typical
cities, which often contain large areas of sub-standard urban form. At
present the property market tends to favour the 'target hardening' of its
own buildings and sites, at the expense of the public realm. Arguably,
a more integrated approach to security in housing needs to extend
the 'locks and bolts' approach to a wider household and community
management regime – an active programme of building mutual respect
and social cohesion between genders, generations, ethnic and racial
groups (Whitzman, 2008).

UN-Habitat runs a Safer Cities Programme, which recognizes that
working with urban youth, sub-cultures and gangs is one of the keys to
stemming the increase in crime. This runs according to four principles
(UN-Habitat, 2007a):

- that local governance and urban planning work together on crime
  prevention;
- that housing and slum improvement policies incorporate active crime
  prevention programmes;
- that public spaces and streets, together with derelict land and facilities,
  should be improved with active crime prevention programmes;
- that civil society and particularly disadvantaged groups, together with
  women and youth, are actively involved.

Environmental crime and justice

The urban environmental agenda is now over-arched by the risks and
hazards of climate change – an uncertain and complex set of changes for
all cities, from catastrophic storms and tidal surges, to drastic changes in
water tables and urban ecologies (Chapter 5). Again, this environmental
agenda revolves around the distribution of power and wealth – where the
poor are more likely to be found in hazardous locations, in substandard
housing, with minimal healthcare or emergency services.

One type of illegal activity which is newly emerging is 'environmental
crime'. Cutting across urban and rural, and national and international
boundaries, this can take many forms:

- deliberate flouting of environmental pollution laws and standards, as in
  factory emissions, or major industrial accidents;
- negligence which causes increased risk and hazard from natural
  disasters: for instance the impacts and loss of life from recent
  earthquakes in Turkey, Pakistan and Iran were greatly magnified as the
  local building regulations were systematically ignored;
- management or extraction of resources in a way which causes external
  environmental impacts: for instance, clear felling of old growth forest
  increases soil erosion and thereby flood risk for cities downriver;
- vandalizing or deliberate degradation of environmental assets: the
  classic example is the burning of the Kuwaiti oil wells by the Iraqi
  army in 1991;
- direct theft from urban infrastructure: for instance in cities in Africa
  and Latin America it is common not only for electricity and water to be
  stolen, but for the wires and pipes themselves to be stolen and sold for
  scrap;
- direct expropriation of common environmental assets: the most
  common problem is the diversion of water or groundwater, often by
  foreign-owned industrial or tourist firms, away from local communities
  and subsistence farmers.

In response, there is a broad scale movement for 'environmental justice'.
At the European Workshop on Environmental Justice 2003, 'injustice'
was defined in broad terms not only as direct impacts, but in terms of
access to information and participation (Westra, 2008). The implications
for the urban environment are potentially far-reaching – almost any
environmental impact can generally be traced to its effects on vulnerable

social groups, and there are legislative provisions at the human rights level which provide for redress (Pellow and Brulle, 2005). Even climate change is now being framed as a justice issue. There are attempts to launch a global class action against high emission countries, and in 2008, climate change campaigners in the UK were acquitted of charges of criminal damage involving the attempted blockade of the Kingsnorth power station in Kent (details on http://www.climatecamp.org).

Critical perspectives on risk and security

The urban risk and security agenda involves each of the critical perspectives in this book, and more. The wider context is recognized by the UNCHS Safer Cities Programme (UN-Habitat, 2007a).

- Security against crime – the conventional agenda of law enforcement and the justice system. As above, the response now extends to a wider engagement with communities and sub-cultures, combined with an urban environmental design agenda.
- Security against natural hazards and disasters – an agenda which is often framed as 'emergency planning' or 'disaster response'. This needs to be linked with the urban development processes which often increase risk and vulnerability (Chapter 5). For instance, the impact of the Indian Ocean tsunami of December 2004 was greatly magnified because large areas of mangrove swamps along the coastline had been cleared to make way for industrialized export crop production.
- Security against evictions and access to land, an even more political and volatile agenda. Even in otherwise enlightened democracies such as India, mass evictions of slum dwellers continue and are justified in the name of urban development, modernization and upholding of property rights (Sharma, 2000).

In this context, urban security is increasingly a globalized commodity for mass consumption. It is increasingly liberalized and privatized, as public policing is replaced by private security firms and technologies. The impacts are concentrated on the poorest and most vulnerable populations, and crime contributes to a crisis of urban governance, where many city administrations are clearly powerless to act against it. There is a trend towards globalization of crime and corruption, in terms of financial fraud, human trafficking and political terrorism. At the same time the aspirations of the sustainable community for security and social cohesion are being taken up in cities around the world. All this poses new challenges for governance of cities and the human urban environment (UN-Habitat, 2007a).

8.7 Education and human resources

Education is seen as the key to development in both 'North' and 'South' – generally framed in terms of economic competitiveness and skills for production. Education is also one of the largest public services in most countries, with huge influence on urban social and cultural life. As part of the wider theme of human resources, skills and innovation, it is also crucial to working with the urban environment. To link this with the human urban environment agenda we review three themes:

- urban environmental themes in formal education;
- planning and management of the urban environment for children and young people;
- mobilizing human resources, skills, innovation and creativity, for the urban environment agenda.

Education for the urban environment

In many countries there is now some form of environment and sustainability programme of classroom education, which is beginning to translate the agenda into the requirements of the formal curriculum (Academy for Sustainable Communities, 2007; Foundation for Environmental Education, 2003). Environmental education is often slotted into 'science' or 'geography' subjects – but in reality, urban studies, current affairs, business and personal/social education each link to the environment and sustainability agenda. The formal education system often carries a hidden agenda of competition and exploitation; this needs to be countered with positive visions for peace, equality and sustainable development, using 'circle sessions', discussion groups, practical projects, and self-learning programmes (Orr, 1992; Hutchinson, 1996).

Arts and media projects are one way to overcome inbuilt cultural and generation gaps – in Manchester for instance, over a thousand pupils built environmental sculptures in the 'Art in Schools' project (Ravetz,

2000). These show the 'hands-on' approach of 'sustainable education' – improved school buildings and grounds, food gardens and wildlife habitats, social enterprise initiatives and arts/media projects. Such projects can be geared to pupils' lifestyles and self-directed learning, and co-ordinated with wider agendas, such as intermediate labour markets for environmental improvements (Chapter 7).

A city for young people

The modern city is an increasingly dangerous place for children, and for all the policies on education, there is little room for street parties or tree-houses (Ward, 1978). The urban environment tends to be structured by hostility and danger, for both young people and for those threatened by them, and the results often show up in crime and drug abuse (Taylor, *et al.*, 1996). Bridging the generation gap needs a new focus on young people's spaces and activities, in the design and management of housing, neighbourhoods and city centres. This might start with reclaiming the streets – for instance, the European concept of 'Home Zone' traffic calming of all residential areas, pedestrian routes, wildlife corridors and water habitats, to enable freedom for play and exploration (Stine, 1997). Economic restructuring for more flexible and localized activity should enable greater contact between children, parents and workers, for work experience and work attitudes.

Yet while the generation gap grows in education, crime and health, many urban youth and pre-school services are cut to the minimum. Pre-school provision is the key to fostering a culture of education, with locally based personal advisors who facilitate parent-child contact. The UK 'Sure Start' programme had many faults but showed a positive alternative to conventional urban regeneration: it is also an exemplar in the follow-up evaluation process (http://www.ness.bbk.ac.uk). For youth work, active out-of-school and work experience projects are needed in collaboration with employers, voluntary sector, sports and media. Such programmes must start from young people's daily experience – if drug use, for instance, is regarded as a danger to society, the majority of young people may be alienated and criminalized at an early age. A holistic youth programme starts from realistic experience, to bridge the 'trust gap' between the generations (Box 8.6).

Box 8.6

Youth and public space in New York City

'Take Back The Park', initiated in 1987, represents a creative departure from previous youth programming in that it was the first project of its kind in New York City that gave high-risk young people a lead role in motivating peers and adults in reclaiming community recreational space from drug dealers. Since its inception, 'Take Back the Park', has every summer been mobilizing one or more New York City neighbourhoods to reclaim a local park that has been taken away from the community by drug dealing, vandalism, and/or substance abuse. The programme mobilizes and trains community coalitions, including youth, police, parks department, community-based agencies and tenants associations. All 'Take Back the Park' efforts remain in action today.

Source: UN-Habitat Best Practices, http://www.bestpractices.org

---

Innovation and creativity in the urban environment

Environmental managers, sustainability assessors, community ecologists, sustainability co-ordinators are new professions which have emerged in recent decades. There are various 'sustainable communities' university programmes which aim to raise commitment for low-impact lifestyles and positive futures (Gough and Scott, 2008). These aim to build active links between teaching, project work, internships and case studies.

However it is clear that many of the necessary skills and resources do not fit easily within subject curricula or professional boundaries. One leading initiative, a national Academy for Sustainable Communities, was formed in the UK following the 'Egan Review' of skills and professions involved in creating and managing sustainable communities (Egan, 2004). This covers environmental issues among social and economic skills, but also recognizes the need for inter-disciplinary and inter-professional working (Roberts, 2007).

This agenda works at many levels beyond that of the urban community. There are emerging environmental technologies such as micro-generation, water management, waste recycling and land remediation: each involves high technology skill levels, but also a practical awareness of business potential, consumer behaviour, political strategy, infrastructure constraints, and urban development opportunities. The conventional technology-focused 'innovation policy' which is seen as more or less essential for competitiveness and growth, is now facing wider challenges:

- social and community innovation – new patterns of citizenship or social enterprise to enable environmental improvements;
- organizational innovation and learning – new ways of working and managing complex institutions;
- market innovation – new types of business models and trading structures, such as emissions trading, or ecosystems service trading;
- political innovation – new types of contract or compact, between market, state and citizens.

All these may be combined in the concept of the 'creative city', and/or the 'learning region' (Cooke and Morgan, 1998; Landry, 2006). It is this wider and more joined up innovation agenda for which skills and human resource development is urgently needed.

8.8 Urban governance

The public realm agendas above – health, security and education – point towards wider questions: how best to run and finance public services; how to work with diverse cultures and networks; and how to put these together for practical improvements in the urban environment? The idea of 'governance' as a collective responsibility for the city, beyond the traditional concerns of elected government, was introduced in Chapters 1 and 2. This section relates the governance debate to social and community issues.

Managing change

If there is one certain feature of cities, it is that of continuous flux and development. Competition between cities and regions has intensified in the global economy; the dominance of advanced capital and knowledge-based industries has increased the speed of change; and large sections of an urban economy may rapidly find themselves obsolete. This raises the agenda of 'change management' for all organizations, across public, private and civic sectors.

Such change management is increasingly focused on creativity and innovation as the driver of competitiveness. The 'creative city' is seen as an agenda for cultural industries, for economic investment, social enterprise, and political support, backed up by analysis of urban history (Hall, 1998; Landry, 2006). Innovation is seen as the driving force for competitiveness, and not only the traditional technology R&D, but a range of social, cultural and political innovations. The creative city agenda emerged alongside that of the 'creative classes', an emerging social and cultural grouping, around knowledge-based cultural industries, professional mobility and global networks (Florida, 2002 and 2004).

There are other dimensions to change management. In many cities of the 'South', rapid and chaotic urbanization is a reality which overtakes conventional urban planning and infrastructure development. Alternative ways of indigenous community-based housing and infrastructure development have begun to emerge from the failures of conventional imported solutions. Likewise, there are many cities and regions which are vulnerable to natural disasters, hazardous locations, economic instability or military conflict. In such conditions, new kinds of civil defence and emergency responses are needed for management of chaotic and traumatic situations.

Meanwhile the mainstream urban change management industry is busier than ever. In areas of rapid restructuring, shrinkage and decline, accelerated growth poles, and in wider patterns of urban agglomeration and metropolitanization, there are myriad policies and projects for regeneration and development. Often at the root of such activity are the choices between economic and social priorities – to invest in the winners or support the losers – and the urban environmental agenda is at the crux of such choices.

Gender and equality in the urban environment

Recent decades have seen much progress in the empowerment of women, ethnic groups, disadvantaged groups and other sub-cultures, but there is a never-ending task in balancing cohesion with diversity, to create a 'city for all' (Beall, 1996). To find out what is wrong in the urban environment we need only to 'ask the women' (Tibaijuka, 2008).

Women actually outnumber men in most cities, but at present their experience of the city is often that of fear, insecurity and exclusion: many women cannot travel at night, have less control over housing, and less access to public space and facilities (Darke, 1996). The majority in the 'South' struggle in substandard or dangerous housing, victims of crime and harassment. The majority in the 'North' live in suburbs, constrained by a social structure based on working men serviced by dependent housewives. Meanwhile as women gradually attain better education, social skills, birth control and life expectancy, former patterns of male dominance are shifting (Peck and Ward, 2002). In each area of public services there are positive responses to the changing gender balance:

- for crime prevention, there are community safety and awareness programmes, domestic violence and victim support networks (Moser, 2005);
- for health, there are well-woman centres, womens' support groups and self-help networks;
- for transport, there are women-only taxi services, accessible design in vehicles and buildings, infant facilities, car sharing clubs and lift networks;

Box 8.7

Women's co-operative housing, Uruguay

The co-operative UFAMA al SUR is a mutual help housing project. It centres on the conversion of a derelict building in central Montevideo into 36 apartments by a cooperative of Afro-Uruguayan women-headed households.

Initiated by the local community-based organisation Mundo Afro, this innovative project addresses the issue of derelict buildings in central areas of the city, allows for the incremental improvement of flats and promotes social integration as well as racial and gender equality.

Source: World Habitat Day Awards, http://www.worldhabitatawards.org

- for employment, there are childcare and after-school projects, positive employment policies, training and business development, support networks, micro-finance networks, and anti-poverty programmes (Yunus, 2008).

Similar approaches work for ethnic, disadvantaged, disabled, cultural and other disadvantaged minorities – in general, the majority of the population who are not in the dominant mode – i.e. male, able-bodied, heterosexual and economically active. In practice many urban 'communities' based on neighbourhoods or local industries, are being replaced by urban 'sub-cultures' based on social and economic networks (Box 8.6).

New approaches in governance

The critical perspectives on governance and change management show many tensions and conflicts (Chapters 1 and 2). By way of demonstration, these can be focused on some topical questions in the urban environment:

- Should environmental policies be produced piecemeal by urban municipal authorities, or standardized across larger areas?
- Should private firms be able to buy their way out of pollution control, in return for bringing jobs and investment?
- What is the most effective and democratic level of governance for larger mega-cities or agglomerations, containing rich and poor areas?
- How can transient, fragmented, informal or illegal urban communities be involved in local environmental improvements?

Such questions, and many more, point towards some of the underlying principles in governance for sustainability – partnership and networking, multi-level and multi-lateral governance, participation and capacity building, and innovation and change management. These are challenging enough in affluent and organized cities in the 'North' (Amin and Thrift, 1995): they are put to much harder tests in the squatter settlements of the 'South' (Hardoy, *et al.*, 2001).

Integrated urban environmental governance

While many public services were created in the nineteenth century by local government itself, the principle of 'partnership' now underlies new modes of urban governance. Tackling environmental problems and mobilizing environmental opportunities generally depends on an integrated approach to urban governance – often lacking at present. For instance:

- improving energy efficiency in housing depends on active collaboration and innovation with many actors – landlords, designers and builders, utilities, developers, citizens groups, unions, training agencies and so on;
- improving urban green spaces depends on active collaboration between municipalities, landowners, leisure enterprises, NGOs, faith groups, health and education services, housing organizations, local community groups and so on.

In many cases the fundamental issue is how to tackle the divide between the civic elite of decision-makers and stakeholders. For this new patterns of public participation and social entrepreneurship are needed – as in the phrase 'one good community entrepreneur is worth a dozen civil servants' (Henderson, 1978). There is a parallel economic agenda for the third sector, and the unlocking of resources for public services. A cultural agenda underlies each of these, encouraging new social movements, multi-cultural alliances, and local-global networks (Borja and Castells, 1997). But there are many obstacles – political, economic and cultural – and as local neighborhood-based 'communities' are shifting to globally networked 'sub-cultures', the problems of exclusion and alienation are as widespread as ever.

There are similar barriers between each of the public services – housing, health, education and crime prevention each tend to tackle a small slice of the symptoms, rather than the underlying causes. The result is that people and communities with compound problems fall through the gaps, and that environmental problems and opportunities are ignored. The increasing trends of civic fragmentation and social exclusion – the so-called 'splintering city' – raises crucial questions on the future of cities (Graham and Marvin, 2001).

Drawing from such challenges, the components can be sketched for a more integrated urban environmental governance:

- multi-sectoral co-ordination: bringing together strategy and action in public services such as housing, unemployment, anti-poverty, health, crime prevention, and education;

- multi-agency co-ordination: bringing together voluntary sectors, churches, unions, employers, business agencies, neighbourhood bodies and others;
- in-depth social audits and assessment, using both technical data and street-level exploration of perceptions, problems and quality of life factors;
- active participation and mobilization of human resources, through the third sector, social economy and social trading systems to enhance public services and social cohesion;
- multi-cultural approach: extending beyond the 'technocratic' paradigm to engage with other generations, sub-cultures and ethnic groups on their own terms.

Box 8.8

Integrated city development strategies

The Urban Management Programme, a UN-Habitat multi-donor programme, in collaboration with the Cities Alliance, pioneered the first set of City Development Strategies (CDS) in seven cities – Bamako (Mali), Cuenca (Ecuador), Colombo (Sri Lanka), Johannesburg (South Africa), Santo Andre (Brazil), Shenyang (China) and Tunis (Tunisia). The basic idea is to develop pro-poor urban governance in cities within the following framework:

- First, inclusive cities provide their residents, especially the poor and the marginalized, with the opportunities and capacities to participate in the decision-making process and to share equitably its social benefits.
- Second, well-governed cities can expect to improve the efficiency with which their scarce resources are allocated.
- A third area of value-added is in expanded productivity, both in the private and public sectors. Cities that understand their competitive position, and move wisely and quickly to capitalize on their comparative advantage, can expect economic returns.
- Finally, cities that plan their strategic moves over decades will waste fewer resources 'catching up' with rapid growth and poorly sited facilities and services.

Source: UN-Habitat, 2005b

These challenges go way beyond the normal remit of local authorities and public services and address a wider web of organizations at neighbourhood, district, regional and national level. There is not necessarily a single point of responsibility for 'making it happen', rather an extended framework for cooperation and co-ordination, as the foundation for urban environment strategy (Box 8.7).

8.9 Conclusions

The 'urban social environment' is a far-reaching agenda, covering the public services of health, education, crime prevention and anti-poverty strategy. It also raises questions of social distribution, citizenship and responsibility, human resources and innovation. And each of these is a complex debate, involving globalization, liberalization, multi-culturalism, risk management and so on.

There is a huge range of experimentation going on, in cities of the 'South', the 'North', and everywhere between. There are new institutions being invented or about to be invented, on the interface between public, private and civic sectors. There are immense new possibilities, hardly touched in this chapter, from the application of ICT, particularly the semantic or second generation Web, and its potential for catalyzing new forms of social networks and capacity building.

Overall, cities seem to be moving towards an urbanizing, warming and increasingly crowded world. It is clear that the previous model of materialistic consumption is unsustainable both environmentally and socially, and that social policy and urban governance will need to find new directions.

Further information

Homes and Communities Academy   http://www.hcacademy.co.uk

Association for Advancement of Sustainability in Higher Education   http://www.aashe.org

EU Sustainable Consumption and Production Programme   http://ec.europa.eu/environment/eussd/escp_en.htm

Foundation for Environmental Education   http://www.eco-schools.org

International Association for Urban Development   http://www.inta-aivn.org/

International Healthy Cities Foundation   http://www.healthycities.org/

Micro-Finance Gateway   http://www.microfinancegateway.org/

New Internationalist   http://www.newint.org/

UN Habitat Safer Cities Programme   http://ww2.unhabitat.org/programmes/
safercities/

UNEP Sustainable Consumption and Production Programme   http://www.unep.
org/themes/consumption

World Health Organization   http://www.who.int

World Social Forum   http://www.forumsocialmundial.org.br/

Further reading

Graham, S. and Marvin, S. (2001) *Splintering Urbanism: Networked
Infrastructures, Technological Mobilities and the Urban Condition*, London:
Routledge.

Klein, N. (2004) *Fences and Windows: Dispatches From the Front Lines of the
Globalization Debate*, London: Picador.

Putnam, R. (2000) *Bowling Alone: The Collapse and Revival of American
Community*, New York: Simon & Schuster.

Sachs, J. (2002) *The End of Poverty: Economic Possibilities for Our Time*, New
York: Simon & Schuster.

Sandercock, L. (2003) *Cosmopolis II: Mongrel Cities of the 21st Century*,
London and New York: Continuum.

Stiglitz, J. (2006) *Making Globalization Work*, New York: W.W. Norton and Co.

UN-Habitat (2004) *The State of the World's Cities 2004/2005: Globalization and
Urban Culture*, Nairobi: UN-Habitat.

UN-Habitat (2007a), *Enhancing Urban Safety and Security: Global Report
on Human Settlements 2007*, Nairobi: United Nations Human Settlements
Programme.

Yunus, M. (2008) *Creating a World Without Poverty: Social Business and the
Future of Capitalism* (co-author: Karl Weber); Public Affairs; 2008

9 Methods and tools for the urban environment

9.1 Introduction

This chapter looks at methods, tools, techniques and actions for effective planning and management of the urban environment. Some of the methods and tools are well established, while others are only just emerging. Tried and tested methods may show new potential when combined with others.

There are also underlying issues which can determine the success of any method or tool, for instance: the capacity of users and organizations; the system of governance; the quality of justice and legal systems; and the commitment to ethics in markets and organisations. Other factors may also be important, such as the quality of environmental data, the transparency of decision making, the presence of corruption.

Some aspects of environmental management may be particularly well suited to a laissez-faire market economy, while others may require a greater degree of government intervention. They all have important roles to play, whatever the form of governance, and whatever the level of development.

9.2 Urban environmental management

The Rio Declaration's number one principle, that 'human beings are at the centre of concerns for sustainable development' (UNCED, 1992, 3), is nowhere truer than in a city. Its built environment is built by people, to suit their own or other people's needs and wants. It is no place for flora and fauna, except for pigeons, parks and pets. The non-human species that live there do so on sufferance. Its environment is a human environment, of buildings that are beautiful or ugly, streets that are clean or filthy, air that is fresh or polluted, and water that is potable or contaminated.

As we saw in Chapters 5, 6, 7 and 8, the social, economic and environmental aspects of a city's development all blend into each other. Apart from their influence on the world outside, the city's environmental policies are a sub-set of its social ones. Even that sub-set is not entirely distinct, as shown in Figure 9.1.

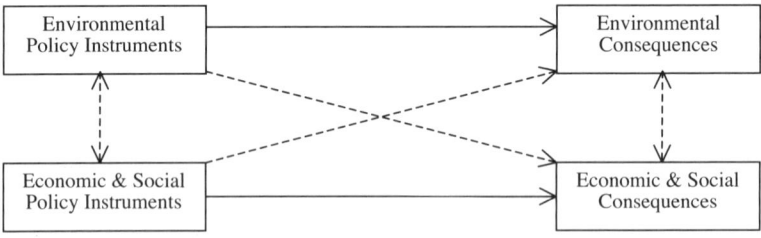

Source: Lee and George, 2000

FIGURE 9.1 *Linkages between policy instruments and their consequences*

Consumption and pollution

The city's economic, social and environmental policies also influence the more natural environment beyond the city's boundaries, as shown in Figure 9.2. Food, water and other natural resources must be brought in, sometimes processed on the way, and sometimes within the city itself. Wastes get sent back out. The inputs and outputs can both be reduced by recycling, to a degree, but every city is a net consumer and polluter of external environmental resources.

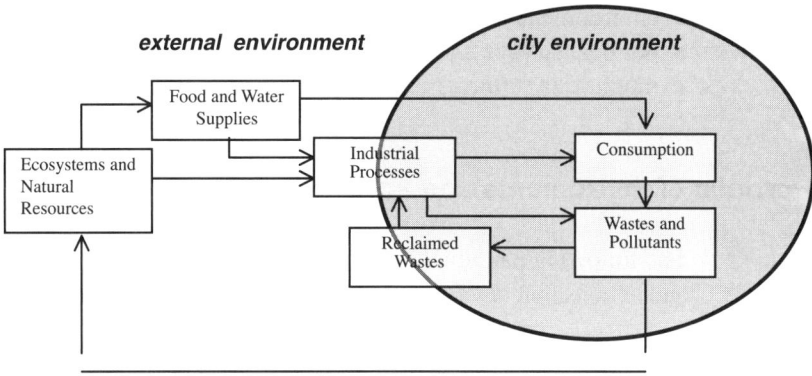

FIGURE 9.2 *Environmental inputs and outputs*

Management instruments

There are two main aspects to the task of managing the city's environment and its influence on the world outside itself. The different needs and wants of different groups of citizens must be understood and reconciled, and the decisions that are made must then be implemented. In the idealized world of economic theory, both aspects can be left to take care of themselves, through the invisible hand of the market (Easterbrook, 1995; Lomborg, 2001). In the real world of market imperfections, irrational actors, non-monetary values, and structures of power and expropriation, intervention is needed in order to identify needs and wants, and to deliver them.

Instruments for urban environmental management may be categorized into five broad groups (Lee and George, 2000):

● direct expenditure;
● command and control instruments;
● planning;
● economic instruments;
● voluntary codes and practices.

All five are dependent on four more groups of tools and actions (see also Chapter 2):

● environmental science;

- environmental economics;
- political science and policy analysis;
- monitoring, information management and ICT.

Portfolio of instruments and actions

The following paragraphs provide a summary of the major instruments and actions that are available to manage the urban environment.

Direct expenditure

Direct expenditure is the principal mechanism through which a city manages the most essential aspects of its environment. This typically takes the form of public sector infrastructure investment, for example in waste disposal, sewage systems, roads and parks. The necessary finance is normally raised directly or indirectly through taxation of the city's own wealth-creation activities.

Command and control

Command and control instruments or regulatory instruments include a wide range of laws and regulations which prohibit certain activities and place constraints on others. In the United Kingdom these include the successors to the Alkali Acts of the 1860s and the Clean Air Act of 1956, whose contributions to improving the environment and public health in the country's cities was dramatic (Davison and Barnes, 1992). Regulations include many that are related to pollution control, and others for the designation of protected species and controls on their exploitation.

Planning

Planning instruments generally complement regulatory command and control to deal in particular with non-pollution types of environmental impact. They include designation of protected areas, land use or spatial planning, sectoral planning (e.g. for minerals and water resources), environmental impact assessment and strategic environmental assessment.

Economic

Economic instruments are used to steer the economic decisions made by companies and individuals towards environmental goals. They include licensing charges for pollution permits, application fees for planning approvals, fines for non-compliance, emission trading schemes, and a variety of taxes and subsidies. Economic instruments also include frameworks which enable and encourage private sector investment in environmental infrastucture.

Voluntary codes

Voluntary codes and practices include cleaner production technology, good housekeeping disciplines, industry sustainability strategies, product labelling, certification schemes for sources of supply, and environmental management systems.

Environmental science

Environmental science is the backbone of every environmental management technique, covering subjects as diverse as atmospheric physics, soil chemistry, ecology, hydrology and epidemiology. Its scope is far too deep and broad for satisfactory treatment in a book of this nature, other than to observe that without a sound scientific basis, any attempt at environmental management is liable to fall short of its objectives or be counter-productive.

Environmental economics

Environmental economics provides an analytical link between the state of the environment and the human behaviour patterns which influence it. It may be used, with varying degrees of rigour, to place monetary values on environmental parameters in the process of addressing market imperfections or market failure. Environmental economics can often be an appropriate tool for increasing the rationality of trade-offs between environmental goals and economic ones, but care must be taken to avoid inappropriate application to non-economic human values.

Political science

Political science and policy analysis enable a better understanding of how decisions are made, how social groups can resolve differences, how organisations work and change, and how policy translates into practice.

Monitoring, evaluation and information

Monitoring and information dissemination provide essential data for professional analysis in the design of any environmental management instrument, and for informed involvement of the public in decision-making. This instrument also includes the use of evaluation methods for assessing the impact of actions and as a learning tool.

9.3 The decision-making framework

These various instruments, tools and actions contribute to making environmental decisions and implementing them for the city as a whole,

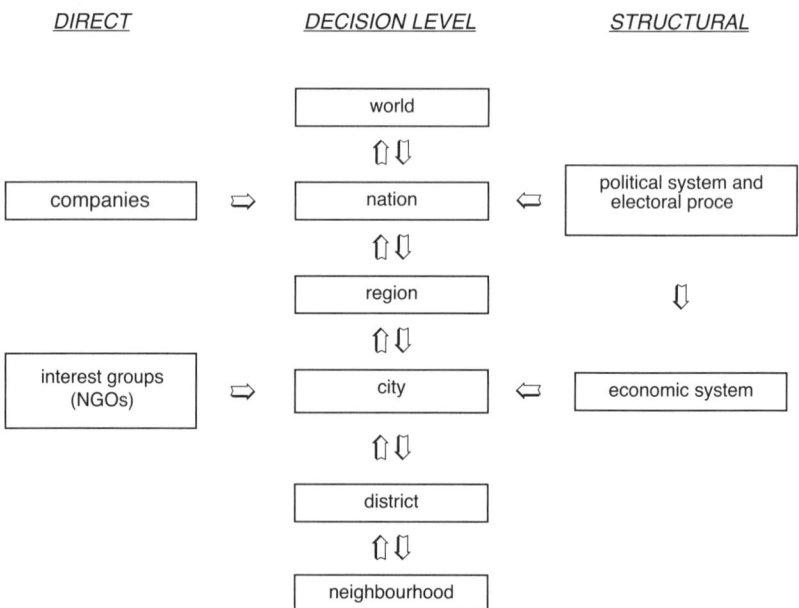

FIGURE 9.3 *Decision-making framework*

for its individual districts and neighbourhoods, for the region and the country of which it is part, and for the surrounding global economic and environmental complex. Decisions made at any level in this spatial hierarchy are influenced by (and influence) decisions at other levels, and are also influenced by prevailing political and economic systems and participatory processes. This overall framework within which environmental decisions are made is summarized in Figure 9.3.

At the top of this spatial hierarchy, environmental decisions include international agreements such as the UN Convention on Biological Diversity and its Framework Convention on Climate Change. At the bottom they may include a plan to set up a wildlife area in a local school.

Influences from above and below

Decisions at each spatial level are framed by those made at higher levels. A regional transport plan must be consistent with national policy on climate change, which in turn must comply with the country's international commitments. Higher level decisions may also limit the ability to make decisions at lower levels, by restricting decision-making authority, finance, or the freedom to raise it.

As well as being constrained from above, decisions at each spatial level are influenced from below. This may be through representative processes, as defined by the political structure and the electoral system, or through participatory processes, in which communities and individuals contribute directly to the debate on a particular issue. A combination of both representative processes and participatory processes may often apply.

In some parts of the spatial hierarchy, where political power is stronger at a lower level than at the one above it, the ability to make a decision is also constrained from below. This applies for instance to the European Union, where the principle of subsidiarity limits decision-making powers to those that are delegated upwards by member states, and to the United Nations, whose powers are strictly limited by the sovereign states that form it.

Subsidiarity is the concept that decisions should be made at the lowest level competent to act on behalf of all those affected, and that the democratic process should only delegate powers up to a higher level for those matters where the common interest over-rides the particular interests of separate groups.

Box 9.1

Spatial hierarchy in energy efficiency

- Household level: typically a matter for individual buildings and owners
- Street or block level: organisation of upgrading contracts
- Neighborhood level: area-based packages for infrastructure
- District level: technical management, finance and administration
- City-region level: for regeneration, utilities, finance packages
- Regional level: co-ordination of utilities, government and development agencies
- National level: overall policy targets and resources

Source: Ravetz (2000)

Box 9.1 gives examples of powers exercised at different levels. The principle may involve conflict and competition, for instance where the problems of traffic reduction are the responsibility of local authorities, without adequate powers and resources to deal with them. In reverse, the concept of residuarity legitimates the exercise of responsibility by lower levels, where higher levels fail to carry out necessary functions. Much activity under the banner of local sustainability assumes residual responsibilities which are sidelined in mainstream institutions. A typical result is that marginal agendas, for example green space, are either pushed out by the centre or pulled in by the 'community', leaving more mainstream agendas such as economic development under central control.

The upshot for the urban environment agenda is that political powers and resources generally come from one of three directions, as a result of pushing, pulling, or redistribution of new roles and functions:

- powers and resources devolved down from the regional or higher level;
- powers and resources devolved up from the local authority level;
- new functions on the interface between different levels, or between the public sector, the private sector and other civil society institutions.

Stakeholder influences

Within the constraints and influences applied from above and below in the central column of Figure 9.3, decisions are made in response to the direct and indirect influences shown laterally in the figure. Key stakeholders exerting direct influence include commercial firms and groups which represent them, such as chambers of commerce and trade associations. The interests of other groups of stakeholders are represented by other interest groups or non-governmental organisations (NGOs). The interests of the public as a whole are represented by the political system and the electoral process. The political system also influences decisions indirectly, through the economic system which it establishes and the public services which it manages or sponsors.

The city's need to export

Whether a decision is made by elected representatives, a government official or a company's board of directors, it is conditioned by the economic system in which the city operates. A city's existence depends on its ability to export goods and/or services, in return for food and other supplies. Some cities, such as Canberra or Washington DC, exist largely on the export of government services, paid for through taxes. Most other cities depend more strongly on exports of commercial services and/or manufactured goods. In order to maximize the standards of living of their citizens, such cities must compete with others in an increasingly international marketplace. Decisions that are made on the city's environment, and its impacts on other environments, are therefore strongly influenced by the nature of the economic processes through which its citizens earn their living.

Economic efficiency

A city's principal trading relationships have traditionally been with its own hinterland, importing food for example, in return for exported services such as the provision of a marketplace. For a capital city, the hinterland is the whole country. The city and its hinterland together, the city-region, trade with others, by exporting goods and services for which they have a comparative advantage, and importing those for which the advantage lies elsewhere. Success in competing economically with other cities depends on the city's own internal economic structures. To compete regionally, nationally and globally, the city has to maximize the commercial competitiveness of its own internal economy. For instance, the distribution of food through small high street retailers and local corner-shops may be less efficient economically than through an out-of-town hypermarket, because the extra travel costs paid by customers may be more than offset by the lower purchase price and transport costs of bulk supplies. Any environmental costs that are felt by the city itself will affect this balance. External ones such as greenhouse gas emissions do not, unless a higher level in the decision-making hierarchy introduces an economic instrument to internalize the cost. This presents difficulties in dealing with global impacts, since only limited powers are delegated upwards to the global level, and these are as yet insufficient to deal with the problem.

In practice, difficulties in environmental managment occur when problems are displaced and transferred from one group to another – from the city's waste water to the fishing industry downstream; or from individual drivers to the children who breathe their exhausts. Economic instruments may in some cases be used to internalize the external costs, while in others each appropriate level of governance may need to plan, regulate, or provide necessary infrastructure.

From government to governance

Government at each spatial level should in principle weigh up all these different interests and influences, in order to make decisions which are judged to be in the best overall interest of the electorate it represents. In doing so, decision-makers have increasingly made use of direct participatory approaches. As we have seen in previous chapters, this increasing engagement of civil society in participatory decision-making is happening in parallel with a smaller role for government within the broader mechanisms of 'governance'.

As discussed in Chapters 1 and 2, this shift from government to governance is associated with the difficulty faced by the state in achieving both social cohesion and economic success in an increasingly competitive international marketplace (SEU, 1998). In order to maximize its ability to compete with other cities and other states, the state is under pressure to relax its constraints on market mechanisms, and to replace state regulation by public-private partnerships. Citizens thus have less influence on decision-making through their democratically elected representatives, and must instead rely on direct participation in individual development decisions.

Within this shift from government to governance, the influence of different social groups on the decisions that are made may increase or decrease, depending on the nature of the participatory processes that are used. In some cases, bureaucratic decisions based on a minimal understanding of public concerns may be replaced by something approaching self-determination on the part of the communities most affected. In others, participation may serve only to persuade the public that it has a say, in decisions that are dominated by individuals or groups that have the greatest influence (Cooke and Kothari, 2001). The extent to which governance then serves the best overall interest of the city's citizens depends on factors similar to those discussed in Chapter 4, including the ideas and ideals of influential people, and the transparency of the decision-making process.

These changes have resulted in a shift of emphasis from the first two environmental management instruments discussed above, command and control and planning, to the third and fourth, i.e. economic instruments and voluntary codes and practices. Nonetheless, all four remain essential to successful environmental management, along with the information and analytical tools which support them, and the involvement of citizens in implementing them.

Participation

While community participation is generally held to be a 'good thing', it has its risks and shortcomings. In any group, large or small, an inner elite circle tends to emerge, which then manipulates the majority. While these relationships may be mediated by the checks and balances of democratic structures, direct participation often aims to extend or to challenge these structures. It is often seen as dangerous by the elite, who may endeavour to capture the process and steer it to its own ends. The Arnstein 'ladder of participation' identifies different levels in this relationship, from 'placation' to 'citizen power' (Arnstein, 1969). Such models tend to assume that 'the community' has a single voice and single agenda – but in practice most communities are diffuse and divided. In a globalizing post-industrial city based on brand images, consumption, and information and communication technologies, the nature of community itself is changing, from one of collective identity to a search for self-identity through the choice of marketed lifestyles (Henley Centre, 1996).

In principle, participation is a guiding theme for almost every aspect of urban decision-making, from strategic planning to neighbourhood regeneration (Box 9.2). In practice there are many problems – strategic issues are complex and dominated by corporate interests, while at the neighbourhood level, most structural decisions have already been taken. Defining the 'community' is a perennial problem, and the need for speed, commercial confidentiality and institutional inertia all serve to distance

---

Box 9.2

Community participation in the built environment

Development stage	Spatial scale	Methods and organization
strategic or regional planning guidance	conurbations, counties	sustainability indicators & forums
local authority development plans	cities, towns, districts	focus groups, alternative plans
neighbourhood/area strategy/programme	neighbourhoods	area partnerships, community planning
site allocations & development briefs	development sites	planning for real, urban design events
building design & mix of uses	larger buildings & sites	design participation, local audits
detailed design, facilities, externals	buildings & surroundings	models, workshops, simulations, visits
construction, access, local employment	larger buildings & surroundings	business/community partnerships
medium–long term use, management & access	buildings, sites, community facilities	community development, trusts & associations

Source: Ravetz J, for the RIBA Community Architecture Group, 1995

the community from the real decision point. Participation processes themselves tend to create a community elite – those who learn to work the system to their own advantage (Jeffrey, 1997). While each local authority has to evolve its own approach, there is again a case for an urban or regional framework of best practice for participation:

● common standards for decentralized decision-making in public services;
● specialist resources such as internet hosts and community technical aid as channels from communities to authorities;
● participation framework for other city-region functions such as transport, health or utilities;
● allocation of a percentage of budgets to participative decision-making.

Local Agenda 21

Local Agenda 21 (LA21), as agreed at the 1992 Rio Earth Summit, advises all local authorities to consult with their citizens to produce a statement on local issues and actions for sustainable development. LA21 has a broad focus on every aspect of the local environment, and also

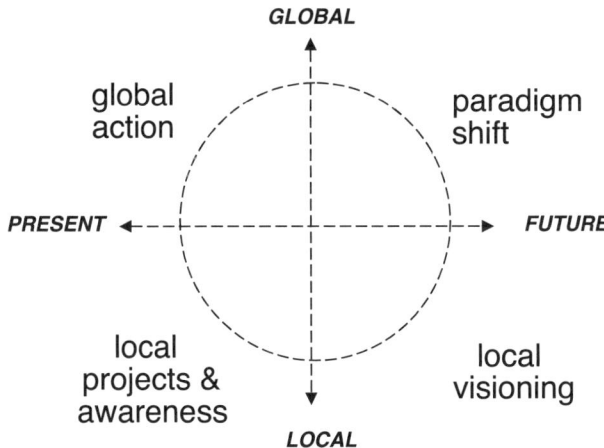

Overview of themes and directions for Local Agenda 21.
Adapted from UN DESA, 2002

FIGURE 9.4 *Local Agenda 21*

addresses complex and challenging social and economic issues. The 'agenda' of LA21 is intended to extend from the local to the global, and from the present to the distant future (Figure 9.4). It also includes several kinds of 'discourse' – from welfare reform to a political programme, and from community services to cultural space.

From the LA21 programmes and results around the world, there are several themes which stand out:

- for 'quality of life', public discussion is essential and can influence political action, but the quality of life agenda can also tend to gloss over harder conflicts and contradictions;
- for the vision aspect, envisioning of future goals and scenarios is an essential step for any community, however, without effective facilitation the participants are often self-selected, with wish-lists disconnected from the real world;
- for direct democracy, LA21 is a potentially valuable feedback channel for the citizenry, and brings groups and networks together in new ways – LA21 forums can also be unfocused and marginal, with little formal legitimacy or accountability;
- for the environmental agenda, LA21 is valuable for networking and open discussion, but information and expertise may be lacking on complex environmental issues, and key players such as utilities and government agencies often show only token interest.

A sceptical view of LA21 might see local authorities as 'playing along' with a public relations exercise, until it challenges the political order (Kitchen, 1997). A more positive view sees local democracy in many countries in a state of flux, and LA21 as a catalyst for new networks with social and environmental agendas (ICLEI, 1996). Initiatives have been successful in some countries and less so in others, with some city governments delegating considerable authority, and others little or none. LA21 appears to be taking the first steps along a very long road towards thinking globally in the development of local action.

9.4 Monitoring and information

Reliable data are essential for sound environmental management. This applies both to general information on environmental quality, and to the specific impacts of individual activities. General environmental information is needed for planning, and specific information for the enforcement of environmental law. However, 'monitoring costs can become astronomical very quickly and this tendency often guides law-makers in their judgements' (Newson, 1992a).

Cost-effective monitoring

Programmes for surveillance of the state of the environment, and for monitoring pollution sources and other impacts, need to be designed in such a way as to focus on the information that is actually needed in order to make decisions, or to prosecute significant offenders. As Newson observes, 'it is essential that monitoring is adequate, even if a certain amount of pure guile is necessary on the part of those policing compliance'. Regulators often work with polluters, helping them as well as goading them to achieve the desired standards. This form of public-private partnership is a long-established conciliatory technique, designed to achieve cheap compliance, and avoid the expensive detailed monitoring required by an adversarial approach. It can also help to achieve potential for more effective results and avoid end-of-pipe thinking.

Monitoring by the public

Conciliatory regulation does not always sit easily alongside public involvement. A complainant may be seen as an irritant to a cosy relationship, rather than as an ally of the regulator. Nonetheless, if the monitoring system is sufficiently transparent, members of the public can serve as an army of unpaid volunteers, focusing the regulator's efforts on issues of concern, and providing information that would otherwise be too expensive to gather. Much of the highly professional data on Britain's bird species comes from members of voluntary organisations, while in developing countries NGOs have made major contributions to environmental monitoring programmes (Eckman, 1996; Rasid, 1996).

State of the environment (SOE) reporting

When a city's development plan is being drawn up or reviewed, the decisions to be made must be based on the best available information on the actual state of the city's environment. A plan based on faulty assumptions may well have the opposite effect to that intended. A state of the environment report will draw the available information together, and may in some cases trigger further data-gathering to fill important gaps.

Indicators

As well as being needed for environmental planning and law-enforcement, reliable data on the state of the environment are necessary for monitoring the achievement of sustainable development goals, and taking corrective action where necessary. As stated in Chapter 40 of Agenda 21 (information for decision-making), 'indicators of sustainable development need to be developed to provide solid bases for decision-making at all levels and to contribute to a self-regulating sustainability of integrated environment and development systems'(UNCED, 1992, 473). Since the UN Conference on Environment and Development (UNCED) in Rio in 1992, considerable effort has been devoted to developing appropriate indicators and associated monitoring systems, as discussed above in Chapter 2.

Many urban authorities are developing new sets of indicators and monitoring and evaluation techniques. This is part of a wider process aiming to implement the principles of sustainable development at every level – global, national, regional and local. At the local level there is a clear need for local sustainable development indicators (SDI) which provide information to enable more informed decision-making.

This has led to a great experimentation with new forms of local SDIs, in the context of raising awareness, forming policy, setting targets, and monitoring conditions and trends. There is now a range of networks which aim to co-ordinate the many individual local efforts in order to promote best practice, and link local to national and international concerns. However there are problems in developing SDI sets which are technically sound, policy relevant, and meaningful to local communities. Therefore the SDI network initiatives have a challenging task.

Box 9.3 describes how sustainable development indicators of this nature may be developed at the local level.

Box 9.3

Local Sustainable Development Indicators (SDIs)

Sustainable development objectives are many and diverse – and therefore the SDIs to measure progress are equally diverse, with a variety of functions and applications. They may be strategic 'top-down', community-driven and 'bottom-up', or a combination of the two. They may be organized by issues or socio-economic sectors or embedded within policy/management structures, to focus on 'urban sustainability', 'local sustainability', or sustainability in general. They may be designed for individual places, or co-ordinated for national or international comparison. These different kinds of functions need to be clear in the design of SDIs, to ensure that they serve the needs of municipalities and the intended users, while remaining consistent with overarching sustainable development principles and objectives. So there are questions which come up at every stage of formulating, testing and maintaining local indicators.

- What are we indicating? For example, is it possible to define accurately 'access to open space'?
- How to indicate it? Is there a clear and practical way to measure this definition?
- How much to measure? Should it show differences, for instance, between neighbourhoods, age-groups or social groups?
- Who is the information for? Management, businesses, residents, consumers, investors, politicians – who needs to know?
- When is it to be used? Agenda setting, policy forming, monitoring or evaluation?
- Where is the link? Is the indicator directly linked to practical action, to planning policies, or to wider debate?
- Why is the theme important? – is the indicator based on a clear concept or objective for 'local sustainability' which is shared by all involved?
- Does the indicator set measure the local contribution to global sustainablity? – or is it just a wish-list of conventional aspirations?

The role of indicators can also be seen as part of a general information flow between different groups – meaning that their limitations can be seen in their context as part of a social learning process.

Access to information

Public access to environmental information was established as a key requirement at the 1992 UN Conference, in the Rio Declaration's Principle 10:

> Environmental issues are best handled with the participation of all concerned citizens, at the relevant level. At the national level, each individual shall have appropriate access to information concerning the environment that is held by public authorities, including information on hazardous materials and activities in their communities, and the opportunity to participate in decision-making processes. States shall facilitate and encourage public awareness and participation by making information widely available. Effective access to judicial and administrative proceedings, including redress and remedy, shall be provided (UNCED, 1992, 3).

These requirements have been reinforced in the UN Economic Commission for Europe's Aarhus Convention (UNECE, 1998). This establishes more detailed rights to information on the environment, to justice in environmental matters, and to participation in decisions affecting the environment.

Environmental education

As was discussed in Chapter 8, effective public participation in environmental decision-making depends not only on reliable information, but on a sound understanding of the issues and their scientific basis. This may not be easy for people without an appropriate formal education, but adult education and distance learning programmes can be a great help for those with a strong interest, while environmental education in schools can lay a solid foundation. For particular issues, such as making choices over alternative siting and technologies for solid waste disposal, specially formed focus groups with expert inputs can facilitate highly productive debate. Such techniques can however be time-consuming and expensive, and may require specialist social science expertise for effective facilitation.

Modelling and scenario-building

Computer simulation can also provide a valuable tool for developing understanding of the relationships between policy decisions and their environmental consequences. Modelling and scenario-building techniques can be combined with interactive interfaces and expressive graphics to give users a powerful impression of the potential outcomes associated with alternative choices (e.g. Rothman, *et al.*, 2002).

In defending the simplicity of the computer simulations that are often used, Rothman *et al.* draw a distinction between 'science-driven and problem-driven approaches to understanding the relationship between human and natural systems', preferring the latter. The distinction can be misleading. Sophisticated models which use more rigorous scientific methods are no less problem-driven than simple ones, and address the same problem, to understand the relationships between human behaviour and the natural environment. Simple models can be as valid for their purpose as complex ones, provided that their users understand the limitations of the science on which they are based.

It is however crucial that the scientists understand the limitations of their science even more fully than the people who will use its outputs. Uncertainties should be identified, and wherever possible, quantified. In cases where the science is inadequate to give reliable results, the Precautionary Principle should be applied. This is one of the key principles of the Rio Declaration, stating that where significant environmental damage may occur, but knowledge on the matter is incomplete, the decisions that are made and the measures implemented should err on the side of caution.

Potential for propaganda

With or without the hidden algorithms and seductive interfaces of a computer simulation, all techniques for facilitating public participation in environmental decision-making are susceptible to being used as propaganda, on one side of an argument or the other. As discussed above in relation to governance issues, participatory techniques may be used to persuade the public that it has a say, when actual decisions are made by dominant individuals or groups. Antonio Gramsci suggested that the participatory institutions of civil society can easily become 'the outer earthworks of the state' (quoted in Kurtz, 1999) through which ruling elites protect their dominance of the state's formal institutions.

Sound environmental science is a key tool for avoiding the misuse of such techniques, along with equally sound social science. As will be seen below, environmental economics links the two, with its own limitations.

9.5 Environmental economics and multi-criteria techniques

Sustainable development requires an appropriate balance to be established between the environmental, social and economic aspects of development. This entails making value judgements about the relative importance of each type of factor. Environmental economics offers one way of doing this.

Economics and ethics

Some environmental qualities have an obvious monetary value. If soil degradation has no significant effect on wildlife, but a large impact on agricultural productivity, the cost is the cost of the lost output. In such circumstances, a rational decision cannot be made without valuing the environmental effect in monetary terms. If on the other hand an endangered species is affected, monetary valuation is more problematic, with difficulties that are both practical and ethical. The less direct the relationship between an environmental quality and economic performance, the greater are the technical difficulties of assigning it a price. Beyond this, even if the Rio Declaration's highly anthropocentric Principle 1 is accepted, placing human beings at the centre of concerns for sustainable development, it is far from universally agreed that human values are exclusively economic. However, provided that other motivations are not excluded from the analysis, the quality of decision-making can often be improved by making the best estimate possible of the economic contribution.

Components of total economic value

Environmental economics generally recognizes five components of an environmental parameter's total economic value (Dixon, *et al.*, 1994; OECD, 1995; Kirkpatrick, 2000):

- direct use value refers to an economic benefit that is obtained directly, such as agricultural production from fertile soil, or tourism in a wildlife park – this is usually the easiest component to measure;
- indirect use value may provide an additional economic benefit through secondary or higher order effects – a wetland for example may act as a water purifier, which improves water quality, and increases fishing yields indirectly;
- option value may yield an economic benefit in the future, although no economic gain is currently made from it – diversity of species may for example enable new agricultural or medicinal products to be developed in response to future diseases;
- existence value is a non-use value, derived from the knowledge that something exists – people may wish to protect a rare species from extinction, even if they will never see it;
- bequest value is a non-use value whose benefit is derived from the knowledge that something exists for future generations, such as archaeological remains that lie buried underground.

Valuation techniques

There are three main types of valuation technique that can be used to estimate each of these components of total economic value. All three give estimates of what people are willing to pay for an environmental quality, or are willing to accept in compensation for its loss.

1 Market price. This approach can be used when a market actually exists for the environmental parameter being studied. Productivity analysis measures direct or indirect use value, from the value of marketed goods and services. Human capital cost valuation measures the value of damage to human health, through healthcare costs, lost output, and other economic losses resulting from sickness or premature death.
2 Benefit transfer. Where an environmental quality is traded in one context and prices are known, the same economic value may be assumed to apply in a different context. The assumption is not necessarily valid, e.g. the cost of damage to human health in different countries.
3 Individuals' preferences. When there is no existing market for the environmental quality being studied, an estimate is made of what the value would be if there were a market. There are three main techniques.

i *Replacement cost or preventive expenditure* is the cost of rectifying environmental damage or preventing it, for example by building a desalination plant to stop an aquifer being depleted. This may over-state or understate the value (e.g. people may not be willing to pay the full cost of a desalination plant, or alternatively, the damage caused by depletion may have a much higher cost).

ii *Surrogate market* approaches use actual prices for related goods or services, which are influenced by the environmental quality being studied. The travel cost method can be used to estimate the recreational value of natural resources, through the costs in money and time paid by visitors. Property value or hedonic price methods make use of the fact that residential and other property prices tend to be higher in areas where environmental quality is high than where it is low. The price difference is used to calculate the value which people put on the environmental differences.

iii *Contingent valuation* is a market survey technique, which is used to estimate an economic value for environmental qualities that have no direct economic influence. A representative sample of people is asked what they would be willing to pay or accept for an environmental quality if it were actually traded. It can be used to estimate all components of total economic value, for any environmental parameter. Uncertainties are associated with survey design and the hypothetical nature of the exercise.

Cost-Benefit Analysis

The use of environmental economics techniques is the least controversial, and the most necessary, for direct and indirect use values where an environmental quality is actually traded, and the market price is known. An environmental good can then be given a money value, which may be incorporated into a Cost-Benefit Analysis (CBA) of alternative options for its conservation or use. However, ethical issues arise even here, for example for depletable natural resources. Conventional financial accounting gives them a higher net present value if used now than if they are preserved for future generations, since the proceeds can be invested to give a higher economic benefit for those same generations.

Discount rate

The difference in value depends on the discount rate that is used to calculate net present value. Conflicting views on whether this rate should be adjusted in favour of future generations are associated with different visions of future lifestyles and technologies. For example, if it is assumed that future generations will value oil for oil's sake, a low discount rate might be chosen, in order to share the resource out over a long period of time (von Amsberg, 1994, 127–51). Alternatively, it may be assumed that future generations will value oil as only one possible source of energy, and that they will make their own decisions on the relative values of energy and other things for their own future lifestyles. In that case, oil may be regarded as a purely economic resource. Provided that the resource is used in such a way as to generate equal or greater economic capital in other sectors of the economy, the appropriate discount rate is then the normal commercial rate. However, other forms of environmental capital are not so readily treated as having only economic value.

Non-economic values

For those components of environmental quality that are not traded, the ethical issues are more immediate, in the assumption that if the quality in question could be bought and sold, it should be. How much are you willing to pay for the air you breathe? How much are you willing to accept for it being denied to your children? If you were allowed to sell your grandmother, how much would you charge? It only takes one person to respond to a contingent valuation question with the answer infinity, and the technique has to invalidate itself by discarding that person's values. 'How can you buy or sell the sky, the warmth and the land?' (Seattle, 1857). As Chief Seattle discovered when he signed the treaty which brought the Indian Wars to a close, you can buy or sell anything, but he did not believe that you should. His values were not economic ones.

Environmental economics is an approach developed by neo-classical economic theory in order to draw environmental decisions into an economic framework. There are other ways of making value judgements than environmental economics. In many circumstances it is an extremely useful tool, provided that the uncertainties in its estimates are fully acknowledged and taken into account. However, the use of the tool is itself a value judgement.

Multi-criteria analysis

Non-monetary techniques for balancing different environmental, social and economic factors against each other include goals analysis, cost-effectiveness analysis and other forms of multi-criteria analysis.

Goals analysis enables an initial screening of alternative proposals, according their ability to achieve all the environmental, social and economic objectives of the policy or other action being designed (Hill, 1968). If only one alternative meets all the goals, no further analysis may be needed. Any which fail to do so may be rejected. If more than one alternative meets all the goals, the cheapest may be the most appropriate (cost-effectiveness analysis). If the policy is likely to have different impacts on different sections of the community, a planning balance sheet may be used to evaluate separate goals for each (Lichfield, 1996). Commonly however, the goals of development cannot be defined so precisely, and each of the alternatives will differ in how closely they meet each of the objectives. Some form of multi-criteria analysis may then be needed which balances the different environmental, social and economic factors against each other.

Pair-wise comparison may enable some alternatives to be rejected, if they perform worse on all counts than at least one other of the available options. Otherwise, multi-criteria techniques generally involve some form of scoring and weighting, through which impacts on different parameters are standardized on a common scale, and weighted for their relative importance. A variety of techniques is available for this (DETR, 2000a), all of which entail making value judgements, notably in the choice of weights. Although monetary valuation is avoided, the problem remains of comparing different values, for example of an environmental loss and an economic gain. The weights that are chosen need to reflect the values of the public as a whole, not those of the experts undertaking the analysis. If a market for the environmental quality exists, environmental economics may yield a more accurate understanding of aggregate public values than a multi-criteria weighting exercise, even when it is highly participatory.

As a technical approach to guiding decisions, multi-criteria analysis can be used to give a pseudo-scientific aura to the values of the experts involved, or to mask the influence of the decision-makers they advise. However, when the techniques are well implemented, they can clarify the value judgements that must be made, and enhance the openness and rationality of the process of making them. 'Where multicriteria evaluation has been effective, it has been applied in a self-aware and transparent fashion. Its goal has not been to provide a demonstration by experts, but to enhance dialogue among stakeholders' (Funtowicz, *et al.*, 2002).

The democratic process

Environmental economics and multi-criteria analysis are both limited in the extent to which they can, in themselves, identify the most appropriate trade-offs between environmental, social and economic factors. Decision-making has to take account of all the different interests and different values of different people. It is of its nature a political process, in which decision-makers may choose to trust their own judgement of the values of the people they represent, or engage them in debate on individual decisions. Either way, these techniques provide valuable tools to inform the democratic process, but cannot substitute for it.

9.6 Economic instruments

Economic instruments are used to steer the economic decisions made by companies and individuals towards environmental goals. These instruments include:

- licensing charges for pollution permits;
- financial liability regimes, e.g. for contaminated land;
- application fees for planning approvals;
- fines for non-compliance with command and control or land use regulations;
- taxes on the use or disposal of polluting products;
- environmental protection subsidies and grants;
- government spending on environmental improvement;
- emission trading and other impact trading schemes.

Direct and indirect economic influences

Some of these instruments are used to intervene directly in the economy, through taxes, subsidies and other mechanisms, in order to change prices and hence consumer preferences. This may apply for example to fertilizer use or water use.

Others operate in conjunction with command and control or planning regulations, which enable the collection of charges and fines. For example, discharges of certain pollutants into the public sewage network are often managed by charging a fee that is at least equal to the cost of removing the pollutants in a common waste water treatment plant. This does not however obviate the need for monitoring by the environmental authorities to ensure that the discharges being paid for are not exceeded, with penalties for abuse. The cost of monitoring can also be recovered through the fees charged. Similarly, the entire costs of the environmental authorities can be recovered through licensing charges, or charges for submitting planning or licensing applications. When the costs are not recovered in this way, they are paid for by the public as a whole, rather than by developers.

The polluter pays principle

Many economic instruments operate on the principle that the polluter pays the polluted, generally as a contribution to the public revenue. In some cases this is used to finance clean-up (e.g. license fees for discharges into the municipal sewage system), while in others the money raised pays the costs of regulation (charges for planning approval). In other cases the revenue is used to reduce the overall tax burden, and thereby compensate the public for the environmental loss which they bear.

However, certain types of subsidies, grants and government spending programmes may run counter to the polluter pays principle. Money may be taken from the affected public, in the form of taxes, and given to polluters as an incentive to stop them polluting. If spent on rectifying the damage, government spending does not act as a disincentive, but changes the cost borne by the public from an environmental one into an economic one.

Impact trading

Impact trading schemes can be an economically efficient way of reducing pollution, by allowing the most efficient producers to sell their emission rights to less efficient ones. The US has used such instruments effectively for the reduction of sulphur dioxide emissions, and the EU's Emissions Trading Scheme applies a similar approach to greenhouse gases. The initial phase of the EU scheme was fairly generous in granting emission rights and had little effect in reducing emissions (Stern, 2006). Future implementation is expected to be more rigorous, with extensions to incorporate the flexibility mechanisms of the Kyoto Protocol. These allow Annex I countries (those with commitments to reduce emissions) to partially offset these commitments through joint implementation projects or international trading between each other, or through projects in developing countries under the Clean Development Mechanism (CDM).

Much of the debate surrounding carbon trading schemes relates to the extent to which they may shift the burden for reducing greenhouse gas emissions from richer to poorer countries (Ott and Sachs, 2000). Unless they are appropriately designed or accompanied by other measures, such schemes may serve to allow high-income countries to maintain high emission levels, with a relatively low level of financial compensation to developing countries for keeping their emissions low.

All impact trading schemes create an artificial market in order to allow market mechanisms to operate on public goods. The market mechanism breaks down for public goods in what is termed market failure. Although such schemes can be an economically efficient means of achieving a regulatory objective, the magnitude of their influence is not determined by the market, but by government decisions in establishing the level of the credits that are granted.

9.7 Regulatory instruments

Regulatory instruments, or command and control instruments, include regulations which cover, for example:

- the maximum quantities and concentrations of different types of pollutant which may be released into the environment;
- types of products and wastes that may be used, produced, sold or disposed of;
- processes, technologies and management techniques which must or must not be used;

- procedures for obtaining permits and licenses;
- protection of wildlife.

Pollution prevention and control

Many pollution control laws and regulations are in principle relatively simple, prohibiting certain polluting activities, or placing tight constraints on them. However, as industry has become more complex, more sophisticated regulations have been added. The fairly straightforward prohibitions and constraints appropriate to small-scale industries and individual households are supported by complex regulatory processes designed to cater for the environmental management of medium- and large-scale industry. As discussed in Chapter 7, the techniques of Industrial Pollution Prevention and Control (IPPC) are similar in all industrialized countries, and are increasingly being used in developing ones.

Three main factors characterize IPPC:

- the integration of protective measures for air, water, land and certain other environmental effects;
- management system requirements, including staff training and supervision, equipment maintenance and record keeping;
- the requirement to use best available techniques (BAT).

As well as exerting pressure on industry to continuously improve the environmental performance of its processes, BAT creates incentives for manufacturers of pollution abatement equipment to develop and market new techniques. A high degree of expertise is required of regulatory staff, as well as sufficient numbers of personnel and strong legal backup (as is the case for all regulatory approaches).

Other command and control instruments

In addition to their role in pollution prevention and control, command and control instruments also contribute to biodiversity conservation, through, for example, laws which prohibit exploitation of endangered species.

Mandatory product labelling may be regarded as both an economic

instrument and a command and control one, using regulatory
requirements to enable market mechanisms to operate.

9.8 Environmental planning

Planning instruments generally complement regulatory command and
control, to deal in particular with non-pollution types of environmental
impact. They include:

- land use or spatial planning;
- land use approval mechanisms;
- sectoral planning (e.g. for transport, energy, minerals, water
  resources);
- environmental impact assessment for major development projects;
- strategic environmental assessment of development policies and
  plans;
- designation of protected areas.

Land use planning

The planning of a city's development is generally based on a land
use plan or master plan, which for example defines those areas of the
city which will be developed for commercial, heavy industrial, light
industrial, residential, recreational and other uses. In segregating these
activities where appropriate, the plan is essentially a plan for the city's
environment, but its design is based strongly on policies and plans for
the social and economic aspects of development. Infrastructural needs
such as transport, water supply, sanitation, waste disposal and energy are
considered within the plan, or in associated sectoral plans.

In a market economy the economic component of the plan is defined in
outline only, to identify those types of industry or commerce the city
will aim to attract, for example through the provision of appropriate
land and infrastructure. Once a plan has been adopted, applications for
individual developments are evaluated for accordance with the plan, and
for compliance with a wide range of laws and regulations which define
specific environmental requirements. 'Environmental planning' then
aims to build in environmental objectives, in considering the location of
development, transport impacts, industry and other infrastructure.

Environmental impact assessment

The planning system and its associated rules and regulations are normally sufficient to control the environmental effects of most development activities, particularly those of a relatively routine nature. However, large or complex developments may have impacts that are specific to their own design, and potentially large. Such projects may therefore be subject to additional checks before they are approved (Lee and George, 2000; Wood, 2003), through a formal process of environmental impact assessment (EIA). This is a time-consuming and relatively costly form of environmental check, but a highly versatile one, whose aim is to:

- identify any potentially significant impacts of a proposed development;
- ensure that they have been properly allowed for in the design;
- evaluate the significance of any residual impacts;
- make this information available to decision-makers.

The developer is normally responsible for undertaking and paying for the EIA study, which may involve detailed scientific and technical investigations by specialist consultants, along with stakeholder consultation. Once the adequacy and validity of the EIA report have been checked and accepted by the competent government authority, its findings are included in the information available to decision-makers in their evaluation of the proposal. Typically, the EIA process results in decision-makers requiring design modifications or other improvements to the proposal as conditions of approval.

Strategic environmental assessment

If a development project complies with a land use plan, it may nonetheless contribute to major environmental impacts, if the plan itself is not environmentally sound. Similarly, if government policies are changed in ways which encourage the use of potential pollutants, such as fertilizers or pesticides, the environmental impacts can be considerable. A change in transport policy may encourage an expansion of motor vehicle use, while energy policies may require power stations to be built. Strategic environmental assessment (SEA) aims to deal with these issues, by subjecting policies, plans and programmes to environmental assessment, and also other strategic decisions such as changes in law (Therivel and Partidario,1996).

The need for SEA is particularly acute when existing policy-making or planning processes cater poorly for environmental considerations. One of the arguments against introducing it systematically is the belief that environmental factors are already well integrated into existing processes.

Sustainability assessment

Sustainability assessment, sustainability appraisal, or sustainability impact assessment, are forms of integrated assessment which combine strategic environmental assessment with a corresponding evaluation of the social and economic impacts of a development policy or plan. The term appraisal is commonly used to denote a less in-depth study than an impact assessment. Integrated assessments can be particularly valuable for policies with specific aims, such as trade policy, where the policy formulation process may not take account of a wide range of impacts in each of the three sustainable development spheres. For policies with broader socio-economic goals, it may be argued that the approach may dilute efforts to strengthen the environmental component of planning (George, 2001), and therefore be less effective in promoting sustainable development than SEA.

Evaluation of programmes and policies

Evaluation or ex-post impact assessment is one of the cornerstones of an evidence-based approach to governance and management. It incorporates appraisal and assessment techniques to evaluate the actual 'outputs and outcomes' in terms of the original objectives behind the programme or policy. For instance the EU's Structural Funds for promoting the development and cohesion of its cities and regions, have a well developed system for evaluation of these very large 5–7 year investment programmes. This set of evaluation procedures also overlaps with the European Commission's own impact assessment procedures which work at the strategic level of policy development (George and Kirkpatrick, 2006). Both approaches are developing new techniques to deal with the challenges of sustainability, beyond the conventional means of economic and environmental assessment (Ravetz, *et al.*, 2004):

● many uncertainties on cause-effect relationships and synergistic effects in complex systems;

- many intangible and qualitative factors, such as 'social capital';
- alternative perspectives on the definition and interpretation of sustainable development.

Protected areas and biodiversity conservation

Conservation of biodiversity within a city's boundaries is generally less important than managing the city's impacts outside itself. Many of these impacts are indirect, related to consumption and waste generation, and are managed through national laws and policies. However, some aspects of a city's development plan may have major direct biodiversity impacts, such as the building or extension of an airport.

One of the principal mechanisms for conserving biological diversity, within the overall arena of land use planning, is the designation of protected areas of natural or semi-natural habitat. Sites can be designated by the relevant authorities at any spatial level, from a local wildlife conservation area at the bottom, to internationally important wetland sites recognized under the Ramsar Convention at the top. In general, the higher the spatial level at which a site is designated, the greater the degree of protection. In the UK, the highest protection is afforded to nationally designated Sites of Special Scientific Interest (SSSIs), but even these are not inviolate.

Laws which define the status of conservation areas generally allow government to override the protection if social or economic considerations are sufficiently important. To cater for this, and at the same time prevent loss of biodiversity, effective measures for habitat restoration or regeneration have to be included in project approvals, and/or in the overall planning process. Use of an ecosystem approach (UNEP/CBD, 2000) in land use planning and associated impact assessments additionally allows for the important contribution to biological diversity that is made by non-protected areas.

9.9 Voluntary codes and practices

Voluntary measures for environmental management include:

- cleaner production technology;

- product labelling (voluntary or mandatory);
- certified sources of supply and ethical standards;
- environmental audit procedures and environmental management systems.

Market incentives

Most of these measures rely on market mechanisms, and in turn on the relationships between economy and ecology which was discussed in Chapter 7. They generally fall into one of three groups:

- economic incentives to develop and use more environmentally efficient technologies;
- consumer demands for a better environment, often associated with increasing affluence (the Environmental Kuznets Curve, as explained in Chapter 3);
- avoidance of penalties and liabilities.

Cleaner production

The first of these market mechanisms applies in particular to raw material consumption, or wastes which are themselves unused raw materials, so that a reduction in waste reduces costs. However, while a reduction in the emission of coal dust from a coal-fired boiler may reduce a company's costs, reducing sulphur emissions may increase them. In such cases, regulatory measures remain necessary. Even then, however, the use of cleaner production methods may still reduce a company's costs, through replacing expensive 'end-of-pipe' solutions by more efficient processes. Appropriate combinations of regulatory measures and frameworks that encourage the adoption of efficient production techniques may be used to stimulate the kinds of market transformation we discussed in Chapter 7.

Among the supportive frameworks which governments and trade associations can introduce to these ends, benchmarking systems allow companies to judge themselves against the standards achieved by others which produce similar products. Because there is a close relationship between environmental efficiency and profitability, and hence a company's success in competion with others in its market, there will always be limits to the extent to which companies are prepared to participate in such approaches. The type of benchmarking undertaken by inspectors within the regulatory framework therefore remains a necessary extension to voluntary schemes.

For similar reasons, voluntary publication of a company's performance for key environmental indicators needs to be supported by mandatory requirements. In general, a company's main incentive for releasing such information is to create a favourable impression for its customers.

Product labelling and certification

The second market mechanism applies for example to forest product certification, associated with voluntary product labelling, where consumer preferences permit a company to charge higher prices for products which cause less environmental damage. However, command and control legislation may still be needed, to avoid false marketing claims.

Environmental Management Systems

The third incentive for use of voluntary measures, the avoidance of penalties, relies on the existence of appropriate, well-enforced command and control legislation. In countries where regulation is strong, one of the principal benefits of an Environmental Management System (EMS) is the avoidance of heavy financial penalties for unintentional breaches of pollution permits.

EMS may also be adopted as a marketing tool, in response to consumer demand. In support of this, companies may require their own suppliers to operate an EMS, certified under the ISO 14000 series of international standards (Welford, 1996). A further benefit of an effective EMS is that it requires cleaner production and good housekeeping measures to be applied in a systematic way, through which environmental damage can be minimized while maximizing profitability.

The ISO 14000 standards can be used by municipalities to manage their own activities. Used in conjunction with appropriate goals, and with management commitment, the standards help improve environmental performance and reduce negative impacts. They provide an objective basis for verifying claims about a local government's environmental performance in its day-to-day operations.

For local governments, the key goals are to be environmentally efficient, and serve as a model for resource-saving and replication. It is important to note that the ISO 14000 series do not themselves specify environmental performance goals. These must be set by the local government or city itself, taking into account the effects it has on the environment, and the views of its stakeholders.

The role of the state

Voluntary codes and practices can have highly significant environmental benefits. However, many types of environmental impact cannot be reduced without cost to the developer (in order to reduce the costs borne by the public). Hence, voluntary approaches can only be an adjunct to regulatory processes and economic instruments, rather than a replacement for them. The principal role of the state in voluntary mechanisms is to establish appropriate frameworks, raise awareness of the commercial benefits, and encourage companies to adopt such measures.

9.10 Conclusions

Between them, command and control instruments, planning instruments, economic instruments and voluntary codes and practices can make a major contribution to achieving the sustainable development of the socio-enviro-economic complex of which a city is part. Difficulties in achieving this goal are associated with shortcomings at each of the spatial levels of this global complex.

At the local level of the city itself, the equity of development for local communities is a key sustainable development principle. Poor public participation and weak democratic processes limit the extent to which this principle is applied in practice.

At the national level, the environmental sustainability of national development is the key issue. The precautionary principle needs to be applied, to identify potentially critical environmental factors that may impede future development, and ensure no net loss. Lack of openness and rigour in decision-making mechanisms limits the extent to which this is achieved. Meanwhile, non-critical environmental factors need to be appropriately balanced against social and economic ones, to ensure no overall net loss of the capital inherited by future generations. Poor integration of environmental, social and economic factors in development decision-making inhibits this.

At the trans-national level, trans-boundary impacts are the main sustainability issue. Within Europe, many of these impacts are handled fairly effectively, through for example the the Large Combustion Plants Directive, which controls transboundary air pollution, and the Espoo convention on EIA in a Transboundary Context. More widely, the ability of environmental management tools to deal with trans-boundary impacts rests largely on bilateral and regional agreements on specific issues, such as control of fisheries and water abstraction. Such agreements are lacking in many regions.

At the global level, impacts such as biodiversity loss and climate change are the prime concern for the sustainability of global development. Again, the ability of environmental management tools to control these effects rests on international agreements for specific impacts. Those made under under Rio's Biodiversity Convention lack sufficient detail to achieve binding and effective commitments. The Kyoto agreement under the Climate Convention applies primarily to high-income countries, has not been ratified by all of them, and is limited in the constraints it imposes. Neither convention fully addresses issues of global equity between high-income countries and low- and middle-income ones, in respect of their past and present contributions to biodiversity loss and climate change. This limits the extent to which binding agreements can be reached, and hence the ability of any environmental management tool to achieve globally sustainable development.

Some of these shortcomings can be and are being addressed by further improvements in environmental management techniques, and strengthening the legal and institutional frameworks within which they operate. Others are characteristic of more general shortcomings in processes of democratic governance, at all levels from the local to the global.

Further information

European Commission Environment   http://ec.europa.eu/environment/index_en.htm

European Environment Agency   http://www.eea.europa.eu/

International Association for Impact Assessment   http://www.iaia.org/

International Institute for Environment and Development   http://www.iied.org/

International Institute for Sustainable Development   http://www.iisd.org/

World Bank Environment   http://go.worldbank.org/B28KB6VQQ0

Further reading

Abaza, H. and Baranzini, A. (eds) *Implementing Sustainable Development: Integrated Assessment and Participatory Decision-making Processes*, Cheltenham: Edward Elgar.

Cooke, W. and Kothari, U. (2001) (eds) *Participation: The New Tyranny?* London: Zed Books.

DETR (2000) *Integrated Pollution Prevention and Control: A Practical Guide*, London: Department of Environment, Transport and the Regions.

DETR (2000a) *Multi-Criteria Analysis: A Manual*, London: Department of Environment, Transport and the Regions.

Farmer, A. (2007) *Handbook of Environmental Protection and Enforcement: Principles and Practice*, London: Earthscan.

Gibson, R.B., Hassan, S., Holtz, S., Tansey, J. and Whitelaw,G. (2005) *Sustainability Assessment Criteria and Processes*, London: Earthscan.

Lee, N. and George, C. (eds) (2000) *Environmental Assessment in Developing and Transitional Countries*, London: John Wiley and Sons.

Newson, M. (ed) (1992a) *Managing the Human Impact on the Natural Environment*, London: Belhaven Press.

OECD (1995) *The Economic Appraisal of Environmental Projects and Policies*, Paris: Organisation for Economic Co-operation and Development.

⑩ Towards sustainable cities and regions

- 10.1 Introduction
- 10.2 Key challenges
- 10.3 Principles for sustainable cities and regions
- 10.4 Half a world away
- 10.5 A future for cities

10.1 Introduction

Although the preceding chapters of this book have presented many varied and different aspects of the 'human urban environment', the most important issues can be summarized in a relatively straightforward question: given that the scientific knowledge and technical means exist to establish sustainable cities, can this knowledge and capability be matched by the political will and social determination to deliver sustainable cities and communities? White (2002) argues that the means have been established that allow for the creation of cities and communities that are sustainable, but also notes, at least in relation to the majority of nations and cities, that 'what we generally lack is the political imagination, political will and personal motivation' (White, 2002, 194) to deliver the necessary changes in terms of approach, technology and practice.

This dichotomy between scientific/technological capability and political/social capacity is often stark, and it reflects the corrosive power of certain of the sources of economic influence that act upon cities, some of which have been discussed in earlier chapters (especially in Chapters 4, 7, and 9). For example, the globalization of economic matters has reduced some cities in the 'South' to the status of low-wage sweatshops serving the almost endless demands of the more affluent cities of the 'North', whilst the cities and nations of the 'North' benefit from the ability to exploit cheap labour, the natural resources present in less-developed nations, and the often weak regulatory environment which prevails in such countries.

In a similar manner, the social justice component of sustainable development can be seen to be treated very differently in the cities of the 'North' and the 'South'. Here the situation is now frequently worse than it was when the Brundtland Report was published 20 years ago (World Commission on Environment and Development, 1987). As was noted earlier (in Chapters 1 and 8), the lack of progress in terms of social justice and inclusion leaves many city dwellers in both the 'North' and, especially, the 'South' in poverty and marginalized from the mainstream of social and political life. However, as has been indicated, there are signs in cities of the 'North' and the 'South' that both the intra-generational and inter-generational goals of the Brundtland Report are now beginning to be taken more seriously and have been acted upon. Curitiba, in the 'South' (see Chapter 4), and the Eldonian community of Liverpool, in the 'North' (see Chapter 1), represent beacons of innovation and inspiration that demonstrate the power of local authorities and local people to plan, develop and manage 'their places'. This 'learning through doing' approach to shaping the future history of cities represents a considerable shift away from the traditional national, top-down approach of the past, which so frequently resulted in things been 'done to' people and their communities, rather than 'with' them. The approach adopted in both these cities took account of the needs and wants of local people, with the result that they now experience a high degree of 'ownership' of the resulting strategies and actions.

The third element of the sustainable development triad – the effective management of the environment and natural resources – also presents a series of difficulties and dilemmas. These difficulties, once again, chiefly reflect a failure of collective will, rather then any innate inability to understand and appreciate what needs to be done. Nowhere is this more evident than in the case of climate change and, especially, the relationship between the burning of fossil fuels and the occurrence and consequences of climate change. The nightmare scenario of continuing pressure on the environment in the form of additional carbon-intensive development, especially through the further unrestricted growth of energy production, car use and air travel, is now fully understood. Responding to these issues is seen to require effective abatement action through enhancing knowledge flows between actors; providing realistic information about the limits to policy, technological and market development; illustrating the political and regulatory risks involved; and outlining the institutional and societal changes needed to deal with the problem (Foxon, 2003). What is even more important than dealing with carbon alone, is the need to explore further the extended consequences of carbon and climate change, including the resulting changes in the wider environment, such as increases in flooding, season change and desertification. In addition, it is essential to consider the resulting effects on economic activities, such as the reduced carrying capacity of agricultural areas, and the implications for social structures, such as increased national and international migration to the cities.

These brief illustrations reinforce lessons and messages that have been evident for many years but which have still to command full support from all sections of the global political, social and economic community. As was discussed in Chapter 9, whilst there are many factors which can determine the successful adoption of change strategies, it is important to recognize the urgent need to learn from each other in order to promote an 'upward spiral' of improved urban management. This mutual learning can be seen as a single unifying task for cities.

This chapter outlines some of the lessons that have been presented earlier in the book. In identifying and presenting these lessons, three guiding principles are used: the importance of demonstrating what is established evidence, what is an emergent issue and what is an appropriate response. As ever, there is no 'one size fits all' solution that can be applied to all cities, and neither is an attempt made to provide such a solution. Indeed, as a number of other authors have observed, such as Hall (2003), the search for a universal approach can be seen as a somewhat pointless exercise akin to the medieval search for the philosophers' stone. Rather than attempting such a search, the remaining sections of this chapter focus on the common challenges which confront cities and city-regions irrespective of their location or stage of development; what Hall (2003, 65) calls a 'portfolio approach' to the planning, development and management of cities. Following the discussion of the challenges, some principles for the establishment of sustainable cities and regions are offered.

10.2 Key challenges

This section outlines a number of key challenges that confront cities in both the 'South' and the 'North'. The intention in presenting these challenges is to identify what needs to be done in order to support and improve the 'human urban environment', rather than attempting to provide a summary of the messages from the earlier chapters of this book.

Managing change

As has been implied in the preceding section, possibly the most significant common challenge confronting cities is the urgent need to convert the established knowledge and understanding related to the problems associated with attempts to create sustainable 'human urban environments', into effective programmes of implementation. In Chapters 1 and 3 evidence and arguments were presented that point to the importance of understanding and promoting managed change and transitions. This challenge is all the more urgent given the increasing pace of unmanaged change and the increasing severity of many of the consequences of such change. Urban population growth is one example. In previous eras of urbanization both the scale and pace of urban population growth were modest, as De Toro (2003, 42) puts it, 'it took Jericho, the oldest city in the world, 7,000 years to change from a small village to a "city" of about 3,000 people strong', whilst today the urbanization growth scenario is staggering with an anticipated 60 per cent of the global population expected to live in cities by 2030.

Because of the presence of these twin drivers of pace and scale, the time available to introduce changes to the ways in which cities are planned, developed and managed is short. Indeed, depending on which of the many forecasts presented in this book you choose to believe, some issues are already beyond the effective control of either existing technologies or current institutions. For example, the managed retreat or abandonment of coastal settlements as a consequence of rising sea level is now a feature of cities in both the 'South' and the 'North' – the only real differences between such cities are the extent to which alternative locations and compensation are offered to the affected population, and the length of time available to implement managed solutions.

In Chapter 9, some of the methods and tools for managing the 'human urban environment' were presented. These methods and tools are, in themselves, proven ways of dealing with many of the issues which confront cities, but, as was admitted, the efficacy and effectiveness of the methods and tools are severely prescribed by the circumstances in which they are used. As was demonstrated, it is not the availability of knowledge or of tools and methods that is the problem, rather the real issue is extent to which governments, the private sector and society as a whole use the available knowledge, tools and methods to promote positive change.

A further consideration with regard to the management of change is the issue of skills and capability. Whilst there is an ever-increasing supply of knowledge about what constitutes good (and bad) practice, and the causes and consequences of such practice, a significant gap exists in terms of the availability of skills and capability. This skills and capability gap is evident in a number of ways: there is an absolute shortage of people with the required skills; there is an inadequate supply of professionals and others who can work across the individual sectors of action (such as housing, planning, education, transport, design, social development, or health) in order to offer a comprehensive, integrated approach to urban management; and inadequate attention is paid to the provision of skills and role models that can be used to establish general community-based capability. This skills, capability and knowledge agenda has generally been ignored in the past, but is now recognized as crucial to the effective management of change in cities of all shapes, sizes and locations (Roberts, 2008).

Economy and employment

Although the above sections may give the appearance of suggesting that the rate and style of economic activity change should be restricted or otherwise controlled, the reality is that this is unlikely to happen in the short-term. It is, for example, difficult to imagine a situation in which the global ownership of economic activities will diminish significantly over the next two decades, or that rapidly rising demand for consumer durables will abate, especially in the cities of the 'South'. Rather than attempting to deal with the problems associated with unsustainable economic growth as a whole, it is more realistic to focus on the redirection of economic drivers and to demonstrate the viability of alternative ways of achieving economic objectives through the utilization of different development pathways and methods. Such an approach, as will be developed later, might involve a move towards local and regional procurement as a way of reducing the high and rising cost of transport, or the promotion of the use of secondary raw materials as a means of cutting dependence upon scarce resources.

Whatever the motivation, and however the particular economic pathway develops in an individual city, the realities of economic development in cities revolve around the twin roles of cities as providers of local support services and as players in the global marketplace. As discussed in

Chapter 7, these roles are chief among the 'layers' of economic activity evident in cities and they change over time in terms of their relationship with the 'human urban environment'. For example, in many cities of the 'North' the decline of manufacturing has brought with it a loss of traditional skilled employment and a consequent search for replacement jobs. This has not been an easy adjustment and has been accompanied by a series of environmental and social repercussions, including the need for the retraining of individuals, the relocation of households and in some extreme cases the virtual abandonment of traditional industrial settlements. In the cities of the 'South' the reverse has often been the case, with the establishment of new manufacturing ventures necessitating the creation of new settlements to house an increasing workforce and, as a consequence, the need for the development of new skills in the urban population. In both cases this process of socio-economic adjustment has had significant consequences and challenges for the environment: in 'Northern' cities the under-use or abandonment of social and economic infrastructure may occur alongside a failure to clean-up derelict brownfield land or polluted watercourses, whilst in the cities of the 'South' the demand for basic resources frequently overwhelms supply, especially with regard to the provision of housing, clean water, sanitation, health and education.

Social and community

In Chapters 1, 4 and 8 the key social issues related to the 'human urban environment' have been outlined and discussed. What emerges from these considerations is a set of interlinked challenges, although they are often treated as separate matters. For example, the social cohesion challenges that are an inevitable accompaniment to the rapid expansion of the urban population through the arrival of migrants are frequently not linked to the desirability of promoting participation and citizen engagement (see section 4.5), nor are they considered as an important aspect of the inculcation of a greater sense of civic 'ownership' and engagement. However, as the lessons of successful practice demonstrate, many solutions to the problems associated with social exclusion and the absence of social cohesion have their roots in the encouragement of citizen participation and engagement in both formal and informal community activities.

Alongside the institutional and political challenges that beset the society of cities in both the 'North' and the 'South', are more specific difficulties that relate to the welfare and life chances of individuals and groups. Irrespective of the location of cities, major dichotomies between social groups are evident in relation to the provision of health, education, welfare, housing and other services. The general indicator used collectively to assess the extent of differences in social provision and opportunity is income, but in many cities, especially in the 'South', this is an inadequate or inappropriate measure of absolute or relative deprivation. For example, as the World Bank (2000, 3) notes, urban poverty defines not only the absence of income or material goods, but has 'a broader meaning of cumulative deprivation, characterized by squalid living conditions; risk to life and health of poor sanitation, air pollution, crime and violence, traffic accidents, and natural disasters; and the breakdown of traditional family and community safety nets'.

It is with regard to the social dimension of the 'human urban environment' that the most severe failures to achieve either inter-generational or intra-generational equity can be observed, and this absence of progress is, in the eyes of some observers, getting worse in some cities (UNCHS, 2001). This failure to close the development gap, with all the associated difficulties that this suggests, is not a new feature of the urban world, but it is one which is likely to become more significant in the future, especially as a new geopolitics of natural resource control develops. As a consequence, it is essential to regard the social dimension of the 'human urban environment' as at least as important as the economic aspect.

Environment and resources

At the same time as the manuscript of this book was completed, the Olympic Games were taking place in Beijing. These games proved to be memorable for far more than the performance of those who participated: the Beijing Games accelerated the introduction of environmental controls in the host region, together with the restriction of car movements on the city's roads. These innovative and brave measures were seen as necessary in order to reduce the level of atmospheric pollution in the city and region; a problem that threatens human health and, ultimately, prosperity in Beijing and in many other major cities in China and elsewhere. This vivid illustration of the fragility of the (un)natural environment of many of the world's major urban centres was heightened by the coincidence of the strong and well-considered actions taken by the Chinese government to reduce pollution in Beijing, with the stark realization that the world's oil reserves are finite and rapidly diminishing.

The Beijing incident and the oil crisis are illustrations of a much deeper malaise that threatens the capacity of the natural environment to continue in the long term to support the world's cities. This deeper malaise takes a variety of forms. First, it reflects the growing divorce of the majority of the urban population from the realities of the natural environment. This was discussed in Chapters 1 and 5, and it provides an important explanation for the lack of awareness and understanding prevalent among many city dwellers that they are contributing massively to the environmental damage that afflicts both urban and rural areas, but which is concentrated particularly heavily in cities. Second, environmental resources of all types, be they stock or flow resources, are infrequently valued at a realistic or true level. The imperfections of the market, as illustrated throughout this book, and especially in Chapter 9, fail to assign the full costs of resource use to those who benefit from their exploitation. This is especially the case in countries in the 'South', such as Nigeria, where resources such as oil are exploited at extraction cost rather than at real cost, which would help prevent pollution and provide meaningful compensation to the host communities (Watts, 2008). Third, resources are increasingly used as weapons, both in an economic and a military sense. This use of resources as weapons implies two things: that resources will be removed from even the imperfect controls exercised by the market and possibly wasted, and that resources will be employed as political bargaining counters irrespective of their inherent or financial value. Fourth, many resources are in the hands of the few, whilst the many depend on the availability of resources for their quality of life or their very survival; the control of water resources illustrates this point. Finally, resources are not valued in any total or constant sense; this can cause the unnecessary squandering of resources that are perceived either as currently plentiful or of little present intrinsic value. This final point is of particular importance in cities, especially in the case of land resources.

Politics, will and engagement

This section of the chapter has examined the main challenges which confront cities in their attempts to follow a sustainable development pathway. As was seen in Chapters 1 and 4, these challenges are unlikely to be met in the absence of appropriate political structures, a genuine cross-sector desire to bring about change, and the full engagement of all stakeholders and residents. If political commitment, civic will and public engagement exist, then it is more likely that the other challenges will be addressed.

The city is the place in which politics was invented. Irrespective of continent or era, cities, with their concentration of people, power and potential, have provided the canvas upon which the future of countries has been painted and, in some cases, have also proved to be the ultimate prizes in the resulting wars. Not surprisingly, the wealth, power and influence concentrated in cities has proved to be the breeding ground of politics of every kind and colour (Reader, 2004).

Through the concentration of political power in cities, it would appear at first sight that cities should be able to resolve many of their problems themselves. However, urbanization has frequently been accompanied by resistance from surrounding rural areas to urban incursions into their territory. This urban–rural contestation of territory is nothing new – such disagreements were reported in ancient Athens and are commonplace today: Socrates is reported to have observed that 'Our healthy state is no longer big enough; its size must be enlarged …. We shall have to cut a slice off our neighbour's territory … and that will lead to war' (Plato, translated by Lee, 1987, 123). Similar debates are frequently held in cities and their surrounding regions in both the 'South' and the 'North': in Jakarta, in Birmingham, in Mexico City, and in New York.

More significantly, the above discussion of the politics of the city-countryside relationship illustrates the importance of managing the challenges which confront cities at an appropriate spatial scale. As was introduced in Chapter 1 and developed elsewhere in this book, cities are no longer small, self-contained urban entities. Few major cities are freestanding places and even fewer can be considered to be fully self-determining in terms of their basic political, economic, social or environmental characteristics and choices. Rather, individual cities now form a wider regional, national and international urban order. The 'conurbation' of the early twentieth century is now simply part of a wider global urban system within which individual cities perform both general urban functions to serve their immediate population and global specialized functions that serve the wider world.

One consequence of the presence of the forces identified above is the need to consider cities in the context of the city-regions of which they represent the core. City-regions, or metropolitan regions as they are more generally known, are 'collections of cities and towns which are spatially, economically, socially and environmentally independent' (Roberts, *et al.*, 1999b, 11). Such areas are meso-scale urban systems which can perform many of the strategic functions that individual cities either cannot perform or find difficult to perform at an acceptable level of efficiency. Metropolitan planning is especially suited to dealing with those issues which are driven by spatial rather than sectoral concerns, such as the overall allocation of land uses, the provision of transport and other social and economic infrastructure and, of particular importance for this book, the management of the environment. Irrespective of the country in which a city is situated, the realities of urban population growth, physical sprawl and the ever-increasing 'ecological footprint' of the city, means that the 'hinterland', 'conurbation', city-region or metropolitan region is often the most appropriate spatial scale for planning and managing the key issues which relate to future of the city.

Another factor constraining the ability of cities to solve their problems, is the continuing unwillingness of some central governments to give city governments the necessary powers and financial resources to tackle the challenges which confront them. In the absence of direct legal powers and fiscal competence, cities either have to rely on the willingness of central governments to enact legislation and grant resources that are sufficient and appropriate to their needs, or they have to attempt to develop alternative ways of creating and delivering policy, what Yuen (2004, 258) describes as 'innovative municipal budgeting and ways of delivering the services'. A common response to the latter challenge is to work in partnership with other public sector actors and with the private and voluntary sectors. However, the partnership model is not without its own particular difficulties, and it is all too often assumed that partnership offers a solution to the failure of the traditional government model. In reality, strong partnership relies on effective elected government for its legitimacy, and only in exceptional circumstances is it possible for a city partnership to function in the absence of strong city government. Thus the presence of the willpower and determination to drive forward city development, and to thereby create a successful 'human urban environment', can be considered to be a product of effective interaction between good government and the other chief actors and stakeholders present in a city.

Mobilizing the wider group of actors and stakeholders is no easy task and, as was discussed in Chapter 4, there are many dimensions to the challenge of engaging the residents and other stakeholders present in a city. The lessons presented in Chapter 4 from Curitiba on how to address environmental issues provide one illustration of the importance of engaging the citizens of a city, whilst other examples of successful practice are evident elsewhere. However, as was established by Arnstein (1969) many years ago, token participation is not enough to ensure either the effective direction and delivery of policy, or the provision of sufficient resources to support delivery. Much more is needed in order to move from tokenism to full engagement. This takes the argument back to issues such as the extent and effectiveness of the powers of city government and the quality of engagement of stakeholders and citizens in general, and through collaborative arrangements, such as partnership. As was stated in Chapter 1, the effective and efficient management of the 'human urban environment' is a task which requires the attention, efforts and resources of everyone.

10.3 Principles for sustainable cities and regions

This section identifies a number of guiding principles that can be used to help to plan, develop and manage cities and city-regions in accord with the objectives of sustainable development. The reader should be aware of the inherent danger of attempting to apply these principles directly to an individual city. Without substantial contextual calibration the principles are of limited technical value and are unlikely to command either political support or the respect of stakeholders and citizens. Equally, copycat or unthinking city strategy 'clones' almost always fail to provide meaningful solutions and, at worst, they can exacerbate the problems which confront a city. For, example, some of the urban policies imported directly into the United Kingdom from the United States of America during the 1980s failed to take root, chiefly because of the very different political, legal and fiscal environments which exist in the two countries.

So, in the context of this book and reflecting the urban problems with which it has been concerned, the principles offered below represent an attempt to provide guidance on the future planning, development and management of cities. These ideas draw upon several sources, including the recent work of Hall (2003), Ruble, *et al.* (1996), Yuen (2004), De Toro (2003), Reader (2004), Roberts (2007) and White (2002). They also reflect some of the enduring realities that have been expounded by authors such as Geddes (1915), Mumford (1938), Jacobs (1961), Glikson (1971) and, of course, the World Commission on Environment and Development (1987). In addition, the principles attempt to 'cut with the grain' of political and other agreements that support the search for better ways of planning and managing the 'human urban environment', such as the Aalborg Charter, Agenda 21 and the Bristol Accord.

The first set of principles is concerned with the component parts of the city and the ways in which these individual elements might be managed to best effect. In offering these components, the reader is reminded of the point discussed in section 1.6 and illustrated at Box 1.5, that is, that it is not sufficient to deliver any or all of these individual components in isolation from each other. Rather, they should be delivered together if a lasting change in urban circumstances is to be secured; this is the 'whole of place – whole of life' approach that has been advocated in a number of chapters in this book.

The second set of principles is associated with the application, management and long-term maintenance of the city. These principles range from matters concerned with the diagnosis of problems and the appropriateness or otherwise of certain technical and professional methods and approaches, to the importance of securing both political and community support for a desired course of action. In one sense this second set of management principles is about the realization of the first set of components, and it can be considered as representing the 'politics and implementation' element of the illustration provided at Box 1.4.

Bringing these two sets of principles together, it is helpful to consider them as a matrix of activities and actions: activities that represent the components, and actions that are aimed at managing the integration of the individual components in comprehensive programmes designed to deliver sustainable places, now and in the future. Box 10.1 illustrates this matrix.

Components

Reflecting the objectives of this book which were stated in the Preface – that the intention is to provide a series of critical perspectives on the 'human urban environment' and to examine the key factors that should be considered in establishing cities as sustainable places – the preceding

Box 10.1

Principles for sustainable cities and regions

Component Activities

	Value resources	Fit housing	Transport and links	Other infrastructure	Economy and employment	Land use	Environmental management	Social cohesion	Governance and community
Vision and strategy									
Partnership									
Citizen engagement									
Spatial framework									
Analysis and assessment									
Programmes and projects									
Skills and knowledge									
Monitoring and evaluation									

Management Actions

chapters have provided assessments of each of the component parts of the city. Although these components do not fully match the elements of the sustainable communities model (Box 1.5), it is, nevertheless, possible to use this model as a general means of structuring the discussion in this section. In Chapter 1 the point was made that the sectoral component parts of a city vary considerably from place to place and, as a consequence, it is difficult to provide a detailed commentary on the particular form and structure of the content and implementation of individual principles. What might be apparent and appropriate in one city is unlikely to be valid elsewhere without considerable modification.

With all the caveats implied in the above observations, it is possible to offer some general principles for the sectoral components of a sustainable city.

- As seen in Chapters 4 and 5, it is essential to place a true value on resources and to ensure, on the other hand, that the city does not sell itself cheap thereby further reducing the value placed on the natural and human resources that it utilizes.
- In Chapter 6 it was argued the city needs to ensure the provision of sufficient fit housing to meet the requirements of its population. The real issue for many cities is how best to achieve this objective given the many resource constraints which they face and the rapid rise in population.
- As also discussed in Chapter 6, it is essential that the city develops and manages an effective internal transport system and a set of linkages to other places. Many cities in both the 'North' and the 'South' are rapidly becoming aware of the inherent vulnerability and high economic, social and environmental costs of dependence on the private car, whilst lacking either the vision or resources to make the necessary investment in public transport.
- In Chapter 5 and elsewhere it was noted that in addition to dealing with the provision of an effective transport system, the city also needs to offer a range of other social and economic infrastructure, including water and drainage, energy and power, health and education services, and other facilities. For many cities of the 'South' this is an immense challenge, but it is one which is also becoming more difficult in cities of the 'North', especially with rising energy costs, water shortages and other supply constraints.
- Chapters 7 and 9 argued that the city should strive to provide a diverse and robust economic structure and match this structure with

an appropriate range of employment opportunities. However, just providing employment opportunities is not enough, cities also need to attend to the need to match local people to the available jobs, and this suggests the need for better education, training and employment access.

- As outlined in Chapters 4, 6 and 8, the city also needs to deal with a number of matters concerned with the availability and condition of land, and the provision of effective land use management. These are not purely physical or environmental considerations, rather they also reflect the consequences of political, legal and economic actions, including the extent to which any particular area of land can be provided with all the necessary services and permissions.

- Chapters 1, 4, 5, and 6 indicated that there are a number of principles of good built and natural environmental management that should be respected and reflected in the planning, development and management of a city. Chief among these are: the need to ensure that a true value is placed on all resources consumed and the cost of controlling or cleaning up contamination and pollution; the desirability of maintaining city space in a manner which respects ecological requirements; the need to ensure that effective environmental assessment procedures and enforcement mechanisms are in place; and the application of the precautionary principle.

- As indicated in Chapters 1, 4 and 8 it is also essential to ensure that cities are safe, socially cohesive and inclusive. These principles are difficult enough to apply in the better managed cities of the 'North', whilst in the cities of the 'South' they frequently remain aspirations rather than easily realizable objectives. Nevertheless these principles are essential elements of sustainable cities and their delivery frequently proves to be the key to creating lasting improvements.

- From the material presented in many of the chapters it is evident that cities need to be governed and managed in an open, accountable and effective manner. Evidence from cities like Curitiba demonstrates the immense value of good political leadership allied to a real sense of community 'ownership' of the policies and actions required to ensure that a city becomes a sustainable place. This has been a constant theme throughout this book, and it is viewed by many observers to be an essential feature of any successful town or city.

Application, management and maintenance

As was noted above, the second group of cross-cutting principles is concerned with the application, management and maintenance of activities across a city, now and over the long-term. One of the key failures of much of previous urban policy and action has been a lack of continuing vigilance, care and maintenance. Cities need to be cared for, and this 'whole of life' approach has been acknowledged in international and national reviews and assessments of urban issues (see for example, Hall, 2003; Ruble, *et al.*, 1996).

Equally, and as argued throughout this book, it is essential that the various component activities are planned, developed, delivered and managed in an integrated manner – this is the 'whole of place' part of the sustainable city package.

Reflecting the discussions presented in the previous chapters and drawing upon the many examples and references referred to in this book, the main management principles and actions are presented below.

- As discussed in Chapters 1, 3, 4 and 9, it is essential to use scenario building and other methods to provide a grounded vision and a clear strategy that can be used to guide the present and future planning, development and management of a city. Even in ideal circumstances and with the full availability of powers, resources and skills, the disconnected delivery of individual component parts of a city will not bring about the desired result. A substantial evidence base exists which demonstrates the considerable added value of strategic thinking and action, especially when it is based on the development and testing of alternative models.
- Chapters 1, 4 and 8 introduced the importance of establishing strong and lasting partnerships which engage and harness the contributions of all of the stakeholders involved in planning and managing a city. In the absence of real partnership it is unlikely that a city will be able to identify, mobilize and direct all of the resources that are required to bring about the delivery of the necessary actions and to ensure lasting change.
- In Chapters 4 and 8 evidence was presented which indicates the importance of involving the citizens of a city in both the development of policy and a range of implementation actions. Evidence from cities of the 'South' and the 'North' highlights the desirability of direct

citizen involvement in order to foster the 'ownership' of policies, practices and places.

- From Chapters 1, 4 and 6 it is evident that as well as dealing with sectoral matters, it is also vital to provide a spatial context and framework for the development and management of individual cities and the wider city-regions in which they are located. As has been discussed, cities are frequently unable to solve the problems which they face within their own boundaries, and, because of the role which they play as regional centres, neither should they be expected to. It is also important to provide a spatial framework for the guidance of the development and management of the city itself; without such a framework it is difficult to provide a coherent basis for improving the physical, social and economic circumstances of a city.

- In most of the chapters of this book reference has been made to the need for excellent analysis, clear understanding and high-quality assessment of the policy options available to a city. The search for a sustainable city – or for a sustainable neighbourhood or city-region – commences with the recognition and acknowledgment of the problems. In the absence of analysis and understanding, policy and action is unlikely to be directed at the real targets and is also unlikely to represent the best use of scarce resources. Equally, it is essential that diagnosis is linked to prognosis; there is hardly ever an automatic direct link between the two, but analysis without action can become a purposeless activity. Providing a link is often best achieved through the development and testing of options for the future of the city or for parts of the city.

- As was discussed in Chapters 4 and 9, irrespective of the chosen strategy or model for the future development and management of a city, it is essential to move beyond vague statements of intent or broad objectives and to develop detailed mechanisms that can provide for effective delivery. Programmes and projects allow for the implementation of specific activities in a logical manner, and they also offer the additional benefit of integrating individual sectoral actions. Put simply, integrated programme management is a proven method for delivering more than the sum of the component parts.

- Chapters 3 and 9 demonstrated the importance of ensuring that all of the actors and stakeholders who are involved in the planning, development and management of a city have access to the necessary skills and knowledge. In cities of both 'North' and 'South' substantial skills and knowledge deficits or gaps are evident. Such deficits can result in either the absence of effective action or the implementation of sub-optimal actions. Cities are now discovering the benefits and merits of better education, training and action learning. The sharing of these new understandings of what are often common problems and solutions offers a means of working smarter. Further skills and knowledge development, coupled with the promotion of skills and knowledge transfer, provides a major resource for cities irrespective of their location or stage of development.

● In Chapters 1, 3, 4 and 9 evidence has been presented which points to the importance of monitoring, evaluating and regularly reviewing the progress of a city. It has also been argued in a number of places in this book that tracking the progress of a city, both overall and in relation to specific aspects of activity, should not rely overly on quantitative measures. The inherent weakness of many standard quantitative measures, such as Gross Domestic Product (GDP) or income per capita, has been demonstrated many times in the past (Roberts, 2006a), whilst many of the attempts to employ comprehensive-rational systems methods of city planning and management have failed to deliver the promised advantages. Nevertheless, it is important to monitor progress and to evaluate performance, and these tasks are best achieved through a mixed portfolio of 'soft' and 'hard' indicators and techniques. Perhaps the most appropriate long-term approach is that suggested by the government of Bhutan, who have proposed the use of a new indicator of progress: the level of Gross National Happiness.

10.4 Half a world away

In bringing together and applying the principles which have been offered in the preceding section, it is also important to appreciate the considerable influence exerted on the evolution of cities by spatial and temporal variation. Much has been made in this book of the differences between the cities of the 'South' and the 'North', but it is also essential to recognize both the major differences which exist between adjacent cities, or even between cities in one country, and the massive variations in quality of life, income and physical condition which are evident within cities. So even in the rich nations of the 'North' it is possible to point to the contrasting fortunes of the 'rustbelt' and 'sunbelt' cities of North America – for example, Detroit and San Francisco – or to the 'rich' and 'poor' cities of Latin America – Rio de Janerio and Belem. Variations within cities are even more prevalent, both in the 'North' and the 'South', with massive contrasts evident in even the most advanced cities.

Although these intra-city variations are often more difficult to measure in quantitative terms, they are the everyday realities of life for city dwellers and they also reflect the underlying dynamics of the city.

This takes the discussion to the issue of temporal variation. Cities change over time, and as many authors have argued, successful cities are the ones that are open and responsive to new ideas, new uses for places and spaces, and the demands of new activities. Monoindustrial, rigid, imposed or tightly-controlled cities are often those that struggle to survive or become trapped in the thinking of the past. But temporal change, like spatial variation, also demonstrates the inherent dynamics of the city. Cities that were successful in the past, but are failing today, may become the leading edge places of the future.

Although all cities experience spatial and temporal variations and changes to a lesser or greater extent, there are also constant elements at work. These constant elements reflect, for example, the particular roles and functions performed by some cities, such as cultural or government activities. Other cities, indeed the majority of cities, are much more exposed to the various processes of transition which have been outlined in earlier chapters.

Irrespective of the vagaries of city evolution, it is evident that the principles outlined in section 10.3 can be used to help to guide, direct and manage the transition of cities and their neighbourhoods. Indeed, these principles can be used to create new communities, to regenerate substandard or failing neighbourhoods, or to support the recovery of places that have been subject to natural or socio-economic disasters and difficulties. The two examples which follow illustrate the point.

The annual World Habitat Awards recognize and disseminate the lessons of successful and sustainable solutions to the problems encountered in cities. In 2004, two Awards were made, one to a project in the 'South', La Paz in El Salvador, and the other – half a world away – in the 'North', the Eldonian Village in Liverpool, UK. As Box 10.2 illustrates, both projects demonstrate what can be achieved under even the most difficult circumstances. In both cases the programmes of reconstruction were guided by a clear vision and strategy, with the full involvement of local people, and were delivered through active partnership and well-managed programmes. In short, even though the cases differ considerably in terms of their heritage, physical form, socio-economic circumstances, political context, and environmental backdrop, they display many common features and offer a range of important lessons (Building and Social Housing Foundation, 2004).

Box 10.2

La Paz and Liverpool

Resources

Feature	La Paz	Liverpool
Challenge	Post-earthquake reconstruction of parts of the city.	Rebuilding community following economic collapse and physical decay.
Style	Municipality-led with extensive citizen participation and broad partnership.	Community-led with high resident engagement and partnership arrangements with public and private agencies.
Activities	Housing design and construction, education and training, business creation, reforestation and pollution management.	Housing provision, creation of social and health facilities, environmental improvement, community businesses established.
Outputs	New and refurbished houses, new jobs and businesses, innovative housing and services, enhanced citizen engagement.	New houses, care and education facilities, community safety and security, brownfield remediation, new jobs and businesses, citizen engagement.
Outcomes	New community skills and confidence, help for vulnerable groups, enhanced business capacity, environmental improvements.	Established community capacity, enhanced quality of neighbourhood, increased citizen confidence, offers a role model for other communities.

10.5 A future for cities

Despite the many difficulties which have confronted them in the past, cities have proved to be highly resilient and there is little sign at present that this quality is on the verge of disappearing. However, the difficulties which now challenge some cities are on an unprecedented scale, and whilst more prosperous or more fortunate cities still display a robust fortitude in the face of mounting adversity, other cities are under considerable strain. Contrast Venice with the coastal cities of Bangladesh. The former city is wealthy, attracts financial and technical support from across the world, and is the subject of a number of pathbreaking technical innovations in the search for a way of reinforcing its coastal defences. The coastal towns and cities of Bangladesh, home to a substantial population, do not share the advantages enjoyed by Venice and some 35 million people are threatened by an imminent rise in sea level. Other contrasts exist in their droves. Former mining settlements, places subject to atmospheric pollution, cities in earthquake zones, urban areas in the hurricane belt, cities subject to drought. Generally, cities in advanced situations are better able to recover, repair or regenerate than those in less favoured situations.

Although the wealthy and advantaged cities have so far been able to avoid many of the difficulties faced by the poorer cities, a new equality of environmental challenge is emerging. Climate change, accompanied by selective and more general resource restrictions, means that even advantaged places now have to take the questions raised in this book more seriously than in the past. It is no longer satisfactory to deal with environmental, social and economic problems through introducing 'end of pipe' solutions. Rather, it is now evident that a more radical approach is necessary, and this will mean placing greater emphasis on precaution, prevention and a new cultural politics of better resource management (Roberts and Gouldson, 2000).

Despite the many challenges which are now evident, it is also the case that cities remain inherently attractive, especially to the rural poor, and for this reason if no other, cities will continue to grow. However, this apparently negative perspective is not without its merits, because cities with their relatively compact settlement form also continue to offer potential scale economies in terms of the use of resources. But such economies will only be realized if the messages contained in this and other similar books are translated into action.

It is now a decade and a half since the publication of *Planning for a Sustainable Environment* (Blowers, 1993) and the warning notes sounded in that book are still not accepted in all cities. However, many of the lessons presented in this book, and especially the principles for successful cities and regions that were offered in section 10.3, reflect both the accepted state of knowledge and understanding in 1993 and the experience of the past fifteen years – a period of time during which the analysis and forecasts of the World Commission on Environment and Development (1987) have become an ever more evident reality. One of the chapters in *Planning for a Sustainable Environment* explored the art and science of place-making and offered a number of important insights into how cities and city-regions can be planned, developed, and managed in order to ensure their future. Chief among these ideas were the need for clear vision, excellent leadership and the extensive engagement of stakeholders and citizens, a comprehensive and integrated programme of implementation, and, above all else, a genuine willingness to change. As John Holliday put it in 1993: 'Land, water, wind, sun, vegetation and minerals: these are the resources that urban populations must learn to use well, in new ways, and without the mind-sets which centuries of a certain kind of urban progress have instilled' (Holliday, 1993, 51). Cities and the people who live in them need to change the ways in which they plan, develop and manage their affairs if they are to survive as the embodiment of the 'human urban environment'.

Further Information

Academy for Sustainable Communities   http://www.ascskills.org.uk/pages/home

United Nations Environment Programme   http://www.unep.org/

Further reading

Girard, L., Forte, B., Cerreta, M., De Torro, P. and Forte, F. (2003) *The Human Sustainable City*, Aldershot: Ashgate.

Gouldson, A. and Roberts, P. (eds) (2000) *Integrating Environment and Economy*, London: Routledge.

Hall, P. (1988) *Cities of Tomorrow*, Oxford: Blackwell.

Reader, J. (2004) *Cities*, London: Heinemann.

Ruble, B., Tulchin, J., Cohen, M. and Garland, A. (1996) *Preparing for the Urban Future*, Washington DC: Johns Hopkins University Press.

White, R. (2002) *Building the Ecological City*, Cambridge: Woodhead Publishing.

Appendices

Appendix 1 Abbreviations

$\mu g\ m^{-3}$	micro gram per cubic metre: $= 10^{-16}\ g/m^3$
ABC	Dutch system of matching land uses to their accessibility profile
ALARA	'as low as reasonable achievable': principle for environmental management
ASC	Academy for Sustainable Communities – UK central government agency
BANANA	'build absolutely nothing anywhere near anyone'
BATNEEC	'best available technology not entailing excessive cost': the application of IPPC
BAU	'business as usual': effect of continuing trends with current policies and commitments
BOD	biochemical oxygen demand: overall indicator of water quality
BPEO	'best practical environmental option': the main principle behind IPPC
BS 7750	British Standard on Environmental Management
CAP	Common Agricultural Policy of the EU: sets subsidies and levies for all food markets
CBA	Cost-Benefit Analysis
CBO	citizen-based organization (voluntary sector)

CDI	City Development Index – aggregated urban health measure
CEC	Commission of the European Community
CFCs	chloro-fluorocarbons: main cause of damage to the ozone layer
CH_4	methane: the 2nd most potent greenhouse gas, mainly from landfills and agriculture
CHP	combined heat and power: local power generation with use of waste heat
CO	carbon monoxide: emitted mainly by diesel engines
CO_2	carbon dioxide: from fossil fuels: responsible for two-thirds of the global warming effect
CSR	Corporate Social Responsibility – sustainability programme in business
DINKY	'double income and no kids' type of household
DPSIR	'driving forces, pressure, state, impact, response': EU indicators system
dw/acre	dwellings per acre: 10 dw/acre = 25 dw/hectare
EEA	European Environment Agency
EIA	Environmental Impact Assessment: at the project level
EMAS	Environmental management and audit system:
ENDS	ENDS Report: environmental journal
EQS	environmental quality standards: statutory pollution limit
ETR	ecological tax reform
EU	European Union
FGD	flue gas desulphurization: removal of SO_2 emissions from fossil fuel combustion
GDP	Gross Domestic Product: annual value of goods and services in the national economy
GHG	'greenhouse gas' – defined as basket of 6 gases by IPCC: CO_2, CH_4, N_2O, CFC, HCFC, SF_6
GIS	geographic information system: computer-based mapping and geographical analysis
GJ	energy unit, 1 gigajoule = 9.5 therms = 278 kWh
GVA	'gross value added' – recent alternative to GDP
GWh	energy unit, 1 gigawatt-hours = 1 million kilowatt-hours (Kwh)
H^+	hydrogen ion (measure of acidity)

ha or hectares	1 hectare area = 10000 m^2 = 2.4 acres = 1/100 of 1 km^2
HFCs	hydro-fluorocarbons: developed as a substitute for CFCs and significant GHGs.
hh	household
HIC	high income countries
HIPC	highly indebted poor countries
IA	integrated assessment – including environmental, social and economic impacts
ICT	information and communication technologies
IPCC	Intergovernmental Panel for the Scientific Assessment of Climate Change
IPPC	'integrated pollution prevention and control': draft EU directive for environmental management
ISEW	'index of sustainable economic welfare': a measure of total benefit to society
ISO 14001	International Standards Organization: standard on environmental management
km^2	1 square kilometre = 100 hectares
kT	1 kilo-tonne = 1000 metric tonnes
LA21	Local Agenda 21: the production of sustainability plans by social consensus
LAC	Latin American countries
LCA	life-cycle analysis: total assessment of impacts for goods and processes
LCP	'least cost planning': method of allocating resources between multiple objectives
LDC	less-developed countries
LETS	local exchange and trading systems: network for non-monetary exchange
LRT	light rapid transit
LZC	low- or zero-carbon development, i.e. with mainly or wholly renewable energy sources.
m^2	1 square metre = 10.76 square feet
MDG	Millennium Development Goals of the UN
MEA	Millenium Ecosystems Assessment of the UNEP
mg m^{-3}	milligrams per cubic metre
MIPS	'material input per unit service': measure of physical impacts of material production
Mt	1 mega-tonne = 1 million metric tonnes
mtC	million tonnes of carbon: 1mtC equivalent to 3.67 million tonnes of CO$_2$

mtoe	energy unit, 1 million tonnes of oil equivalent = 12700 GWh primary energy
MW	power unit, 1 mega-watt = 1000 kW (kilowatt) = 1000 kW per hour (kWh)
N_2O	nitrous oxide, significant GHG
ng m^{-3}	nanogram per cubic metre = 10^{-19} g/m^3
NGO	non-governmental organization
NIMBY	'not in my back yard' attitude of residents to development
NIS	Newly Independent States of the Soviet Union
NO_2	nitrogen dioxide, the most common oxide of nitrogen
NO_x	nitrous oxides incl. nitrogen dioxide and nitrous oxide, the largest pollutant from transport
O_3	tropospheric or low-level ozone
OECD	Organization for Economic Co-operation and Development:
ONS	Office of National Statistics: UK central government agency
p.a.	per annum or per year
p.c.	per capita or per person
pH	chemical measure of acidity
PM_{10} or $PM_{2.5}$	particulate matter less than 10μm or 2.5μm diameter
ppb	parts per billion, by volume: pollution measure
pph	persons per hectare: planning measure for urban density
PV	photo-voltaic panels which generate power from sunlight
R&D	research and development of new technologies
SA	'sustainability appraisal' – method for comprehensive assessment
SD	'sustainable development' shorthand
SEA	Strategic Environmental Assessment of policies and plans
SME	small and medium enterprise (firm with up to 500 employees)
SMR	Standardized Mortality Rate: statistical death rate allowing for age distribution
SO_2	sulphur dioxide, the main component of acid rain
SRI	Socially Responsible Investment – index measuring sustainability of finance.
TCPA	Town and Country Planning Association

therm	energy unit for gas, 1 therm = 29.3 kWh = 0.105 GJ = 100 000 BTU
TTWA	'Travel to Work Area': area with>70% containment in local labour and employment
UNCHS	United Nations Centre for Human Settlements, now titled UN Habitat
UNDP	United Nations Development Programme
UNECE	United Nations Economic Commission for Europe
UNEP	United Nations Environment Programme
USEPA	United States Environmental Protection Agency
VOCs	volatile organic compounds: large class of hydrocarbons
vpd	transport unit, vehicles per day in both directions
WCED	World Commission on Environment and Development: produced Brundtland Report
WHO	World Health Organization: agency of United Nations
WTE	waste-to-energy: incineration with energy recovery
WTO	World Trade Organization – regulates international trade
WWT	waste water treatment plant and infrastructure

Appendix 2 Glossary

Agglomeration	see urban agglomeration.
Brownfield	an area of land or property which has previously been used, may also be polluted, vacant and/or derelict.
City-region	see metropolitan area or region.
Death rate (or crude death rate)	the number of deaths per 1,000 population in a given year.
Deforestation	the loss of trees due to overcutting of forests. One consequence of deforestation is soil erosion, which results in the loss of protective soil cover and the water-holding capacity of the soil.
Desertification	the process of grasslands being converted to desert mainly as a result of deforestation, overgrazing, and erosion due to poor land management.

Externalities   problems or impacts which are caused beyond the boundary of the location or the activity being considered: for instance climate change impacts are an externality of a city's use of fossil fuels.

Food insecurity   a situation that exists when people lack secure access to sufficient amounts of safe and nutritious food for normal growth and development and an active and healthy life. Food insecurity may be chronic, seasonal, or transitory.

Fossil fuel   a group of primary energy sources created from the incomplete biological decomposition of dead organic matter. The fossil fuels include oil, coal and natural gas and account for about 90 per cent of all the energy consumed in the world.

Irrigation   the practice of supplying land with water artificially by means of ditches, pipes or streams.

Less developed countries   less developed countries include all countries in Africa, Asia (excluding Japan), Latin America and the Caribbean, and the regions of Melanesia, Micronesia, and Polynesia.

Mega-city   a city with a population of 10 million or more.

Metropolitan area or region   a large concentration of population, usually an area with 100,000 or more people. The area typically includes an important city with 50,000 or more inhabitants, and the administrative areas bordering the city that are socially and economically integrated with it. Also called a conurbation or **city-region**.

More developed countries   more developed countries include all countries in Europe, North America, Australia, New Zealand, and Japan.

OECD   an international co-ordination agency. Its 30 current members include 20 countries from the EU, plus Turkey, Norway, Iceland, USA, Canada, Mexico, Australia, New Zealand, Japan and South Korea.

Rate of natural increase   the rate at which a population is increasing (or decreasing) in a given year due to a surplus (or deficit) of births over deaths, expressed as a percentage of the base population.

Renewable water   the surface water runoff from local precipitation, the inflow from other regions, and the groundwater recharge that replenishes aquifers.

Subsistence agriculture	farming at a level at which only enough food is produced to meet immediate local needs.
Urban	countries differ in the way they classify population as 'urban' or 'rural'. Typically, a community or settlement with a population of 2,000 or more is considered urban. A listing of country definitions is published annually in the United Nations Demographic Yearbook.
Urban agglomeration	urban agglomerations are areas of 1 million population or more. The concept of agglomeration defines the population contained within the contours of contiguous territory inhabited at urban levels of residential density without regard to administrative boundaries: also called conurbation or **city-region**.
Urbanization	growth in the proportion of a population living in urban areas.

Appendix 3 Millennium Development Goals

The Millennium Development Goals (MDGs) are the world's time-bound and quantified targets for addressing extreme poverty in its many dimensions – income poverty, hunger, disease, lack of adequate shelter and exclusion – while promoting gender equality, education and environmental sustainability. They are also basic human rights, i.e. the rights of each person on the planet to health, education, shelter and security.

The world has made significant progress in achieving many of the MDGs. Between 1990 and 2002 average overall incomes increased by approximately 21 per cent. The number of people in extreme poverty declined by an estimated 130 million. Child mortality rates fell from 103 deaths per 1,000 live births a year to 88. Life expectancy rose from 63 years to nearly 65 years. An additional 8 per cent of people in the developing world received access to water. And an additional 15 per cent acquired access to improved sanitation services.

But progress has been far from uniform across the world, or across the MDGs. There are huge disparities across and within countries. Within countries, poverty is greatest for rural areas, though urban poverty is also extensive, growing, and underreported by traditional indicators.

Sub-Saharan Africa is the epicentre of crisis, with continuing political instability, food insecurity, a rise in extreme poverty, high child and maternal mortality, large numbers of people living in slums, and a widespread shortfall for most of the MDGs. Asia is the continent with the fastest progress, but hundreds of millions of people still remain in extreme poverty, and even its fast-growing countries fail to achieve some of the non-income MDGs. Other parts of the world have mixed records, notably Latin America, the transition economies, the Middle East and North Africa, often with slow or no progress on some of the MDGs, and persistent inequalities undermining progress on others.

Targets 10 and 11 relate most closely to the human urban environment agenda. There is a strong debate as to whether these are fair and equitable (why only 100 million slum dwellers?) and how far they are measurable (what is 'significant improvement'?) (see UN-Habitat, 2005b).

Goals and targets

Goal 1 Eradicate extreme poverty and hunger

- Target 1. Halve, between 1990 and 2015, the proportion of people whose income is less than $1 a day.
- Target 2. Halve, between 1990 and 2015, the proportion of people who suffer from hunger.

Goal 2 Achieve universal primary education

- Target 3. Ensure that, by 2015, children everywhere, boys and girls alike, will be able to complete a full course of primary schooling.

Goal 3 Promote gender equality and empower women

- Target 4. Eliminate gender disparity in primary and secondary education, preferably by 2005, and in all levels of education no later than 2015.

Goal 4 Reduce child mortality

- Target 5. Reduce by two-thirds, between 1990 and 2015, the under-five mortality rate.

Goal 5 Improve maternal health

- Target 6. Reduce by three-quarters, between 1990 and 2015, the maternal mortality ratio.

Goal 6 Combat HIV/AIDS, malaria, and other diseases

- Target 7. Have halted by 2015 and begun to reverse the spread of HIV/AIDS.
- Target 8. Have halted by 2015 and begun to reverse the incidence of malaria and other major diseases.

Goal 7 Ensure environmental sustainability

- Target 9. Integrate the principles of sustainable development into country policies and programs and reverse the loss of environmental resources.
- Target 10. Halve, by 2015, the proportion of people without sustainable access to safe drinking water and basic sanitation.
- Target 11. Have achieved by 2020 a significant improvement in the lives of at least 100 million slum dwellers.

Goal 8 Develop a global partnership for development

- Target 12. Develop further an open, rule-based, predictable, nondiscriminatory trading and financial system (includes a commitment to good governance, development, and poverty reduction, both nationally and internationally).
- Target 13. Address the special needs of the Least Developed Countries (includes tariff- and quota-free access for Least Developed Countries' exports, enhanced program of debt relief for Heavily Indebted Poor Countries [HIPCs] and cancellation of official bilateral debt, and more generous official development assistance for countries committed to poverty reduction).
- Target 14. Address the special needs of landlocked developing countries and small island developing states (through the Program of Action for the Sustainable Development of Small Island Developing States and 22nd General Assembly provisions).
- Target 15. Deal comprehensively with the debt problems of developing countries through national and international measures in order to make debt sustainable in the long term.

Some of the indicators listed below are monitored separately for the least developed countries, Africa, landlocked developing countries and small island developing states.

- Target 16. In cooperation with developing countries, develop and implement strategies for decent and productive work for youth.
- Target 17. In cooperation with pharmaceutical companies, provide access to affordable essential drugs in developing countries.
- Target 18. In cooperation with the private sector, make available the benefits of new technologies, especially information and communications technology.

Source: United Nations, 2002

Appendix 4 City Development Index

What does the CDI measure?

There are a number of concepts relating to cities and urban development that, although complex and multifaceted, are meaningful and desirable to measure. These include: development level, liveability, sustainability, relative disadvantage or poverty, congestion and inclusiveness. These multidimensional ideas cannot be encompassed by a single indicator but require a combination of different indicators – corresponding to different aspects of development or city performance – to form an index. The two most useful urban indices discovered to date have been the city product per person, which is analogous to the GDP at the city level, to give the economic output of the city, and the City Development Index (CDI).

The CDI is defined at the city level and could also be taken as a measure of average well-being and access to urban facilities by individuals. The high statistical significance and usefulness of the index indicates that it is actually measuring something real. It appears that the CDI is actually a measure of depreciated total expenditure over time on human and physical urban services and infrastructure, and it is a proxy for the human and physical capital assets of the city.

There is some support for this idea, in that more expensive services such as water treatment tend to be more heavily weighted. This cannot be confirmed with the present data, but it seems likely that a monetary cost can be associated with lifting the CDI by a percentage point.

The City Development Index was developed as a prototype for Habitat II to rank cities according to their level of development. It is used in this report as a benchmark for comparative display of several of the key indicators from the UNCHS (Habitat) Global Urban Indicators Database. The CDI is, to date, the best single measure of the level of development in cities.

Calculating the CDI

Index	Formula
Infrastructure	25 × Water connections + 25 × Sewerage + 25 × Electricity + 25 × Telephone
Waste	Wastewater treated × 50 + Formal solid waste disposal × 50
Health	(Life expectancy – 25) × 50/60 + (32 – Child mortality) × 50/31.92
Education	Literacy × 25 + Combined enrolment × 25
Product	(log City Product – 4.61) × 100/5.99
City development	(Infrastructure index + Waste index + Education index + Health index + City Product index)/5

CDI components by region

Region	CDI	City Product	sub-indices			
			Infrastructure	Waste	Health	Education
Africa	42.85	49.69	36.17	26.04	50.39	51.96
Arab States	64.55	66.52	69.79	45.87	77.18	63.39
Asia-Pacific	65.35	62.90	67.75	44.40	78.27	73.43
HIC	96.23	90.60	99.21	100.00	94.26	97.10
LAC	66.25	62.93	70.42	39.50	82.71	75.68
Transition countries	78.59	71.62	90.64	55.93	85.80	88.94

CDI components by selected cities

Region	CDI	City Product	Infrastructure	Waste	Health	Education
			sub-indices			
Stockholm	97.4	93.5	99.5	100.0	94.0	99.8
Melbourne	95.5	90.0	99.8	100.0	93.7	94.1
Singapore	94.5	91.6	99.5	100.0	92.7	88.6
Hong Kong	92.0	89.4	99.3	99.0	90.9	81.3
Moscow	89.9	81.0	98.7	86.8	83.8	99.3
Seoul	86.0	65.3	98.4	100.0	88.7	77.7
Rio de Janeiro	79.4	82.3	86.2	62.6	81.9	84.3
Sofia	79.1	70.9	93.7	58.5	86.2	86.3
Hanoi	74.2	59.6	72.0	90.0	80.6	69.0
Havana	71.0	65.0	74.8	50.0	80.7	84.7
Jakarta	69.2	66.2	57.3	46.7	80.2	95.7
Ulaanbaatar	68.4	53.7	59.0	90.0	72.5	66.7
Lahore	61.1	71.1	78.5	50.0	64.9	40.8
Colombo	58.4	46.9	68.6	45.0	86.2	45.3
Bangalore	58.0	51.1	82.7	31.3	76.5	48.5
Dhaka	48.4	55.6	45.3	27.5	64.6	48.7
Vientiane	47.1	44.0	58.0	0.0	62.3	71.3
Accra	46.6	49.4	50.0	0.0	71.4	62.0
Phnom Penh	43.5	40.2	33.0	27.0	47.2	69.9
Port Moresby	39.3	69.0	18.1	10.0	59.1	40.2
Lagos	29.3	42.1	29.5	2.0	44.0	29.1
Niamey	21.7	40.0	22.0	0.0	78.3	14.9

Source: UNCHS, 2001

Appendix 5 Urban environment statistics

Table App 5.1 *The world's ten largest metropolitan areas since 1900 – population (millions)*

1000 AD	Pop.	1900	Pop.	2000	Pop.	2015 (projected)	Pop.
Cordova, Spain	0.45	London	6.5	Tokyo	26.4	Tokyo	26.4
Kaifeng, China	0.4	New York	4.2	Mexico City	18.1	Mumbai	26.1
Istanbul, Turkey	0.3	Paris	3.3	Mumbai	18.1	Lagos	23.2
Angkor, Cambodia	0.2	Berlin	2.7	São Paulo	17.8	Dhaka	21.1
Kyoto, Japan	0.18	Chicago	1.7	New York	16.6	São Paulo	20.4
Cairo, Egypt	0.14	Vienna	1.7	Lagos	13.4	Karachi	19.2
Baghdad, Iraq	0.13	Tokyo	1.5	Los Angeles	13.1	Mexico City	19.2
Nishapur, Iran	0.13	St. Petersburg	1.4	Calcutta	12.9	New York	17.4
Al-Hasa, Saudi	0.11	Manchester	1.4	Shanghai	12.9	Jakarta	17.3
Pata, India	0.1	Philadelphia	1.4	Buenos Aires	12.6	Calcutta	17.3

Source: World Bank 2001b (from O'Meara 1999 and United Nations Population Division 2000)

Table App 5.2 *The world's biggest cities, 1980–2010, millions*

	1980		1990		2000		2010	
1	21.9	Tokyo	25.1	Tokyo	26.4	Tokyo	26.4	Tokyo
2	15.6	New York	16.1	New York	18.1	Mexico City	23.6	Bombay
3	13.9	Mexico City	15.1	Mexico City	18.1	Bombay	20.2	Lagos
4	12.5	São Paulo	15.1	São Paulo	17.8	São Paulo	19.7	São Paulo
5	11.7	Shanghai	13.3	Shanghai	16.6	New York	18.7	Mexico City
6	10	Osaka	12.2	Bombay	13.4	Lagos	18.4	Dhaka
7	9.9	Buenos Aires	11.5	Los Angeles	13.1	Los Angeles	17.2	New York
8	9.5	Los Angeles	11.2	Buenos Aires	12.9	Calcutta	16.6	Karachi
9	9	Calcutta	11	Osaka	12.9	Shanghai	15.6	Calcutta
10	9	Beijing	10.9	Calcutta	12.6	Buenos Aires	15.3	Jakarta
11	8.9	Paris	10.8	Beijing	12.3	Dhaka	15.1	Delhi
12	8.7	Rio de Janeiro	10.5	Seoul	11.8	Karachi	13.9	Los Angeles
13	8.3	Seoul	9.7	Rio de Janeiro	11.7	Delhi	13.9	Metro Manila
14	8.1	Moscow	9.3	Paris	11	Jakarta	13.7	Buenos Aires
15	8.1	Bombay	9	Moscow	11	Osaka	13.7	Shanghai
16	7.7	London	8.8	Tianjin	10.9	Metro Manila	12.7	Cairo

(*continued*)

	1980			1990			2000			2010	
17	7.3	Tianjin	8.6	Cairo		10.8	Beijing		11.8	Istanbul	
18	6.9	Cairo	8.2	Delhi		10.6	Rio de Janeiro		11.5	Beijing	
19	6.8	Chicago	8	Metro Manila		10.6	Cairo		11.5	Rio de Janeiro	
20	6.3	Essen	7.9	Karachi		9.9	Seoul		11	Osaka	
21	6	Jakarta	7.7	Lagos		9.6	Paris		10	Tianjin	
22	6	Metro Manila	7.7	London		9.5	Istanbul		9.9	Seoul	
23	5.6	Delhi	7.7	Jakarta		9.3	Moscow		9.7	Paris	
24	5.3	Milan	6.8	Chicago		9.2	Tianjin		9.4	Hyderabad	
25	5.1	Teheran	6.6	Dhaka		7.6	London		9.4	Moscow	
26	5	Karachi	6.5	Istanbul		7.4	Lima		9	Bangkok	
27	4.7	Bangkok	6.4	Teheran		7.3	Bangkok		8.8	Lima	
28	4.6	Saint Petersburg	6.4	Essen		7.2	Teheran		8.6	Lahore	
29	4.6	Hong Kong	5.9	Bangkok		7	Chicago		8.2	Madras	
30	4.4	Lima	5.8	Lima		6.9	Hong Kong		8.1	Teheran	

Source: UNCHS, 2001 (from World Urbanization Prospects 1999)

Table App 5.3 *World urbanization rates*

	Total population (millions)			Urban population (as % of total)			Total population annual growth rate (%)		Urban population annual growth rate (%)	
	1975	1999	2015	1975	1999	2015	1975–99	1999–2015	1975–99	1999–2015
Least developed countries	327	609	892	14.3	25.4	35.1	2.6	2.4	5.1	3.0
All developing countries	2898	4610	5759	25.9	38.9	47.6	1.9	1.4	3.7	1.8
Transitional countries	354	398	383	57.7	65.9	69.6	0.5	−0.2	1.1	0.1
Developed countries	746	873	928	75	78.7	82.2	0.7	0.4	0.9	0.4
World	3987	5863	7048	37.8	46.5	53.2	1.6	1.2	2.5	1.3

Derived from UNDP, 2004

References

Abaza, H. and Baranzini, A. (eds) (2002) *Implementing Sustainable Development: Integrated Assessment and Participatory Decision-making Processes*, Cheltenham: Edward Elgar.

Abbott, J. (1996) *Sharing the City: Community Participation in Urban Management*, London: Earthscan.

Academy for Sustainable Communities (2007) *Making Places: Creating Sustainable Communities*, Leeds: Academy for Sustainable Communities.

Adams, J. (1996) *Risk*, London: University College Press.

Adeyemi, A.S., Olorunfemi J.F. and Adewoye T.O. (2001) Waste scavenging in third world cities: a case study in Illorin, Nigeria, *Environmentalist*, 21(2): 93–6.

Agyeman, J., Bullard, R.D., Evans, R. (2003) *Just Sustainabilities: Development in an Unequal World*, Boston, MA: MIT Press.

Allen, A. and Dávila, J. (2002) Mind the gap! Bridging the rural-urban divide, *ID21 Insights*, 41(May) (available on http://www.id21.org/insights/insights41/index.html).

Alonso, W. (1971) The economics of urban size, *Papers of the Regional Science Association*, 26(1): 67–83.

Amin, A. (ed.) (1994) *Post-Fordism: A Reader*, Oxford: Blackwell.

Amin, A. and Thrift, N (1995) Globalization, institutional thickness and the local economy, in Healey, P., Cameron, S., Davoudi, S., Graham, S. and Madani-Pour, Alı (eds) *Managing Cities: the New Urban Context*, Chichester: Wiley.

Amin, A. and Thrift, N. (2002) *Cities: Reimagining the Urban*, Cambridge: Polity Press.

Anderson, K., Bows, A. and Upham, P. (2008) *Aviation and Climate Change: Lessons From European Policy*, London and New York: Routledge.

Angel, S., Sheppard, S.C. and Civco, D.L. (2005) *The Dynamics of Global Urban Expansion*, Washington DC: World Bank, Transport and Urban Development Department (available on http://www.citiesalliance.org/publications/homepage-features/feb-06/urban-expansion.html).

Appleyard, D. and Lintell, M. (1972) Environmental quality of city streets: a residents' viewpoint, *Journal of American Institute of Planners*, 38(2): 84–101.

Arend, M., Bruns, A. and McCurry, J.W. (2004) The 2004 global infrastructure report, *Site Selection*, 49 (September): 580–90 (available on www.siteselection.com).

Arnstein, S.R. (1969) A ladder of public participation, *Journal of American Institute of Planners*, 35(4): 216–24.

Athanasiou, T. (1997) *Slow Reckoning: The Ecology of a Dying Planet*, London: Secker and Warburg.

Auty, R.M. (2000) How natural resources affect economic development, *Development Policy Review*, 18(4): 347–64.

Ayres, R. and Simonis, U. (eds) (1997) *Industrial Metabolism: Restructuring for Sustainable Development*, New York: United Nations University Press.

Badcock, B. (2002) *Making Sense of Cities: A Geographical Survey*, London: Arnold; New York: Oxford University Press.

Bai, X. and Imura, H. (2000) A comparative study of urban environment in East Asia: a stage model of urban environmental evolution, *International Review for Environmental Strategies* 1(1): 135–58.

Bailey, P. (1997) IEA: a new methodology for environmental policy, *Environmental Impact Assessment Review*, 17: 221–6.

Banister, D. (2005) *Unsustainable Transport: City Transport in the New Century*, London and New York: Routledge.

Barnett, V. (1998) *Kondratiev and the Dynamics of Economic Development*, London: Macmillan.

Barrett, J., Ravetz, J. and Bond, S. (2006) *Counting Consumption: CO_2 Emissions, Material Flows and Ecological Footprint of the UK by Region and Devolved Country*, Surrey: WWF-UK (available on http://www.ecologicalbudget.org.uk).

Barter, P., Kenworthy, J. and Laube, F. (2002) Lessons for Asia on sustainable transport, in Low, N.P. and Gleeson, B.J. (eds) *Making Urban Transport Sustainable*, Basingstoke: Palgrave (Macmillan).

Bartol, C.R. and Bartol, A.M. (2006) *Current Perspectives in Forensic Psychology and Criminal Justice*, New York: Sage.

Bartone, C., Bernstein, J., Leitmann, J. and Eigen, J. (1994) *Toward Environmental Strategies for Cities: Policy Considerations for Urban Environmental Management in Developing Countries*, Washington DC: World Bank.

Batty, M. (1995) Cities and complexity: implications for modelling sustainability, in Brotchie, M., Batty, M., Blakely, E., Hall, P. and Newton, P. (eds) *Cities in Competition: Productive and Sustainable Cities for the 21st Century*, Melbourne: Longman Australia.

Batty, M. and Longley, P. (1994) *Fractal Cities: A Geometry of Form and Function*, London: Academic Press.

Baue, W. (2007) Investing for sustainability, in Gardner, G. and Prugh, T. (eds) *State of the World 2008: Ideas and Opportunities for Sustainable Economies*, Washington DC: Worldwatch Institute.

Beall, J (ed.) (1996) *A City for All: Valuing Difference and Working with Diversity*, London: Earthscan.

Beatley, T. (2000) *Green Urbanism: Learning from Cities in Europe*, Washington DC: Island Press.

Beck, U. (1995) *Ecological Politics in an Age of Risk*, Bristol: Polity Press.

Bell, D. and Jayne, M. (2006) *Small Cities: Urban Experience Beyond the Metropolis*, London and New York: Routledge.

Benton-Short, L. and Short, J.R. (2007) *Cities and Nature*, London and New York: Routledge.

Bigio, A.G. and Dahiya, B. (2004) *Urban Environment and Infrastructure: Toward Livable Cities*, Washington DC: World Bank.

Birkeland, J. (ed.) (2002) *Design for Sustainability*, London: Earthscan.

Black, M. and Fawcett, B. (2008) *The Last Taboo: Opening The Door On The Global Sanitation Crisis*, London: Earthscan.

Bleischwitz, R. and Proske, A. (2006) Urban waste management, in von Weizsacker, E.U., Young, O.R., Finger, M. with Beisheim, M. (eds) *Limits to Privatization: How to Avoid Too Much of a Good Thing: A Report to the Club of Rome*, London: Earthscan.

Blowers, A. (ed.) (1993) *Planning for a Sustainable Environment*, London: Earthscan.

Boardman, B., Darby, S., Killip, G., Hinnells, M., Jardine, C.N., Palmer, J. and Sinden, G. (2005) *The 40% House*, Oxford: Environmental Change Institute (available on http://www.40percent.org.uk/).

Bookchin, M. (1997) *The Politics of Social Ecology: Libertarian Municipalism*, Montreal: Black Rose Books.

Bontje, M. (2001) Dealing with deconcentration: population deconcentration and planning response in polynucleated urban regions in North-west Europe, *Urban Studies*, 38(4): 769–85.

Borja, J. and Castells, M. (1997) *Local and Global: The Management of Cities in the Information Age*, London: Earthscan.

Bosch, J., van der Meijden, J., Nio, M., Nijenhuis, W. and de Vries, N. (1999) *Eating Brasil*, Rotterdam: MVRDV (Office of Architecture and Urbanism).

Bostoen, K., Kolsky, P., and Hunt, C. (2006) Improving urban water and sanitation services: health, access and boundaries, in Granahan, G. and Marcotullio, P. (eds) *Scaling Urban Environmental Challenges: From Local to Global and Back*, London: Earthscan with the International Institute for Environment and Development.

Bourg, D. and Erkman, S. (2003) *Perspectives on Industrial Ecology*, Sheffield: Greenleaf Publishing.

Braczyck, H., Cooke, P. and Heidenreich, M. (eds) (1998) *Regional Innovation Systems*, London: UCL Press.

Brand, P. with Thomas, M.J. (2005) *Urban Environmentalism: Global Change and the Mediation of Local Conflict*, New York and Oxford: Routledge.

Brand, S. (1994) *How Buildings Learn: What Happens After They're Built*, New York: Viking.

Brandon, P. and Lombardi, P. (2005) *Evaluating Sustainable Development in the Built Environment*, Oxford: Blackwell Publishing.

Brantingham, P.J. and Brantingham, P.L. (1991) *Environmental Criminology*, Prospect Heights, IL: Waveland Press.

Breheny, M. (1992) The contradictions of the compact city: a review, in Breheny, M. (ed.) *Sustainable Development and Urban Form*, London: Pion.

Breheny, M., Gent, T. and Lock, D. (1994) *Alternative Development Patterns: New Settlements*, London: DOE Planning Research Programme.

Breheny, M. and Rookwood, R. (1993) Planning the sustainable city region, in Blowers, A. (ed.) *Planning for a Sustainable Environment*, London: Earthscan.

Brenner, N. and Theodore, P. (2002) *Spaces of Neo-liberalization*, Oxford: Blackwell.

Briffett, C. and Mackie, J. (2002) Environmental assessment in Singapore: an enigma wrapped up in a mystery!, *Impact Assessment and Project Appraisal*, 20(2): 113–25.

Brown, L.R. (1996) *State of the World 1996*, London: Earthscan.

Bruegmann, R. (2008) *Sprawl: A Compact History*, Chicago: University of Chicago Press.

Buck, N., Gordon, I., Harding, A. and Turok, I. (2005) *Rethinking Urban Competitiveness, Cohesion and Governance*, Oxford: Blackwell.

Building and Social Housing Foundation (2004) *World Habitat Awards*, Coalville: Building and Social Housing Foundation.

Bulkeley, H., Betsill, M.M. and Merrill, M. (2003) *Cities and Climate Change: Urban Sustainability and Global Environmental Governance*, Abingdon: Routledge.

Burgess, E. (1925) The growth of the city, in Park, R. and Burgess, E. (eds) *The City*, Chicago, Chicago University Press.

Burwood, S. and Roberts, P. (2002) *Learning from Experience*, London: Office of the Deputy Prime Minister (ODPM) and British Urban Regeneration Association.

Byrne, D.S., Harrisson, S., Keithley, J. and McCarthy, P. (1986) *Housing and Health*, Aldershot: Gower.

Caldwell, J.C., Caldwell, B.K., McDonald, P.F., Schindlmayr, T. (2006) *Demographic Transition Theory*, Dordrecht, The Netherlands: Kluwer.

Calhoun, C. (1995) *Critical Social Theory: Culture, History, and the Challenge of Difference*, Oxford: Blackwell.

Calthorpe, P. (1993) *The Next American Metropolis*, New York: Princeton Architectural Press.

Camagni, R., Capello, R. and Nijkamp, P. (1998) Towards sustainable city policy: an economy-environment technology nexus, *Ecological Economics*, 24(1): 103–18.

Cambridge Econometrics and AEA TEchnology (2003) *The Benefits of Greener Business*, Bristol: Environment Agency.

Capello, R. and Camagni, R. (2000) Beyond optimal city size: an evaluation of alternative urban growth patterns, *Urban Studies*, 37(9): 1479–96.

Cardoso, F.H. (2001) *Charting a New Course: The Politics of Globalization and Social Transformation*, Lanham, MD: Rowman and Littlefield.

Carley, M. and Spapens, P. (1997) *Sharing the World: Sustainable Living and Global Equity in the 21st Century*, London: Earthscan.

Carr-Hill, R. and Lintott, J. (2002) *Consumption, Jobs and the Environment*, London: Macmillan.

Castells, M. (1983) *The City and the Grassroots: A Cross-cultural Theory of Urban Social Movements*, Berkeley: University of California Press.

Castells, M. (1996) *The Rise of the Network Society*, Blackwell: Oxford.

Castells, M. and Hall, P. (1994) *Technopoles of the World: The Making of 21st Century Industrial Complexes*, London: Routledge.

CEC (Commission of the European Communities) (1990) *Green Paper on the Urban Environment*, Brussels: CEC.

CEC (1996) Council directive 96/61/EC concerning integrated pollution prevention and control, *Official Journal L257/28*, Brussels: CEC.

CEC (2003) *European Energy and Transport Trends to 2030*, Luxembourg: CEC.

Cervero, R. (1998) *The Transit Metropolis: A Global Inquiry*, Washington DC: Island Press.

Chambers, N., Simmons, C. and Wackernagel, M. (2000) *Sharing Nature's Interest: Ecological Footprints as an Indicator of Sustainability*, London: Earthscan.

Champion, A. (1999) Urbanization, suburbanization, counterurbanization and re-urbanization, in Paddison, R. and Lever, W. (eds) *Handbook of Urban Studies*, Beverley Hills, CA: Sage.

Chang, Ha-Joon (2002) *Kicking Away the Ladder: Development Strategy in Historical Perspective*, London: Anthem.

Chatterton, P. and Style, S. (2001) Putting sustainable development into practice? The role of local policy partnership networks, *Local Environment*, 6(4): 439–52.

Christaller, W. (1966) *Central Places in Southern Germany*, Englewood Cliffs, NJ: Prentice Hall.

Clayton, A. and Radcliffe, N. (1996) *Sustainability: A Systems Approach*, London: Earthscan.

Cohen, M. (1993) Megacities and the environment, *Finance and Development*, 30: 40–7.

Cohen, M.J. (1996) Risk society and ecological modernization: alternative visions for post-industrial nations, *Futures*, 29(2): 105–19.

Cohen, M.J. (1998) Stock and flow: making better use of metropolitan resources, *Brookings Review*, 16(4): 37–8

Common, M. and Perrings, C. (1992) Towards an ecological economics of sustainability, *Ecological Economics*, 6(1): 7–34

Communities and Local Government (CLG) (2007) *Building a Greener Future: Policy Statement*, London: CLG (available on www.communities.gov.uk/ publications/planningandbuilding/building-a-greener).

Connolly, P. (2006) Fixing environmental agendas in Mexico, in McGranahan, G. and Marcotullio, P. (eds) *Scaling Urban Environmental Challenges: From Local to Global and Back*, London: Earthscan with the International Institute for Environment and Development.

Cook, A. and Ng, M.K. (2001) *Building Sustainable Communities*, Hong Kong: University of Hong Kong.

Cooke, P., Heidenreich, M. and Braczyk, H.-J. (eds) (2004) *Regional Innovation Systems: The Role of Governance in a Globalized World*, 2nd edition, London: Routledge.

Cooke, W. and Kothari, U. (eds) (2001) *Participation: the New Tyranny?*, London: Zed Books.

Cooke, P. and Morgan, K. (1998) *The Associational Economy: Firms, Regions and Innovation*, Oxford: Oxford University Press.

Costanza, R., d'Arge, R., de Groot, R., Farberk, S., Grasso, M., Hannon, B., Limburg, K., Naeem, S., O'Neill, R.V., Paruelo, J., Raskin, R.G., Sutton, P. and van den Belt, M. (1997) The value of the world's ecosystem services and natural capital, *Nature*, 387(15): 253–60.

Curwell, S., March, C. and Venables, R. (eds) (1990) *Buildings and Health: The Rosehaugh Guide*, London: RIBA Publications.

Daily, G. and Ellison, K. (2003) *The New Economy of Nature*, New York: Island Press.

Darke, J. (1996) The man-shaped city, in Booth, C., Darke, J. and Yeandle, S. (eds) *Changing Places: Women's Lives in the City*, London: Paul Chapman Publishing.

Davis, J. and Kelly, M. (1993) *Healthy Cities: Research and Practice*, London: Routledge.

Davis, M. (1998) *The Ecology of Fear: Los Angeles and the Imagination of Disaster*, New York: Metropolitan Press.

Davis, M. (2005) *Planet of Slums*, London: Verso.

Davis, S., Lukomnick, J. and Pitt-Watson, D. (2006) *The New Capitalists: How Citizen Investors are Reshaping the Corporate Agenda*, Boston: Harvard Business School Publishing.

Davison, A. and Barnes, J. (1992) Patterns of air pollution: critical loads and abatement strategies, in Newson, M. (ed.) *Managing the Human Impact on the Natural Environment*, London: Belhaven Press.

de Botton, A. (2006) *The Architecture of Happiness*, London: Pantheon Books.

de Groot, R.S., Wilson, M. and Boumans, R. (2002) A typology for the description, classification and valuation of ecosystem functions, goods and services, *Ecological Economics*, 41(3): 393–408.

de Toro, P. (2003) Introduction to Part I, in Girard, L., Forte, B., Cerreta, M., De Toro, P. and Forte, F. (eds) *The Human Sustainable City*, Aldershot: Ashgate.

Deffeyes, K. (2005) *Beyond Oil: The View from Hubbert's Peak*, New York: Hill and Wang.

DETR (Department of Environment, Transport and the Regions) (2000a) *Multi-criteria Analysis: A Manual*, London: DETR.

DETR (2000b) *Integrated Pollution Prevention and Control: A Practical Guide*, London: DETR.

DFID-UN (UK Department for International Development and UN-Habitat) (2002) *Sustainable Urbanisation: Achieving Agenda 21*, London and Nairobi: United Nations.

Dickinson, R. (1967) *The City Region in Western Europe*, London: Routledge.

Dixon, J.A., Scura, L.F., Carpenter, R.A. and Sherman, P.B. (1994) *Economic Analysis of Environmental Impacts*, London: Earthscan.

Dobbs, L. and Moore, C. (2002) Engaging communities in area-based regeneration: the role of participatory evaluation, *Policy Studies*, 23(1–2): 157–71.

Douglas, I. (1983) *The Urban Environment*, London: Edward Arnold.

Douglass, M. (2000) Mega-urban regions and world city formation: globalisation, the economic crisis and urban policy issues in Pacific Asia, *Urban Studies*, 37(12): 2315–36.

Douthwaite, R. (1995) *Short Circuit*, Hartland: Green Books.

Dreborg, K.H. (1996) Essence of backcasting, *Futures*, 28(9): 813–28.

Drewe, P., Klein, J. and Hulsbergen, E. (2007) *The Challenge of Social Innovation in Urban Revitalization*, Amsterdam: Techne Press.

Dreze, J. and Sen, A. (1989) *Hunger and Public Policy*, Oxford: Clarendon Press.

Droege, P. (ed.) (2008) *Urban Energy Transition: From Fossil Fuels to Renewable Power*, New York: Elsevier.

Duany, A., Plater-Zyberk, E. and Alminana, R. (2003) *The New Civic Art: Elements of Town Planning*, New York: Rizzoli International Publications.

Duany, A., Plater-Zyberk, E. and Speck, J. (2000) *Suburban Nation: The Rise of Sprawl and the Decline of the American Dream*, New York: North Point Press.

Dwivedi, V. and Vajpeyi, D.K. (1995) *Environmental Policies in the Third World*, Westport, CT: Greenwood.

Easterbrook, G. (1995) *A Moment on the Earth: The Coming Age of Environmental Optimism*, New York: Penguin.

Eckman, K. (1996) How NGOs monitor projects for impacts: results of recent research, *Impact Assessment*, 14(September): 241–68.

ECOTEC (1994) Sustainability, employment and growth: the employment impact of environmental policies, Working Paper No. 1, Birmingham: ECOTEC.

Edwards, C. (1992) Industrialisation in South Korea, in Hewitt, T., Johnson, H. and Wield, D. *Industrialisation and Development*, Oxford: Oxford University Press.

Edwards, P. and Tsouros, A. (2006) *Promoting Physical Activity and Active Living in Urban Environments: The Role of Local Governments*, Copenhagen: WHO Regional Office for Europe.

Edwards-Jones, G., Davies, B. and Hussain, S. (2000) *Ecological Economics: An Introduction*, Oxford: Blackwell Science.

Egan, J. (2004) *Skills for Sustainable Communities: The Egan Review*, London: Communities and Local Government.

Ekins, P. (1994) A Four Capital Mode of Wealth Creation, in Ekins, P. and Max-Neef, M. (eds) *Real Life Economics: Understanding Wealth Creation*, London: Routledge.

Ekins, P (1997) The Kuznets Curve for the environment and economic growth: examining the evidence, *Environment and Planning A*, 29: 805–3.

Eldredge, H.W. (1967) *Taming Megalopolis*, New York: Anchor Books.

Elkington, J. (1997) *Cannibals with Forks: The Triple Bottom Line of 21st Century Business*, Oxford: Capstone Publishing.

Ellerman, A.D., Joskow, P.L., Montero, J., Schmalensee, E. and Bailey, M. (2000) *Markets for Clean Air: The US Acid Rain Programme*, Cambridge and New York: Cambridge University Press.

Elson, M., Walker, S., MacDonald, R. and Edge, J. (1993) *The Effectiveness of Green Belts*, London: DOE, HMSO.

Engels, F. (1845) *The Condition of the Working Class in England*, 1958 edition, Oxford: Blackwell.

Etzioni, A. (1996) *Spirit of Community: Rights, Responsibilities and the Communitarian Agenda*, New York: Harper and Row.

European Commission (2008) *Action Plan on Sustainable Production and Consumption and Sustainable Industrial Policy*, *Memo/08/507*, Brussels: European Commission.

European Commission Expert Group on Urban Environment (1998) *Response to the Communication: Towards an Urban Agenda in the European Union*, Luxembourg: Office for Official Publications of the European Communities.

European Environment Agency (1995) *Europe's Environment: The Dobris Assessment*, Copenhagen: European Environment Agency.

European Environment Agency (1996) DPSIR online resource pack, Copenhagen: European Environment Agency (available on http://ia.ew.eea.europa.eu/Resources/slides/dpsir-folder).

European Environment Agency (2000) *Cloudy Crystal Balls: An Assessment of Recent European and Global Scenario Studies and Models*, Environmental Issue Report No. 17, Copenhagen: European Environment Agency.

European Environment Agency (2001) *Participatory Integrated Assessment Methods: An Assessment of their Usefulness to the EEA*, Technical Report 64, Copenhagen: European Environment Agency.

Eurostat (2000) *Economy-wide Material Flow Accounts and Derived Indicators: A Methodological Guide*, Luxembourg: Office for Official Publications of the European Community.

Farmer, A. (2007) *Handbook of Environmental Protection and Enforcement: Principles and Practice*, London: Earthscan.

Field, S. (1990) *Trends in Crime and their Interpretation*, Home Office Research and Planning Unit report, London: HMSO.

Financial Times (2008) Motorists must pay more and drive less, 8 July.

Florida, R. (2002) *The Rise of the Creative Class: And How It's Transforming Work, Leisure, Community and Everyday Life*, New York: Basic Books.

Florida, R. (2004) *The Creative Talent: Cities and the Creative Class*, New York: Routledge.

Forman R.T.T. (1995) *Land Mosaics: The Ecology of Landscapes and Regions*, Cambridge: Cambridge University Press.

Foster, S., Morris, B., Lawrence, A. and Chilton, J. (1999) Groundwater impacts and issues in developing cities: an introductory review, in Chilton, J. (ed) *Groundwater in the Urban Environment: Selected City Profiles*, Rotterdam: Balkema.

Foucault, M. (1991) Governmentality, in Burchell, G., Gordon, C. and Miller, P. (eds) *The Foucault Effect: Studies in Governmentality*, Chicago: Chicago University Press.

Foundation for Environmental Education (2003) *Eco-schools: Contributing to Local Agenda 21*, Lisbon: Foundation for Environmental Education (available on http://www.eco-schools.org).

Foxon, T. (2003) *Inducing Innovation for a Low-carbon Future: Drivers, Barriers and Policies*, London: The Carbon Trust.

Friedmann, J. (1992) *Empowerment: The Politics of Alternative Development*, Oxford: Blackwell.

Friedmann, J. (1995) The world city hypothesis, in Knox, P.L. and Taylor, P.J. (eds) *World Cities in a World System*, Cambridge: Cambridge University Press.

Friedmann, J. (2005) Civil society revisited: travels in Latin America and China, in Keiner, M., Koll-Schretzenmayr, M. and Schmid, W.A. (eds) *Managing Urban Futures: Sustainability and Urban Growth in Developing Countries*, Aldershot: Ashgate.

Fricdmann J. and Weaver, C. (1979) *Territory and Function*, London: Edward Arnold.

Froebel, F., Heinrichs, J. and Kreye, K. (1980) *The New International Division of Labour*, Cambridge: Cambridge University Press.

Fujita, M., Krugman, P. and Venables, A. (1999) *The Spatial Economy: Cities, Regions and International Trade*, Boston MA: MIT Press.

Funtowicz, S.O., Martinez-Alier, J., Munda, G. and Ravetz, J. (2002) Multi-criteria-based environmental policy, in Abaza, H. and Baranzini, A. (eds) *Implementing Sustainable Development: Integrated Assessment and Participatory Decision-making Processes*, Cheltenham: Edward Elgar.

Funtowicz, S.O. and Ravetz, J.R. (1993) Science for the post-normal age, *Futures*, 25(7): 739–55.

Gallie, D. and White, M. (1997) *Employee Commitment and the Skills Revolution*, London: Policy Studies Institute.

Gandy, M. (1994) *The Politics of Urban Waste*, London: Earthscan.

Gandy, M. (2002) Recycling the past: the dilemmas of ecological urbanism, in Hewitt, M. and Hagen, S. (eds) *City Fights: Debates on Urban Sustainability*, London: James and James.

Garreau, J. (1991) *Edge City: Life on the New Frontier*, New York: Doubleday.

Geddes, P. (1915) *Cities in Evolution: An Introduction to the Town Planning Movement and to the Study of Cities*, 1968 edition, London: Benn.

George, C. (2001) Sustainability appraisal for sustainable development: integrating everything from jobs to climate change, *Impact Assessment and Project Appraisal*, 19(2): 95–106.

George, C. (2007) Sustainable development and global governance, *Journal of Environment and Development*, 16(1): 102–25.

George, C. and Kirkpatrick, C. (2006) Methodological issues in the impact assessment of trade policy: experience from the European Commission's Sustainability Impact Assessment (SIA) programme, *Impact Assessment and Project Appraisal*, 24(4): 325–4.

George, C. and Kirkpatrick, C. (eds) (2007) *Impact Assessment and Sustainable Development: European Practice and Experience*, Cheltenham, UK and Northampton, MA: Edward Elgar.

George, C. and Kirkpatrick, C. (2008) Sustainability Impact Assessment of trade agreements: from public dialogue to international governance, *Journal of Environmental Assessment, Policy and Management*, 10(1): 67–89.

Geyer, H.S. (ed.) (2002) *International Handbook of Urban Systems: Studies of Urbanisation and Migration in Advanced and Developing Countries*, Oxford: Blackwell.

Giaoutzi, M. and Nijkamp, P. (eds) (2006) *Tourism and Regional Development: New Pathways*, Cheltenham, UK and Northampton, MA: Edward Elgar.

Gibson, R.B., Hassan, S., Holtz, S., Tansey, J. and Whitelaw, G. (2005) *Sustainability Assessment Criteria and Processes*, London: Earthscan.

Gilbert, R., Stevenson, D., Girardet, H. and Stren, R. (1996) *Making Cities Work*, London: Earthscan.

Gilbertson, T., Holland, N., Semino, S., and Smith, K. (2008) *Paving the Way for Agrofuels: EU Policy, Sustainability Criteria, and Climate Calculations*, Washington DC: Transnational Institute (available on http://www.tni.org).

Girard, L., Forte, B., Cerreta, M., de Toro, P. and Forte, F. (2003) *The Human Sustainable City*, Aldershot: Ashgate.

Girard, L.F. and Nijkamp, P. (eds) (2009) *Cultural Tourism and Sustainable Local Development*, New Directions in Tourism Analysis, Aldershot: Ashgate.

Glasze, G., Webster, C. and Frantz, K. (2005) *Private Cities: Local and Global Perspectives*, Abingdon: Routledge.

Glikson, A. (1971) *The Ecological Basis of Planning*, The Hague: Martinus Nijhoff.

Goldenstein, L. (2007) Economia creativa: nova oportunidade, *Jornal O Estado de S. Paulo*, 27 April.

Gorz, A. (1980) *Ecology as Politics*, Boston: South End Press

Gough, S. and Scott, W. (2008) *Higher Education and Sustainable Development: Paradox and Possibility*, London and New York: Routledge.

Gouldson, A. and Murphy, J. (1998) Economy-innovation-environment: towards an evolutionary framework of the micro-economic response to the environment, in Gouldson, A. and Murphy, J. *Regulatory Realities*, London: Earthscan.

Graham, P. (2004) *Building Ecology: First Principles for a Sustainable Built Environment*, Oxford: Blackwell Publishing.

Graham, S. and Marvin, S. (2001) *Splintering Urbanism: Networked Infrastructures, Technological Mobilities and the Urban Condition*, London: Routledge.

Green, N. (1999) Art and complexity in London's East End, *Complexity*, 4(6): 14–21.

Grogan, P. and Proscio, A. (2000) *Comeback Cities: A Blueprint for Urban Neighborhood Revival*, Boulder, CO: Westview Press.

Grossman, P. (2002) The effects of free trade on development, democracy and environmental protection, *Sociological Enquiry*, 72(1): 131–50.

Gudgin, G. (1996) Prosperity and growth in UK regions, *Local Economy*, 11(1): 7–26.

Gugler, J. (1988) *The Urbanization of the Third World*, Oxford: Oxford University Press.

Guy, S. and Marvin, S. (1999) Understanding sustainable cities: competing urban futures, *European Urban and Regional Studies*, 6(3): 268–75.

Guy, S., Marvin, S. and Moss, T. (eds) (2001) *Urban Infrastructure in Transition: Networks, Buildings, Plans*, London: Earthscan.

Guy, S. and Shove, E. (2000) *A Sociology of Energy, Buildings and the Environment: Constructing Knowledge, Designing Practice*, Abingdon: Routledge.

Gwilliam, M., Bourne, C., Swain, C. and Prat, A. (1999) *Sustainable Renewal of Suburban Areas*, York: York Publishing Services.

Haase, D., Holzkämper, A. and Seppelt, R. (2009) Rethinking urban development: residential vacancy and demolition in Eastern Germany: conceptual model and spatially explicit results, *Urban Studies* (forthcoming).

Habermas, J. (1992) Further reflections on the public sphere, in Calhoun, C. (ed.) *Habermas and the Public Sphere*, Cambridge, MA: MIT Press.

Hajer, M.A. (2003) *The Politics of Environmental Discourse: Ecological Modernization and the Policy Process*, Oxford: Oxford Scholarship Online.

Hall, P. (1988) *Cities of Tomorrow: An Intellectual History of Urban Planning and Design in the Twentieth Century*, Oxford: Blackwell.

Hall, P. (1998) *Cities in Civilization: Culture, Innovation and Urban Order*, London: Weidenfeld and Nicolson.

Hall, P. (2003) The sustainable city in an age of globalization, in Girard, L., Forte, B., Cerreta, M., de Toro, P. and Forte, F. (eds) *The Human Sustainable City*, Aldershot: Ashgate.

Hall, P. and Pain, K. (2006) *The Polycentric Metropolis Learning from Mega-city Regions in Europe*, London: Earthscan.

Hall, P. and Pfeiffer, U. (2000) *Urban Future 21: A Global Agenda for 21st Century Cities*, London: E. and F.N. Spon.

Hall, P. and Ward, C. (1998) *Sociable Cities: The Legacy of Ebenezer Howard*, Chichester: Wiley.

Hamdi, N. (2004) *Small Change: About the Art of Practice and the Limits of Planning in Cities*, London: Earthscan.

Handmer, J. and Dovers, S. (2007) *The Handbook of Disaster and Emergency Policies and Institutions*, London: Earthscan.

Hannerz, U. (1980) *Exploring The City: Inquiries Towards An Urban Anthropology*, New York: Columbia University Press.

Hansen, J., Sato, M., Kharecha, P., Russell, G., Lea, D.W. and Siddall, M. (2007) Climate change and trace gases, *Philosophical Transactions of the Royal Society America*, 365: 1925–54.

Hardin, G. (1968) The tragedy of the commons, *Science*, 162: 1243–8, re-published in Hardin, G. (1969) *Population, Evolution and Birth Control*, San Francisco: W.H. Freeman.

Hardoy, J., Mitlin, D. and Satterthwaite, D. (2001) *Environmental Problems in an Urbanizing World*, London: Earthscan.

Harvey, D. (1985) *The Urbanization of Capital: Studies in the History and Theory of Capitalist Urbanization*, Baltimore: Johns Hopkins University Press.

Harvey, D. (1995) *The Condition of Post-modernity: An Enquiry into the Conditions of Cultural Change*, Oxford: Blackwell.

Harvey, D. (2005) *A Brief History of Neo-liberalism*, Oxford: Oxford University Press.

Haughton, G. (1997) Developing sustainable urban development models, *Cities*, 14(4): 189–95.

Hawken, P., Lovins, A. and Lovins, L.H. (2005) *Natural Capitalism: The Next Industrial Revolution*, London: Earthscan.

Healey, P. (1997a) *Collaborative Planning: Shaping Places in a Fragmented Society*, London: Macmillan.

Healey, P. (1997b) The revival of spatial planning in Europe, in Healey, P., Khakee, A., Motte, A. and Needham, B. (eds) *Making Strategic Spatial Plans*, London: UCL Press.

Healey, P. (2007) *Urban Complexity and Spatial Strategies: Towards a Relational Planning for Our Times*, London: Routledge.

Henderson, H. (1978) *Creating Alternative Futures: The End of Economics*, New York: Berkley Windhover.

Henderson, J.V., Shalizi, Z. and Venables, A.J. (2001) Geography and development, World Bank working paper, Washington DC: World Bank.

Henley Centre (1996) *Planning for Local Change*, 3 vols, London: The Henley Centre for Economic Forecasting.

Hertz, N. (2005) *The Debt Threat: How Debt Is Destroying the Developing World*, New York: Harper Business.

Heynen, N.C., Kaika, M. and Swyngedouw, E. (2006) *In the Nature of Cities: Urban Political Ecology and the Politics of Urban Metabolism*, London: Routledge.

Hildering, A. (2004) *International Law, Sustainable Development and Water Management*, Delft: Eburon Publishers.

Hill, C.P. (1985) *British Economic and Social History 1700–1982*, 5th edition, London: Edward Arnold.

Hill, M. (1968) A goals-achievement matrix for evaluating alternative plans, *Journal of the American Institute of Planners*, 34(1): 19–28.

Hills, P. and Roberts, P. (2001) Political integration, transboundary pollution and sustainability: challenges for environmental policy in the Pearl River Delta region, *Journal of Environmental Planning and Management*, 44(4): 455–73.

Hitchens, D. (1997) Environmental policy and implications for competitiveness in the regions of the EU, *Regional Studies*, 31(8): 813–19.

Hobsbawm, E.J. (1969) *Industry and Empire: From 1750 to the Present Day*, Harmondsworth: Penguin Books.

Holliday, C.O., Schmidheiny, S., Watts, P. (2002) *Walking the Talk: The Business Case for Sustainable Development*, Geneva: World Business Council for Sustainable Development.

Holliday, J. (1993) Ecosystems and natural resources, in Blowers, A. (ed.) *Planning for a Sustainable Environment*, London: Earthscan.

Hough, M. (1984) *City Form and Natural Processes*, London: Routledge.

Howard, E. (1898) *Garden Cities of To-morrow: A Peaceful Path to Real Reform*, reprinted 1985, Eastbourne: Attic Press.

Howes, R., Skea, J. and Whelan, B. (1998) *Clean and Competitive: Motivating Environmental Performance in Industry*, London: Earthscan.

Hoyt, H. (1939) *The Structure and Growth of Residential Neighbourhoods in American Cities*, Washington DC: Federal Housing Administration.

HRW (Human Rights Watch) (1999) *The Price of Oil: Corporate Responsibility and Human Rights Violations in Nigeria's Oil-producing Communities*, New York: HRW.

Hu, H. (2003) Globalization and huge urban projects: the case of Shanghai, in Delft University, *Globalization and Large Urban Projects*, Delft: Delft University Press.

Hudson-Smith, A. (2008) *Digital Geography: Geographic Visualization for Urban Environments*, London: UCL Centre for Advanced Spatial Analysis.

Hutchinson, F. (1996) *Educating Beyond Violent Futures*, London: Routledge.

ICLEI (International Council for Local Environmental Initiatives) (1996) *The Local Agenda 21 Planning Guide: An Introduction to Sustainable*

Development Planning, Toronto: ICLEI (available on: http://www.idrc.ca/ openebooks/448–2/).

ICLEI (2002) *Curitiba: Orienting Urban Planning to Sustainability ICLEI Case Study 77*, Toronto: ICLEI.

Institute for Local Self-Reliance (2005) *Beyond 40 per cent: Record Setting Recycling and Composting Programmes*, Washington DC: Island Press.

International Energy Agency (IEA) (2007) *World Energy Outlook 2007*, Paris: IEA.

International Labour Organization (2002) *Women and Men in the Informal Economy: A Statistical Picture*, Geneva: International Labour Organization.

International Rivers Network (2006) *Spreading the Water Wealth: Making Water Infrastructure Work for the Poor*, IRN Dams, Rivers and People Report. Berkeley, CA: IRN.

IPCC (Intergovernmental Panel on Climate Change) (2007) Assessment Report 4: Summary for Policy Makers, New York: UN Environment Programme.

IPCC Working Group III (2000) *IPCC Special Report on Emissions Scenarios (SRES)*, New York: UN Environment Programme.

Jackson, T. and Michaelis, L. (2003) *Policies for Sustainable Consumption*, London: Sustainable Development Commission (available on www.sd-commission.gov.uk).

Jacobs, J. (1961) *The Death and Life of Great American Cities*, New York: Vintage Books.

Jacobs, J. (1986) *Cities and the Wealth of Nations*, Harmondsworth: Penguin.

Jacobs, M. (1994) The limits to neo-classicism: towards an institutional environmental economics, in Redclift, M. and Benton, E. (eds) *Social Theory and the Global Environment*, London: Routledge.

James P., Ashley, J. and Evans, A. (2000) Ecological networks: connecting environmental, economic and social systems?, *Landscape Research*, 25(3): 345–53.

Jeffrey, B. (1997) Creating participatory structures in local government, *Local Government Policy Making*, 23(4): 25–31.

Jenks, M., Burton, E. and Williams, K. (eds) (1999) *The Compact City: A Sustainable Urban Form?*, London: E. and F.N. Spon.

Jolly, R. (2002) The history of development policy, in Kirkpatrick, C., Clarke, R. and Polidano, C. (eds) *Handbook of Development Policy and Management*, Cheltenham: Edward Elgar.

Jones, A., Williams, L., Lee, N., Coats, D. and Cowling, M. (2006) *Ideopolis: Knowledge City-regions: A Report to the Work Foundation*, London: Work Foundation.

Junde, Liu and Zaide, Peng (1996) Rural-urban transition and urban growth in Shanghai, *Asian Geographer*, 15(1–2): 106–13.

Kaika, M. (2005) *City of Flows: Modernity, Nature, and the City*, New York: Routledge.

Kaku, M. (1998) *Visions: How Science Will Revolutionize the 21st Century*, Oxford: Oxford University Press.

Kasarda, J.D. (2004) Asia's emerging airport cities, *Urban Land Asia*. Washington, D.C.: Urban Land Institute.

Kats, G., Alevantis, L., Berman, A., Mills, E., and Perlman, J. (2003) *The Costs and Financial Benefits of Green Buildings: A Report to California's Sustainable Building Task Force*, report, Washington, DC: United States Green Building Council.

Kay, J.J. and Schneider, E.D. (1994) Complexity and thermodynamics: towards a new ecology, *Futures*, 26(6): 626–47.

Keil, R. (1995) The environmental problematic in world cities, in Knox, P.L. and Taylor, P.J. (eds) (1995) *World Cities in a World System*, Cambridge: Cambridge University Press.

Kenworthy, J. and Townsend, C. (2006) A comparative perspective on urban transport and emerging environmental problems in middle income cities, in McGranahan, G. and Marcotullio, P. (eds) *Scaling Urban Environmental Challenges: From Local to Global and Back*, London: Earthscan with the International Institute for Environment and Development.

King, J. (2007) *The King Review of Low-carbon Cars. Part I: The Potential for CO_2 Reduction*, London: HM Treasury.

Kirkpatrick, C. (2000) Economic valuation of environmental impacts, in Lee, N. and George, C. (eds) *Environmental Assessment in Developing and Transitional Countries*, London: John Wiley and Sons.

Kitchen, T. (1997) *People, Politics, Policies and Plans: The City Planning Process in Contemporary Britain*, London: Paul Chapman Publishing.

Kitzes, J., Galli, A., Bagliani, M., Barrett, J., Dige, G., Ede, S., Erb, K., Giljum, S., Haberl, H., Hails, C., Jolia-Ferrier, L., Jungwirth, S., Lenzen, M., Lewis, K., Loh, J., Marchettini, N., Messinger, H., Milne, K., Moles, R., Monfreda, C., Mora, D., Nakano, K., Pyhälät, A., Rees, W., Simmons, C., Wackernagel, M., Wada, Y., Walsh, C., and Wiedmann, T. (2008) A research agenda for improving national Ecological Footprint accounts, *Ecological Economics* (forthcoming, corrected proof available on: doi:10.1016/j. ecolecon.2008.06.022).

Klein, N. (2001) *No Logo*, London: Flamingo.

Klein, N. (2004) *Fences and Windows: Dispatches From the Front Lines of the Globalization Debate*, London: Picador.

Knox, P.L. and Taylor, P.J. (1995) *World Cities in a World-system*, New York: Cambridge University Press.

Kobus, D. (ed.) (2003) *Practical Guidebook on Strategic Planning in Municipal Waste Management: An Output of the Cities for Change*, Washington DC: World Bank and Bertelsmann Stiftung.

Kok, M., Vermeulen, Walter, Faaij, Andre and de Jager, David (eds) (2002) *Global Warming and Social Innovation: The Challenge of a Climate Neutral Society*, London: Earthscan.

Konig, K.W. (1999) Rainwater in cities: a note on ecology and practice, in: Inoguchi, T., Newman, E. and Paoletto, G. (eds) *Cities and the Environment: New Approaches for Eco-societies*, New York: United Nations Press.

Kostof, S. (1999) *The City Shaped: Urban Patterns and Meanings Through History*, New York: Thames and Hudson.

Krishna, A. (2002) *Active Social Capital: Tracing the Roots of Development and Democracy*, New York: Columbia University Press

Kumar, K. (1978) *Prophesy and Progress*, Harmondsworth: Penguin.

Kumar, S. and Chambers, R. (2002) *Methods for Community Participation*, London: Intermediate Technology Publications.

Kundu, A. (2006) Dynamics of growth and process of degenerated peripheralization in Delhi: an analysis of socio-economic segmentation and differentiation in the micro environment, in McGranahan, G. and Marcotullio, P. (eds) *Scaling Urban Environmental Challenges: From Local to Global and Back*, London: Earthscan.

Kunstler, J.H. (2006) *The Long Emergency: Surviving the Converging Catastrophes of the Twenty-first Century*, New York: Atlantic Monthly Press.

Kurtz, L. (ed.) (1999) *Encyclopaedia of Violence, Peace and Conflict*, San Diego: Academic Press.

Kuznets, S. (1955) Economic growth and income inequality, *American Economic Review* 45(1): 1–28.

Lacquian, A. (2006) *The Planning and Governance of Asia's Mega-urban Regions*, Vancouver: University of British Columbia.

Landry, C. (2006) *The Art of City Making*, London: Earthscan.

Lash, J. and Urry, J. (1993) *Economies of Signs and Space*, London: Sage.

Lawn, P. (2008) *Employment and the Environment: A Reconciliation*, London and New York: Routledge.

Leach, R. and Barnett, K. (1998) New public management and the local government review, *Local Government Studies*, 23(1): 37–54.

Lee, N. and George, C. (eds) (2000) *Environmental Assessment in Developing and Transitional Countries*, London: John Wiley and Sons.

Lee, Y.F. (2006) Motorization in rapidly developing cities, in McGranahan, G. and Marcotullio, P. (eds) *Scaling Urban Environmental Challenges: From Local to Global and Back*, London: Earthscan, with the International Institute for Environment and Development.

Lefebvre, H. (1991) *The Production of Space*, first published in French in 1973, Oxford: Blackwell.

Leitmann, J. (1999) Integrating the environment in urban development: Singapore as a model of good practice, in *Sustaining Cities: Environmental Planning and Management in Urban Design*, New York: McGraw-Hill.

Lichfield, N. (1996) *Community Impact Evaluation*, London: UCL Press.

Liddle, B. (2001) Free trade and the environment-development system, *Ecological Economics*, 39(1): 21–36.

Ling, A. (2007) *Introducing Goldman Sachs SUSTAIN*, New York: Goldman Sachs Investment Research.

Lipietz, A. (1995) *Green Hopes: The Future of Political Ecology*, Cambridge: Polity Press.

Logan, J. and Molotch, H. (1987) *Urban Fortunes: The Political Economy of Place*, Los Angeles: University of California Press.

Lohman, L. (ed.) (2006) *Carbon Trading: A Critical Conversation on Climate Change, Privatization and Power*, Stockholm: Dag Hammarskjöld Centre.

Lomborg, B. (2001) *The Skeptical Environmentalist*, Cambridge: Cambridge University Press.

Lowe, R. (2007) Technical options and strategies for decarbonizing UK housing, *Building Research and Information*, 35(4): 399–411.

Lu, S. (2001) Environmental management in the People's Republic of China, unpubished masters dissertation, Institute for Development Policy and Management, Manchester: University of Manchester.

Lundqvist, L.J. and Biel, A. (eds) (2007) *From Kyoto to the Town Hall: Making International and National Climate Policy Work at the Local Level*, London: Earthscan.

Lupton, D. (1999) *Risk*, Abingdon: Routledge.

Lynas, M. (2006) *Six Degrees: Our Future on a Hotter Planet*, New York: Fourth Estate, HarperCollins.

Manahan, M.A., Yamamoto, N. and Hoedeman, O. (eds) (2007) *Water Democracy: Reclaiming Public Water in Asia*, report to the Reclaiming Public Water Network, Washington DC: Transnational Institute (available on http://www.tni.org).

Mander, Ü., Wiggering, H. and Helming, K. (eds) (2007) *Multifunctional Land Use: Meeting Future Demands for Landscape Goods and Services*, Berlin: Springer.

Mara, D. (2003) *Domestic Wastewater Treatment in Developing Countries*, London: Earthscan.

Marcotullio, P.J. (2001) Asian urban sustainability in the era of globalization, *Habitat International*, 25(4): 577–98.

Marshall, R. and Macfarlane, R. (2000) *The Intermediate Labour Market: A Tool for Tackling Long-term Unemployment*, York: Joseph Rowntree Foundation.

Martin, J. (2006) *The Meaning of the 21st Century: A Vital Blueprint for Ensuring Our Future*, New York: Penguin.

Mayer, M. (1995) Urban governance in the post-Fordist city, in Healey, P., *et al.* (eds) *Managing Cities: The New Urban Context*, London: Wiley.

Mayo, E. (1997) *Community Banking: A Review of the International Policy and Practice in Social Lending*, London: New Economics Foundation.

Mbiba, B. (2002) The primacy of land conflicts, *ID21 Insights*, 41(May) (available on http://www.id21.org/insights/insights41/insights-issu01-art11.html).

McBane, J. (2008) *The Rebirth of Liverpool*, Liverpool: Liverpool University Press.

McCully, P. (2006) *Spreading the Water Wealth: Making Water Infrastructure Work for the Poor*, Berkeley, CA: International Rivers (available on http://www.internationalrivers.org).

McDonough, W. and Braungart, M. (2002) *Cradle to Cradle: Remaking the Way We Make Things*, New York: North Point Press.

McGranahan, G. (2006) Urban transitions and the spatial displacement of environmental burdens, in McGranahan, G. and Marcotullio, P. (eds) *Scaling Urban Environmental Challenges: From Local to Global and Back*, London: Earthscan with the International Institute for Environment and Development.

McGranahan, G. and Marcotullio, P. (2006) (eds) *Scaling Urban Environmental Challenges: From Local to Global and Back*, London: Earthscan with the International Institute for Environment and Development.

McGranahan, G. and Murray, F. (2003) *Air Pollution and Health in Rapidly Developing Countries*, London: Earthscan.

McGranahan, G., Satterthwaite, D. and Tacoli, C. (2004) Rural–urban change, boundary problems and environmental burdens, IIED Working Paper 10, London: IIED.

McGranahan, G., Songore, J. and Kjellen, M. (1996) Sustainability, poverty and urban environmental transitions, in Pugh, C. (ed.) *Sustainability, Environment and Urbanization*, London: Earthscan.

McKinsey Global Institute (2006) *Mapping the Global Capital Market 2006*, San Francisco: McKinsey Global Institute.

McKinsey Global Institute (2007) *Curbing Global Energy Demand Growth: The Energy Productivity Opportunity*, San Francisco: McKinsey Global Institute.

McKinsey Global Institute (2008a) *Preparing for China's Urban Billion*, San Francisco: McKinsey Global Institute.

McKinsey Global Institute (2008b) *The Carbon Productivity Challenge: Curbing Climate Change and Sustaining Economic Growth*, San Francisco: McKinsey Global Institute.

McMichael, A.J. (2000) The urban environment and health in a world of increasing globalisation: issues for developing countries, *Bulletin of the World Health Organization*, 78(9): 1117–26.

Meadows, D. (1995) The city of first priorities, *Whole Earth Review*, Spring.

Meadows, D., Meadows, D. and Randers, J. (1992) *Beyond the Limits: Confronting Global Collapse and Envisioning a Sustainable Future*, Post Mills, VT: Chelsea Green.

Mega, V. (2000) Cities inventing the civilisation of sustainability: an odyssey in the urban archipelago of the European Union, *Cities*, 17(3): 227–36.

Millennium Ecosystem Assessment (2005a) *Living Beyond Our Means: Natural Assets and Human Wellbeing: A Statement from the Board*, New York: UNEP (available on www.millenniumassessment.org).

Millennium Ecosystem Assessment (2005b) Urban systems, chapter 27 in *Ecosystems and Human Well-being: Current State and Trends: Findings of the Condition and Trends Working Group*, Millennium Ecosystem Assessment Series Vol. 1, Washington DC: Island Press (available on www. milleniumassessment.org).

Millennium Ecosystem Assessment (2005c) *Ecosystems and Human Well-being: A Framework for Assessment*, Washington DC: Island Press (available on www.milleniumassessment.org).

Mitchell, W.J. (1996) *City of Bits: Space, Place, and the Infobahn*, Boston: MIT Press.

Mitlin, D. (2001) Housing and urban poverty: a consideration of the criteria of affordability, diversity and inclusion, *Housing Studies*, 16(4): 509–22.

Monbiot, G. (2005) *The Age of Consent: A Manifesto for a New World Order*, New York: HarperCollins.

Morgan, K. (1997) The learning region: institutions, innovation and regional renewal, *Regional Studies*, 31(5): 491–504.

Moser, C. (2005) City violence and the poor, *In Focus: Poverty and the City*, August, Brasilia: UNDP and International Poverty Centre, pp. 10–12.

Moser, C. and Dani, A.A. (eds) (2008) *Assets, Livelihoods, and Social Policy*, Washington DC: World Bank.

Moulaert, F., Swyngedouw, E. and Rodriguez, A. (2003) *The Globalized City: Economic Restructuring and Social Polarization in European Cities*, Oxford: Oxford University Press.

Mumford, L. (1938) *The Culture of Cities*, London: Secker and Warburg.

Mumford, L. (1961) *The City in History: Its Origins, Its Transformations, and Its Prospects*, Orlando: Harcourt.

Murray, F. and McGranahan, G. (2003) Air pollution and health in developing countries: the context, in McGranahan, G., and Murray, F. (eds) *Air Pollution and Health in Rapidly Developing Countries*, London: Earthscan.

Murray, R. (2002) *Zero Waste*, London: Greenpeace Environmental Trust.

Najam, A. and Robins, N. (2001) Seizing the future: the South, sustainable development and international trade, *International Affairs*, 77(1): 49–68.

Neuwirth, Robert (2005) *Shadow Cities: A Billion Squatters, A New Urban World*, London and New York: Routledge.

Newman, O. (1981) *Community of Interest*, New York: Anchor/Doubleday.

Newman, P. and Kenworthy, J. (1999) *Sustainability and Cities: Overcoming Automobile Dependence*, Washington DC: Island Press.

Newson, M. (1992a) The geography of pollution, in Newson, M. (ed.) *Managing the Human Impact on the Natural Environment*, London: Belhaven Press.

Newson, M, (1992b) Water and sustainable development: the 'turn-around decade', *Journal of Environmental Planning and Management*, 35(2): 175–84.

Nicholson-Lord, D. (1987) *The Greening of the Cities*, London: Routledge.

Niemczynowicz, J. (1996) Megacities from a water perspective, *Water International*, 21(4): 198–205.

Nijkamp, P. and Perrels, A. (1994) *Sustainable Cities in Europe*, London: Earthscan.

ODPM (Office of the Deputy Prime Minister) (2005) *The Bristol Accord*, London: ODPM.

ODPM (2006) *UK Presidency EU Ministerial Informal on Sustainable Communities*, London: ODPM (available on http://www.communities. gov.uk/documents/citiesandregions/pdf/143108.pdf).

OECD (Organisation for Economic Co-operation and Development) (1995) *The Economic Appraisal of Environmental Projects and Policies*, Paris: Organisation for Economic Co-operation and Development.

OECD (1997) *Sustainable Development: OECD Policy Approaches for the 21st Century*, Paris: OECD.

OECD (2008) *OECD Environmental Outlook to 2030*, Paris: OECD.

Offe, C. and Heinze, R.G. (1992) *Beyond Employment: Time, Work and the Informal Economy*, translated by Braley, A., Cambridge: Polity Press.

O'Riordan, T. (1996) Eco-taxation and the sustainability transition, in O'Riordan, T. (ed.) *Ecotaxation*, London: Earthscan.

Ormerod, P. (1994) *The Death of Economics*, London: Faber.

Orr, D.W. (1992) *Ecological Literacy: Education and the Transition to a Post Modern World*, Albany, NY: State of New York Press.

Ortúzar, J.D. and Willumsen, L.G. (2001) *Modelling Transport*, 3rd edition, Oxford: Wiley.

Ott, H.E. and Sachs, W. (2000) *Ethical Aspects of Emissions Trading*, Wuppertal Papers No. 110, Wuppertal: Wuppertal Institute.

Paavola, J. and Adger, W.N. (2005) Institutional ecological economics, *Ecological Economics*, 53(3): 353–68.

Pacione, M. (2001) *Urban Geography: A Global Perspective*, London and New York: Routledge.

Painter, J. (1995) *Politics, Geography and 'Political Geography': A Critical Perspective*, London: Arnold.

Payne, G. and Majale, M. (2004) *The Urban Housing Manual: Making Regulatory Frameworks Work for the Poor*, London: Earthscan.

Pearce, D. (1996) *The True Costs of Road Transport (Blueprint 5)*, London: Earthscan.

Pearce, D. and Barbier, E.B. (2000) *Blueprint for a Sustainable Economy*, London: Earthscan.

Pearce, D. and Moran, D. (1994) *The Economic Value of Biodiversity*, London: Earthscan.

Peck, J. and Ward, K. (eds) (2002) *City of Revolution: Restructuring Manchester*, Manchester: Manchester University Press.

Pellow, D. and Brulle, R.J. (2005) *Power, Justice, and the Environment: A Critical Appraisal of the Environmental Justice Movement*, Cambridge, MA: MIT Press.

Pigou, A.C. (1952) *The Economics of Welfare*, London: Macmillan.

Plato (1955) *The Republic*, translated by Lee, H.D.P., Harmondsworth: Penguin.

Ponting, C. (1992) *A Green History of the World*, Harmondsworth: Penguin.

Porter, M. (1990) *The Competitive Advantage of Nations*, New York, Simon and Schuster.

Portugali, J. (2000) *Self-organization and the City*, Berlin and London: Springer.

Poyner, B. and Webb, B. (1991) *Crime Free Housing*, Oxford: Butterworth Heinemann.

Pucher, J., Korattyswaropam, N., Mittal, N. and Itteyerah, N. (2005) Urban transport crisis in India, *Transport Policy*, 12(3): 185–98.

Putnam, R. (2000) *Bowling Alone: The Collapse and Revival of American Community*, New York: Simon and Schuster.

Rakodi, C., Nunan, F., Mitlin, D., Grant, U., Beall, J., Devas, N. and Amis, P. (2004) *Urban Governance, Voice and Poverty in the Developing World*, London: Earthscan.

Randles, S. and Green, K. (eds) (2006) *Industrial Ecology and Spaces of Innovation*, Aldershot: Ashgate.

Rapaport, A. (1977) *Human Aspects of Urban Form: Towards a Man-environment Approach to Urban Form and Design*, Oxford: Pergamon.

Rasid, H. (1996) Impact assessments from survey of floodplain residents: the case of the Dhaka-Narayangan-Demra (DND) project, Bangladesh, *Impact Assessment*, 14(2) 115–32.

Raskin, P. Banuri, T., Gallopín, G., Gutman, P., Hammond, P., Kates, R.W. and Swart, R. (2003) *Great Transitions: The Promise and Lure of the Times Ahead*, Boston: Tellus Institute (available on http://www.tellus.org).

Ravetz, J. (1995) *Feasibility Studies for Community Projects: A Guide to Good Practice*, for the RIBA Community Architecture Group, London: RIBA Publications.

Ravetz, J. (1999a) Citizen participation for integrated assessment: new pathways in complex systems, *International Journal of Environment and Pollution*, 11(3): 331–50.

Ravetz, J. (1999b) Economy, environment and the sustainable city: notes from Greater Manchester, in Roberts, P. and Gouldson, A. (eds) *Integrating Environment and Economy: Local and Regional Strategies*, London: Routledge.

Ravetz, J. (1999c) Urban form and the sustainability of urban systems: theory and practice in a northern conurbation, in Jenks, M., Burton, E. and Williams, K. (eds) *Achieving Sustainable Urban Form*, London: E. and F. Spon.

Ravetz, J. (2000) *City-region 2020: Integrated Planning for a Sustainable Environment*, with a foreword by the UK Secretary of State for the Environment, London: Earthscan (Chinese language version, translated by Jian-Cheng Lin and Tian-Tian Hu, Taipei: Chan's Publishing).

Ravetz, J. (2006a) Environment in transition in an industrial city-region: analysis and experience, in McGranahan, G. and Marcotullio, P. (eds) *Scaling Urban*

Environmental Challenges: From Local to Global and Back, London: Earthscan with the International Institute for Environment and Development.

Ravetz, J. (2006b) Regional innovation and resource productivity: new approaches to analysis and communication, in Randles, S. and Green, K. (eds) *Industrial Ecology and Spaces of Innovation*, Aldershot: Ashgate.

Ravetz, J. (2008a) Resource flow analysis for sustainable construction: metrics for an integrated supply chain approach, *Proceedings of the Institute of Civil Engineers: Waste and Resource Management*, 161(WR2): 51–66.

Ravetz, J. (2008b) Integration and inter-dependency in governance and markets: a spatial ecology approach, CURE Working Paper 2008/1 (available on http://www.manchester.ac.uk/cure).

Ravetz, J. (2009) *Pathways Towards a One Planet Economy: A Prospectus*, Godalming, Surrey: WWF-UK (available on http://www.ecologicalbudget.org.uk).

Ravetz, A. with Turkington, R. (1995) *The Place of Home: English Domestic Environments 1914–2000*, London: E. and F. Spon.

Ravetz, J., Coccossis, H., Schleicher-Tappeser, R. and Steele, P. (2004) Evaluation of regional sustainable development: transitions and prospects, *Journal of Environmental Assessment Planning and Management*, 6(4): 585–619.

Reader, J. (2004) *Cities*, London: Heinemann.

Redclift, M.R. (ed.) (2005) *New Developments in Environmental Sociology*, Cheltenham: Edward Elgar.

Redman, J. (2008) *World Bank: Climate Profiteer*, Sustainable Energy and Economy Network report to the Institute of Policy Studies, Washington DC (available on http:www.ips-dc.org/reports/#292).

Reed, D. (1996) *Structural Adjustment, the Environment and Sustainable Development*, London: Earthscan.

Rees, W. and Wackernagel, M. (1995) *Our Ecological Footprint: Reducing Human Impact on the Earth*, Gabriola Island, British Columbia: New Society Publishers.

Rifkin, J. (1995) *The End of Work: The Decline of the Global Labor Force and the Dawn of the Post-market Era*, New York: Putnam.

Ringland, G. (2002) *Scenarios in Public Policy*, Chichester: Wiley.

Roaf, S. (2004) *Closing the Loop: Benchmarks for Sustainable Buildings*, London: Construction Books.

Roberts, P. (1995) *Environmentally Sustainable Business: A Local and Regional Perspective*, London: Paul Chapman Publishing.

Roberts, P. (2000) *The New Territorial Governance*, London: Town and Country Planning Association.

Roberts, P. (2001) Social justice through spatial policy, *Regional Contact*, 15: 59–71.

Roberts, P. (2003) Economic restructuring, regional development and the environment: ecological modernisation and the European Union's Structural

Funds, *International Journal of Environment and Sustainable Development*, 2(3): 267–83.

Roberts, P. (2006a) Evaluating regional sustainable development: approaches, methods and the politics of analysis, *Journal of Environmental Planning and Management*, 49(4): 515–32.

Roberts, P. (2006b) Wealth from waste: local and regional economic development and the environment, *Geographical Journal*, 170(2): 126–35.

Roberts, P. (2007) Social innovation, spatial transformation and sustainable communities: Liverpool and the Eldonians, in Drewe, P., Klein, J. and Hulsbergen, E. (eds) *The Challenge of Social Innovation in Urban Revitalization*, Amsterdam: Techne Press.

Roberts, P. (2008) Sustainable communities: policy, practice and professional development: a model for Europe, in Cooper, I. and Symes, M. (eds) *Sustainable Urban Development: Changing Professional Practice*, London: Routledge.

Roberts, P. and Colwell, A. (2007) Moving the environment to centre stage: a new approach to planning and development at European and regional levels, *Local Environment*, 6(4): 421–37.

Roberts, P. and Gouldson, A. (2000) Retrospect and prospect, in Gouldson, A. and Roberts, P. (eds) *Integrating Environment and Economy*, London: Routledge.

Roberts, P., Jackson, T. and Lloyd, G. (1999a) *Environmental Taxation*, London: Local Government Association.

Roberts, P. and Sykes, H. (eds) (2000) *Urban Regeneration: A Handbook*, London: Sage.

Roberts, P., Thomas, K. and Williams, G. (1999b) *Metropolitan Planning in Britain*, London: Jessica Kingsley.

Rothman, D.S., Robinson, J.B. and Biggs, D. (2002) Signs of life: linking indicators and models in the context of QUEST, in Abaza, H. and Baranzini, A. (eds) *Implementing Sustainable Development: Integrated Assessment and Participatory Decision-making Processes*, Cheltenham: Edward Elgar.

Rotmans, J. and van Assaelt, M. (1996) Integrated assessment: a growing child on its way to maturity, *Climatic Change*, 34(3–4): 327–6.

Roy, R. and Caird, S. (2001) Household ecological footprints: moving towards sustainability, *Town and Country Planning*, 70(10): 277–79.

Royal Commission on Environmental Pollution (RCEP) (1994) *18th Report: Transport and the Environment*, London: HMSO.

Roszak, T. (1993) *Eco-psychology*, British Columbia: Island Press.

Ruble, B., Tulchin, J. and Garland, A. (1996) Introduction: globalism and local realities, in Ruble, B., Tulchin, J., Cohen, M. and Garland, A. (eds) *Preparing for the Urban Future*, Washington DC: Johns Hopkins University Press.

Rudlin, D. and Falk, N. (1999) *Building the 21st Century Home: The Sustainable Urban Neighbourhood*, Oxford: Architectural Press.

Rydin, Y. (1995) The greening of the housing market, in Bhatti, M., Brooke, J. and Gibson, M. (eds) *Housing and the Environment: A New Agenda*, London: Chartered Institute of Housing.

Sachs, A. (1999) Virtual ecology: a brief environmental history of Silicon Valley, *Worldwatch Magazine*, January–February 1999 (available on http://www.worldwatch.org/pubs/mag/1999/121).

Sachs, J. (2002) *The End of Poverty: Economic Possibilities for Our Time*, New York: Simon and Schuster.

Sachs, J. and Warner, A. (1995) *Natural Resource Abundance and Economic Growth*, Cambridge, MA: Harvard Institute for International Development.

Said, E. (1978) *Orientalism*, New York: Vintage Books.

Sale, K. (1985) *Dwellers in the Land: The Bioregional Vision*, San Francisco: Sierra Club.

Salet, W., Thornley, A. and Kreukels, A. (2003) *Metropolitan Governance and Spatial Planning: Comparative Case Studies of European City-regions*, London: Routledge and Spon.

Sandercock, L. (2003) *Cosmopolis II: Mongrel Cities of the 21st Century*, London and New York: Continuum.

Sardar, Z. and Ravetz, J.R. (1996) *Cyberfutures: Culture and Politics on the Information Super-highway*, London: Pluto Press.

Sassen, S. (1994) *Cities in a World Economy*, Thousand Oaks, CA: Pine Forge Press.

Sassen, S. (2001) *The Global City: New York, London, Tokyo*, 2nd edition, Princeton, NJ: Princeton University Press.

Sassen, S. (2006) *Cities in a World Economy*, 3rd edition, Thousand Oaks, CA: Pine Forge Press.

Satterthwaite, D. (1996) *An Urbanising World: The Second Global Report on Human Settlements*, Oxford: Oxford University Press.

Satterthwaite, D. (2006) *Outside the Large Cities: The Demographic Importance of Small Urban Centres and Large Villages in Africa, Asia and Latin America*, Human Settlements Discussion Paper Urban 3, London: IIED.

Satterthwaite, D., Huq, S., Pelling, M., Reid, H. and Lankao, P.R. (2007) Adapting to climate change in urban areas: the possibilities and constraints in low- and middle-income nations, IIED working paper (available on http://www.iied.org/pubs/).

Savage, M. and Ward, A. (1993) *Cities and Uneven Development*, London: Macmillan.

Scheumann, W., Neubert, S. and Kipping, M. (eds) (2008) *Water Politics and Development Cooperation: Local Power Plays and Global Governance*, Berlin: Springer.

Schumpeter, J.A. (1939) *Business Cycles: A Theoretical, Historical and Statistical Analysis of the Capitalist Process*, New York: McGraw-Hill.

Schwartz, P. (1991) *The Art of the Long View*, New York: Doubleday.

Scott, A.J. (2000) *The Cultural Economy of Cities: Essays on the Geography of Image-producing Industries*, New York: Sage.

Seattle (1857) Speech by Chief Seattle of the Duwamish and Suquamish tribes of North American Indians at the signing of the Port Elliott treaty, quoted by Henry Smith in the *Seattle Sunday Star*, 29 October.

Selya, R.M. (1999) Taiwan as a service economy, in Bryson, J., Henry, N., Keeble, D. and Martin, R., *The Economic Geography Reader*, Chichester: John Wiley and Sons.

SEU (Social Exclusion Unit) (1998) *Bringing Britain Together: A National Strategy for Neighbourhood Renewal*, London: Cabinet Office.

Sharma, K. (2000) *Rediscovering Dharavi: Stories From Asia's Largest Slum*, Harmondsworth: Penguin.

Sharpe, W. and Hodgson, A. (2006) *Intelligent Infrastructure Futures Technology Forward Look: Towards a Cyber-urban Ecology*, London: Foresight Directorate, Office of Science and Innovation.

Simpson, E.S. (1994) *The Developing World*, Harlow: Longman.

Slater, R., Blore, I. and Devas, N. (2002) *Municipalities and Finance: A Sourcebook for Capacity Building*, London: Earthscan.

Smith, M., Whitelegg, J. and Williams, N. (1998) *Greening the Built Environment*, London: Earthscan.

Social Investment Forum (2006) *Report on Socially Responsible Investing Trends in the United States*, Washington DC: Social Investment Forum.

Soja, E. (2000) *Postmetropolis: Critical Studies of Cities and Regions*, Malden, MA and Oxford: Blackwell.

Solow, R.M. (1970) *Growth Theory: An Exposition*, Oxford: Clarendon.

Spellerberg, I.F. (1992) *Evaluation and Assessment for Conservation: Ecological Guidelines for Determining Priorities for Nature Conservation*, London: Chapman and Hall.

Sperling, J. (2004) *The Great Divide: Retro vs. Metro America*, New York: Polipoint.

Stern, N. (2006) *Stern Review on the Economics of Climate Change*, chapter 15, Cambridge: Cambridge University Press.

Stiglitz, J. (2002) *Globalization and its Discontents*, London: Penguin.

Stiglitz, J. (2006) *Making Globalization Work*, New York: W.W. Norton and Co.

Stine, S. (1997) *Landscapes for Learning: Creating Outdoor Environments for Children and Youth*, Chichester: Wiley.

Strategy Unit (2001) *A Futurist's Toolbox: Methodologies in Futures Work*, London: Strategic Futures Team, Cabinet Office (available on www.cabinetoffice.gov.uk/media/cabinetoffice/strategy/assets/toolbox.pdf).

Strategy Unit (2002) *The Future and How To Think About It*, London: Cabinet Office (available on http://www.strategy.gov.uk).

Stren, R., White, R. and Whitney, J. (1992) *Sustainable Cities: Urbanization and the Environment in International Perspective*, Oxford: Westview Press.

Sustainable Development Commission (2005) *I Will if You Will: Towards*

Sustainable Consumption, London: Sustainable Development Commission (available on www.sustainable.gov.uk).

Swyngedouw, E. (2004) *Social Power and the Urbanization of Water: Flows of Power*, Oxford: Oxford University Press.

Tacioli, C. (2002) Livelihood opportunities? *ID21 Insights*, 41(May) (available on http://www.id21.org/insights/insights41/index.html).

Taniguchi, C. (2001) *Transported to the Future, Transport and Energy*, Geneva: UNEP (available on http://www.unep.org/OurPlanet/imgversn/121/tanig.html).

Tapscott, D. and Williams, A.D. (2006) *Wikinomics: How Mass Collaboration Changes Everything*, New York: Penguin.

Taylor, I., Evans, K. and Fraser, P. (1996) *A Tale of Two Cities: Global Change, Local Feeling and Everyday Life in the North of England*, London: Routledge.

Taylor, I., Walton, P. and Young, J. (1973) *The New Criminology*, London: Routledge and Kegan Paul.

Taylor, J.B. (2002) *New Policies for Economic Development*, presentation by John B. Taylor, Under Secretary for International Affairs, United States Treasury, Annual Bank Conference on Development Economics, 30 April, Washington DC: World Bank.

Teaford, R. (2007) *The American Suburb: The Basics*, New York: Routledge.

Therivel, R. and Partidario, M.R. (eds) (1996) *The Practice of Strategic Environmental Assessment*, London: Earthscan.

Tibaijuka, A. (2008) *Housing and Economic Development*, London: Earthscan.

Townsend, P., Davidson, N., Whitehead, M. (1988) *Inequalitites in Health: The Black Report and the Health Divide*, London: Pelican.

Tuan, Y.-F. (1979) *Landscapes of Fear*, Oxford: Blackwell.

UN (United Nations) (2002) *Implementation of the United Nations Millennium Declaration A/57/270*, New York: United Nations.

UN (2005) *World Urbanisation Prospects: The 2005 Revision*, New York: United Nations.

UNCED (United Nations Conference on Environment and Development) (1992) *Report of the United Nations Conference on Environment and Development*, UNCED Report A/CONF.151/5/Rev.1, New York: United Nations.

UNCHS (United Nations Centre for Human Settlements) (2001) *The State of the World's Cities Report 2001*, Nairobi: UNCHS.

UNDP (United Nations Development Programme) (2004) *Human Development Report 2004*, New York: United Nations Development Programme.

UNDP (2005) *Arab Human Development Report 2004*, UNDP Regional Bureau for Arab States, New York: United Nations.

UNECE (United Nations Economic Commission for Europe) (1998) *Convention on Access to Information, Public Participation in Decision-making and Access to Justice in Environmental Matters*, Geneva: United Nations.

UNEP (United Nations Environment Programme) (2001) *Consumption Opportunities: Strategies for Change*, Geneva: UNEP.

UNEP (2002a) *Capacity Building for Sustainable Development: An Overview of UNEP Environmental Capacity Development Initiatives*, New York: UNEP (available on http://www.unep.org/Pdf/Capacity_building.pdf).

UNEP (2002b), *Vital Water Graphics: An Overview of the State of the World's Fresh and Marine Waters*, Nairobi: UNEP.

UNEP Finance Initiative (2004) *The Materiality of Social, Environmental and Corporate Governance Issues to Equity Prices*, Geneva: UNEP.

UNEP/CBD (2000) *Ecosystem Approach: Description, Principles and Operational Guidance*, COP V/6 Annexes, UNEP/CBD/COP/5/23, United Nations Environment Programme/Convention on Biological Diversity, Geneva: United Nations.

UNEP/IISD (2005) *Environment and Trade: A Handbook*, 2nd edition, Geneva and London: United Nations Environment Programme and International Institute for Sustainable Development.

UNFPA (United Nations Population Fund) (2007) *State of the World Population 2007: Unleasing the Potential of Urban Growth*, New York: UNFPA.

UN-Habitat (United Nations Human Settlements Programme) (2001) *Cities in A Globalizing World: Global Report on Human Settlements 2001*, Nairobi: UN-Habitat (available on www.unhabitat.org/categories.asp?catid=555).

UN-Habitat (2003) *The Challenge of Slums: Global Report on Human Settlements 2003*, Nairobi: UN-Habitat (available on www.unhabitat.org/categories.asp?catid=555).

UN-Habitat (2004) *State of the World's Cities 2004/2005: Globalization and Urban Change*, London: Earthscan.

UN-Habitat (2005a) *Financing Urban Shelter: Global Report on Human Settlements 2005*, Nairobi: UN-Habitat (available on www.unhabitat.org/categories.asp?catid=555).

UN-Habitat (2005b) *Habitat Debate: The MDGs and the City*, September 2005, vol. 11, no. 3.

UN-Habitat (2006) *The State of the World's Cities Report 2006/2007: The Millennium Development Goals and Urban Sustainability: 30 Years of Shaping the Habitat Agenda*, Nairobi: UN Human Settlements Programme, London: Earthscan.

UN-Habitat (2007a) *Enhancing Urban Safety and Security: Global Report on Human Settlements 2007*, Nairobi: UN Human Settlements Programme (available on www.unhabitat.org/categories.asp?catid=555).

UN-Habitat (2007b) *Global Report on Human Settlements: Crime and Violence Statistics*, Nairobi: UN-Habitat.

UN-Habitat, (2003) *The Challenge of Slums: Global Report on Human Settlements 2003*, Nairobi: UN-Habitat (available on www.unhabitat.org/categories.asp?catid=555).

UN-Habitat (2004) *State of the World's Cities 2004/2005: Globalization and Urban Change*, Nairobi: UNCHS; London: Earthscan.

UN-Habitat, (2006) *The State of the World's Cities Report 2006/2007: The Millennium Development Goals and Urban Sustainability: 30 Years of Shaping the Habitat Agenda*, Nairobi: UN-Habitat; London: Earthscan.

UN-Habitat (2007a) *Enhancing Urban Safety and Security: Global Report on Human Settlements 2007*, Nairobi: UN-Habitat.

UN-Habitat (2007b) *Crime and Violence*, Global report on human settlements, Nairobi: UN-Habitat.

UN DESA, Population Division (2006) *World Urbanization Prospects: The 2005 Revision*, New York: UN (available on http://www.un.org/esa/population/publications/WUP2005/2005WUPHighlights_Final_Report.pdf).

UN DESA (2002) *Second Local Agenda 21 Survey: Background Paper 15*, report submitted by International Council of Local Environmental Initiatives, New York: United Nations (available on http://www.iclei.org/documents/Global/final_document.pdf).

United States Building Council (2006) *Foundations of the Leadership in Energy and Environmental Design, Environmental Rating System, A Tool for Market Transformation*, August, Washington DC: United States Building Council (available on http://www.usgbc.org/ShowFile.aspx?DocumentID=2040).

URBED and Newbury King (1998) *Valuing the Value-added: The Role of Housing Plus in Creating Sustainable Communities*, Working Paper 3, London: Housing Corporation.

US Climate Action Partnership (2009) *A Blueprint for Legislative Action*, Washington DC: US Climate Action Partnership (available on http://www.us-cap.org/blueprint).

US Green Building Council, LEED (2003) *Green Building Rating System for Core and Shell Development*, September 2003 (available on http://www.usgreenbuildingcouncil.com/Docs/LEEDdocs/LEED-CS%20Pilot%20Rating%20System.pdf, accessed 30 April 2008).

van der Berg, L., Braun, E. van der Meer, J., van der Meer, E. and European Institute for Comparative Urban Research (1997) *Metropolitan Organising Capacity*, Aldershot: Ashgate.

van Liemt, G. (2001) *Some Social and Welfare Aspects of International Trade*, Brussels: EU-LDC Network (available on http://www.eu-ldc.org/downloads/themes/social/sew_backpaper.doc.).

Vasconcellos, E. (2001) *Urban Transport, Environment and Equity: The Case for Developing Countries*, London: Earthscan.

Veblen, T. (1899) *The Theory of the Leisure Class: An Economic Study of Institutions*, Macmillan: New York.

Vogler, J. (2000) *The Global Commons*, Chichester: Wiley.

von Amsberg, J. (1994) The sustainable supply rule for economic evaluation of natural capital depletion, in Goodland, R. and Edmundson, V. (eds) *Environmental Assessment and Development*, Washington DC: World Bank.

von Weizsacker, E.U. (2005) Environmental impacts of privatization, in von Weizsacker, E.U., Young, O.R., Finger, M. with Beisheim, M. (eds) *Limits to Privatization: How to Avoid Too Much of a Good Thing: A Report to the Club of Rome*, London: Earthscan.

von Weizsacker, E., Lovins, A. and Lovins, L.II. (1997) *Factor of Four: Doubling Wealth, Halving Resource Use*, London: Earthscan.

Wackernagel, M. and Rees, W. (1996) *Our Ecological Footprint*, Gabriola Island, British Columbia: New Society Publishers.

Wallerstein, I.M. (2004) *World-systems Analysis: An Introduction*, Durham, NC: Duke University Press.

Ward, B. (1988) *Progress for a Small Planet*, London: Earthscan.

Ward, C. (1978) *The Child in the City*, New York: Pantheon.

Watts, M. (2008) *Curse of the Black Gold: 50 Years of Oil in the Niger Delta*, London: Powerhouse Books.

WBCSD (World Business Council for Sustainable Development) (1998) *Exploring Sutainable Development: Global Scenarios 2000–2050*, London: WBCSD.

WBCSD (2004) *Mobility 2030: Meeting the Challenges to Sustainability*, Geneva and Washington DC: WBCSD.

WBCSD (2007a) *Energy Efficiency in Buildings: Business Realities and Opportunities*, Geneva and Washington DC: WBCSD.

WBCSD (2007b) *Geneva Dialogue: The Role of Business in Tomorrow's Society*, Geneva and Washington DC: WBCSD.

WBCSD (2007c) *Investing in a Low-carbon Energy Future in the Developing World*, Geneva and Washington DC: WBCSD.

WBCSD (2007d) *Markets for Ecosystem Services: New Challenges and Opportunities for Business and the Environment*, Geneva and Washington DC: WBCSD.

WBCSD (2007e) *Promoting Small and Medium Enterprises for Sustainable Development*, Geneva and Washington DC: WBCSD.

Weber, M. (1922) *Economy and Society*, reprinted 1978, Berkeley: University of California Press.

Webster, C.J. and Lai, W.C.L. (2004) *Property Rights, Planning and Markets: Managing Spontaneous Cities*, Cheltenham, UK and Northampton, MA: Edward Elgar.

WEC (World Energy Council) (2006) *Alleviating Urban Energy Poverty in Latin America*, Geneva: WEC.

Welford, R. (1996) *Corporate Environmental Management*, London: Earthscan.

Westra, L. (2008) *Environmental Justice and the Rights of Indigenous Peoples International and Domestic Legal Perspectives*, London: Earthscan.

White, R. (2002) *Building the Ecological City*, Cambridge: Woodhead Publishing.

White, R. and Whitney, J. (1992) Cities and the Environment: An Overview, in Stren, R., White, R. and Whitney, J. (eds) *Sustainable Cities: Urbanization and the Environment in International Perspective*, Boulder, CO: Westview Press.

White, R.R. (1994) *Urban Environmental Management*, London and New York: Routledge.

Whitzman, C. (2008) *The Handbook of Community Safety, Gender and Violence Prevention Practical Planning Tools*, London: Earthscan.

WHO (World Health Organization) (1992) *Twenty Steps to Developing a Healthy Cities Project*, New York: WHO.

WHO (2004) *World Report on Road Traffic Injury Prevention*, Geneva: WHO.

WHO and UNICEF (2000) *Global Water Supply and Sanitation Assessment 2000 Report*, Geneva: WHO.

Williams, J. (2005) Designing neighbourhoods for social interaction: the case of cohousing, *Journal of Urban Design*, 10(2): 195–227.

Williamson, O.E. (1985) *The Economic Institutions of Capitalism*, London: Collier Macmillan.

Willis, K. and Garrod, G.D. (1993) Valuing landscape: a contingent valuation approach, *Journal of Environmental Management*, 37(1): 1–22.

Wolanski, E. (ed.) (2005) *The Environment in Asia-Pacific Harbours*, Berlin: Springer.

Wood, C. (2003) *Environmental Impact Assessment: A Comparative Review*, Harlow: Longman.

Wood, R. and Ravetz, J. (2000) Recasting the urban fringe, *Landscape Design*, 294(October): 13–16.

World Bank (2000) *Cities in Transition: A Strategic View of Urban and Local Government Issues*, Washington: World Bank.

World Bank (2001a) *Urban Environmental Priorities*, Urban Development Division, Washington DC: World Bank.

World Bank (2001b) *World Development Indicators 2001*, Washington DC: World Bank.

World Bank (2001c) *China: Air, Land and Water*, Washington DC: World Bank.

World Bank (2002) *Empowerment and Poverty Reduction: A Sourcebook*, Washington DC: World Bank.

World Bank (2005) *World Development Report 2005*, Washington DC: World Bank.

World Commission on Environment and Development (1987) *Our Common Future*, Oxford: Oxford University Press.

World Resources Institute, *et al.* (1997) *Resource Flows: The Material Basis of Industrial Economies*, Washington DC: World Resources Institute.

World Wildlife Fund (WWF) (2004) *Living Planet Report*, New York: WWF (available on http://www.panda.org/news_facts/publications).

World Wildlife Fund (2006) *Living Planet Report*, New York: WWF (available on http://www.panda.org/news_facts/publications).

World Wildlife Fund and Global Footprint Network (2005) *Europe 2005: The Ecological Footprint*, Godalming: WWF International.

Worldwatch Institute (2007a) *Biofuels for Transport: Global Potential and Implications for Sustainable Energy and Agriculture*, Washington DC: Worldwatch Institute.

Worldwatch Institute (2007b) *The State of the World: Our Urban Future*, Washington DC: Worldwatch Institute.

Worldwatch Institute (2008) *State of the World 2008: Ideas and Opportunities for Sustainable Economies*, London: Earthscan.

Worthington, J. (2005) *Re-inventing the Workplace*, Oxford: Architectural Press.

Yeang, K. (1995) *Designing with Nature: The Ecological Basis for Architectural Design*, New York: McGraw-Hill.

Yeung, H.W. (2004) *Entrepreneurship and the Internationalisation of Asian Firms*, Cheltenham: Edward Elgar.

Yuen, B. (2004) Resume: lessons learned, in Freire, M. and Yuen, B. (eds) *Enhancing Urban Management in East Asia*, Aldershot: Ashgate.

Yunus, M. (2008) *Creating a World Without Poverty: Social Business and the Future of Capitalism*, New York: Public Affairs.

Zhang, X.Q. (2000) High-rise and high-density urban form, in Jenks, M. and Burgess, R. (eds) *Compact Cities: Sustainable Urban Forms for Developing Countries*, London: Spon Press.

Zukin, S. (1998) Urban lifestyles: diversity and standardisation in spaces of consumption, *Urban Studies*, 35(5–6): 825–39.

Index